Nguyen Phung Quang · Jörg-Andreas Dittrich

Vector Control of Three-Phase AC Machines

System Development in the Practice

With 230 Figures

Springer

Prof. Dr. Nguyen Phung Quang
Hanoi University of Technology
Department of Automatic Control
01 Dai Co Viet Road
Hanoi
Vietnam
quangnp-ac@mail.hut.edu.vn

Dr. Jörg-Andreas Dittrich
Neeserweg 31
8048 Zürich
Switzerland
andreas_d@swissonline.ch

ISBN: 978-3-540-79028-0   e-ISBN: 978-3-540-79029-7

Power Systems ISSN: 1612-1287

Library of Congress Control Number: 2008925606

Cover design: deblik, Berlin

Printed on acid-free paper

9 8 7 6 5 4 3 2 1

springer.com

Power Systems

Nguyen Phung Quang and Jörg-Andreas Dittrich

Vector Control of Three-Phase AC Machines

Dedicated to my parents, my wife and my son
Nguyen Phung Quang

To my mother in grateful memory
Jörg-Andreas Dittrich

Formula symbols and abbreviations

A	System matrix
B	Input matrix
C	Output matrix
dq	Field synchronous or rotor flux orientated coordinate system
E	Sensitivity function
E_I, E_R	Imaginary, real part of the sensitivity function
$\mathbf{e}_N$	Vector of grid voltage
f	General analytical vector function
f_p, f_r, f_s	Pulse, rotor, stator frequency
G	Transfer function
G_{fe}	Iron loss conductance
H, h	Input matrix, input vector of discrete system
h	General analytical vector function
$\mathbf{i}_\mu$	Vector of magnetizing current running through L_m
$\mathbf{i}_m$	Vector of magnetizing current
i_{md}, i_{mq}	dq components of the magnetizing current
$\mathbf{i}_N, \mathbf{i}_T, \mathbf{i}_F$	Vectors of grid, transformer and filter current
$\mathbf{i}_s, \mathbf{i}_r$	Vector of stator, rotor current
$i_{sd}, i_{sq}, i_{rd}, i_{rq}$	dq components of the stator, rotor current
$i_{s\alpha}, i_{s\beta}$	$\alpha\beta$ components of the stator current
i_{su}, i_{sv}, i_{sw}	Stator current of phases u, v, w
J	Torque of inertia
K	Feedback matrix, state feedback matrix
$L_f g$	Lie derivation of the scalar function $g(\mathbf{x})$ along the trajectory $\mathbf{f}(\mathbf{x})$
L_m, L_r, L_s	Mutual, rotor, stator inductance
L_{sd}, L_{sq}	d axis, q axis inductance
$L_{\sigma r}, L_{\sigma s}$	Rotor-side, stator-side leakage inductance
m_M, m_G	Motor torque, generator torque
N	Nonlinear coupling matrix
p_{Cu}	Copper loss
p_v	Total loss
$p_{v,fe}, p_{Fe}$	Iron loss

$\mathbf{R}_I$, $\mathbf{R}_{IN}$	Two-dimensional current controller
R_F, R_D	Filter resistance, inductor resistance
R_{fe}	Iron loss resistance
R_r, R_s	Rotor, stator resistance
R_ψ	Flux controller
$\mathbf{r}$	Vector of relative difference orders
r	Relative difference order
$\mathbf{s}$	Complex power
S	Loss function
s	Slip
T	Sampling period
T_p	Pulse period
T_r, T_s	Rotor, stator time constant
T_{sd}, T_{sq}	d axis, q axis time constant
t_D	Protection time
t_{on}, t_{off}	Turn-on, turn-off time
t_r	Transfer ratio
t_u, t_v, t_w	Switching time of inverter leg IGBT's
$\mathbf{u}$	Input vector
$\mathbf{u}_0$, $\mathbf{u}_1$, ... , $\mathbf{u}_7$	Standard voltage vectors of inverter
U_{DC}	DC link voltage
$\mathbf{u}_N$	Vector of grid voltage
$\mathbf{u}_s$, $\mathbf{u}_r$	Vector of stator, rotor voltage
u_{sd}, u_{sq}, u_{rd}, u_{rq}	dq components of the stator, rotor voltage
$u_{s\alpha}$, $u_{s\beta}$	$\alpha\beta$ components of the stator voltage
$\mathbf{V}$	Pre-filter matrix
$\mathbf{w}$	Input vector
$\mathbf{x}$	State vector
$\mathbf{x}_w$	Control error or control difference
$\mathbf{z}$	State vector
z_p	Number of pole pair
$\underline{Z}_s$	Complex resistance or impedance
$\alpha\beta$	Stator-fixed coordinate system
$\mathbf{\Phi}$	Transition or system matrix of discrete system
ψ_μ	Main flux linkage
ψ_p, ψ_r, ψ_s	Vector of pole, rotor, stator flux
ψ_r', ψ_s'	Vector of rotor, stator flux in terms of L_m
ψ_{sd}, ψ_{sq}, ψ_{rd}, ψ_{rq}	dq components of the stator, rotor flux

$\psi'_{rd}, \psi'_{rq}, \psi'_{r\alpha}, \psi'_{r\beta}$	Components of ψ'_r
ψ'_{sd}, ψ'_{sq}	Components of ψ'_s
λ_i	Eigen value
$\omega, \omega_r, \omega_s$	Mechanical rotor velocity, rotor and stator circuit velocity
ϑ, ϑ_s	Rotor angle, angle of flux orientated coordinate system
σ	Total leakage factor
φ	Angle between vectors of stator or grid voltage and stator current

ADC	Analog to Digital Converter
CAPCOM	Capture/Compare register
DFIM	Doubly-Fed Induction Machine
DSP	Digital Signal Processor
DTC	Direct Torque Control
EKF	Extended Kalman Filter
FAT	Finite Adjustment Time
GC	Grid-side Converter or Front-end Converter
IE	Incremental Encoder
IGBT	Insulated Gate Bipolar Transistor
IM	Induction Machine
KF	Kalman Filter
MIMO	Multi Input – Multi Output
MISO	Multi Input – Single Output
MRAS	Model Reference Adaptive System
NFO	Natural Field Orientation
PLL	Phase Locked Loop
PMSM	Permanent Magnet Excited Synchronous Machine
PWM	Pulse Width Modulation
SISO	Single Input – Single Output
VFC	Voltage to Frequency Converter
VSI	Voltage Source Inverter
µC, µP	Microcontroller, microprocessor

Table of Contents

B   Three-Phase AC Drives with IM and PMSM

1 Principles of vector orientation and vector orientated control structures for systems using three-phase AC machines

From the principles of electrical engineering it is known that the 3-phase quantities of the 3-phase AC machines can be summarized to complex vectors. These vectors can be represented in Cartesian coordinate systems, which are particularly chosen to suitable render the physical relations of the machines. These are the field-orientated coordinate system for the 3-phase AC drive technology or the grid voltage orientated coordinate system for generator systems. The orientation on a certain vector for modelling and design of the feedback control loops is generally called vector orientation.

1.1 Formation of the space vectors and its vector orientated philosophy

The three sinusoidal phase currents i_{su}, i_{sv} and i_{sw} of a neutral point isolated 3-phase AC machine fulfill the following relation:

$$i_{su}(t) + i_{sv}(t) + i_{sw}(t) = 0 \qquad (1.1)$$

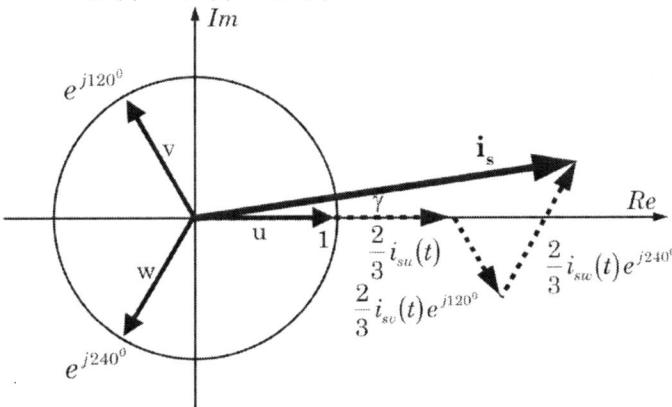

Fig. 1.1 Formation of the stator current vector from the phase currents

These currents can be combined to a vector $\mathbf{i}_s(t)$ circulating with the stator frequency f_s (see fig. 1.1).

$$\mathbf{i}_s = \frac{2}{3}\left[i_{su}(t) + i_{sv}(t)e^{j\gamma} + i_{sw}(t)e^{j2\gamma}\right] \text{ with } \gamma = 2\pi/3 \qquad (1.2)$$

The three phase currents now represent the projections of the vector $\mathbf{i}_s$ on the accompanying winding axes. Using this idea to combine other 3-phase quantities, complex vectors of stator and rotor voltages $\mathbf{u}_s$, $\mathbf{u}_r$ and stator and rotor flux linkages ψ_s, ψ_r are obtained. All vectors circulate with the angular speed ω_s.

In the next step, a Cartesian coordinate system with dq axes, which circulates synchronously with all vectors, will be introduced. In this system, the currents, voltage and flux vectors can be described in two components d and q.

$$\mathbf{u}_s = u_{sd} + ju_{sq}; \mathbf{u}_r = u_{rd} + ju_{rq}$$

$$\mathbf{i}_s = i_{sd} + ji_{sq}; \mathbf{i}_r = i_{rd} + ji_{rq} \qquad (1.3)$$

$$\psi_r = \psi_{rd} + j\psi_{rq}; \psi_s = \psi_{sd} + j\psi_{sq}$$

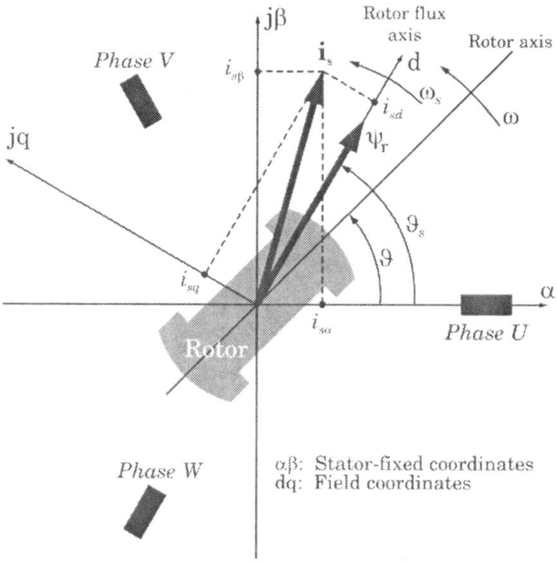

Fig. 1.2 Vector of the stator currents of IM in stator-fixed and field coordinates

Now, typical electrical drive systems shall be looked at more closely. If the real axis d of the coordinate system (see fig. 1.2) is identical with the direction of the rotor flux ψ_r (case IM) or of the pole flux ψ_p (case

PMSM), the quadrature component (q component) of the flux disappears and a physically easily comprehensible representation of the relations between torque, flux and current components is obtained. This representation can be immediately expressed in the following formulae.

- The induction motor with squirrel-cage rotor:

$$\psi_{rd}(s) = \frac{L_m}{1 + sT_r} i_{sd} ; \quad m_M = \frac{3}{2} \frac{L_m}{L_r} z_p \psi_{rd} i_{sq} \tag{1.4}$$

- The permanentmagnet-excited synchronous motor:

$$m_M = \frac{3}{2} z_p \psi_p i_{sq} \tag{1.5}$$

In the equations (1.4) and (1.5), the following symbols are used:

m_M	Motor torque
z_p	Number of pole pairs
$\psi_{rd}, \psi_p = \psi_p$	Rotor and pole flux (IM, PMSM)
i_{sd}, i_{sq}	Direct and quadrature components of stator current
L_m, L_r	Mutual and rotor inductance
	with $L_r = L_m + L_{\sigma r}$ ($L_{\sigma r}$: rotor leakage inductance)
T_r	Rotor time constant with $T_r = L_r/R_r$ (R_r : rotor resistance)
s	Laplace operator

The equations (1.4), (1.5) show that the component i_{sd} of the stator current can be used as a control quantity for the rotor flux ψ_{rd}. If the rotor flux can be kept constant with the help of i_{sd}, then the cross component i_{sq} plays the role of a control variable for the torque m_M.

The linear relation between torque m_M and quadrature component i_{sq} is easily recognizable for the two machine types. If the rotor flux ψ_{rd} is constant (this is actually the case for the PMSM), i_{sq} represents the motor torque m_M so that the output quantity of the speed controller can be directly used as a set point for the quadrature component i_{sq}^*. For the case of the IM, the rotor flux ψ_{rd} may be regarded as nearly constant because of its slow variability in respect to the inner control loop of the stator current. Or, it can really be kept constant when the control scheme contains an outer flux control loop. This philosophy is justified in the formula (1.4) by the fact that the rotor flux ψ_{rd} can only be influenced by the direct component i_{sd} with a delay in the range of the rotor time constant T_r, which is many times greater than the sampling period of the current control loop. Thus, the set point i_{sd}^* of this field-forming component can be provided by the output quantity of the flux controller. For PMSM the pole flux ψ_p is maintained permanently unlike for the IM. Therefore the PMSM must be controlled such that the direct component i_{sd} has the value zero. Fig. 1.2 illustrates the relations described so far.

If the real axis d of the Cartesian dq coordinate system is chosen identical with one of the three winding axes, e.g. with the axis of winding u (fig. 1.2), it is renamed into $\alpha\beta$ coordinate system. A stator-fixed coordinate system is now obtained. The three-winding system of a 3-phase AC machine is a fixed system by nature. Therefore, a transformation is imaginable from the three-winding system into a two-winding system with α and β windings for the currents $i_{s\alpha}$ and $i_{s\beta}$.

$$\begin{cases} i_{s\alpha} = i_{su} \\ i_{s\beta} = \dfrac{1}{\sqrt{3}}\left(i_{su} + 2\,i_{sv}\right) \end{cases} \tag{1.6}$$

In the formula (1.6) the third phase current i_{sw} is not needed because of the (by definition) open neutral-point of the motor.

Figure 1.2 shows two Cartesian coordinate systems with a common origin, of which the system with $\alpha\beta$ coordinates is fixed and the system with dq coordinates circulates with the angular speed $w_s = d\vartheta_s/dt$. The current $\mathbf{i}_s$ can be represented in the two coordinate systems as follows.

- In $\alpha\beta$ coordinates: $\quad\mathbf{i}_s^s = i_{s\alpha} + j\,i_{s\beta}$

- In dq coordinates: $\quad\mathbf{i}_s^f = i_{sd} + j\,i_{sq}$

(Indices: s - stator-fixed, f - field synchronous coordinates)

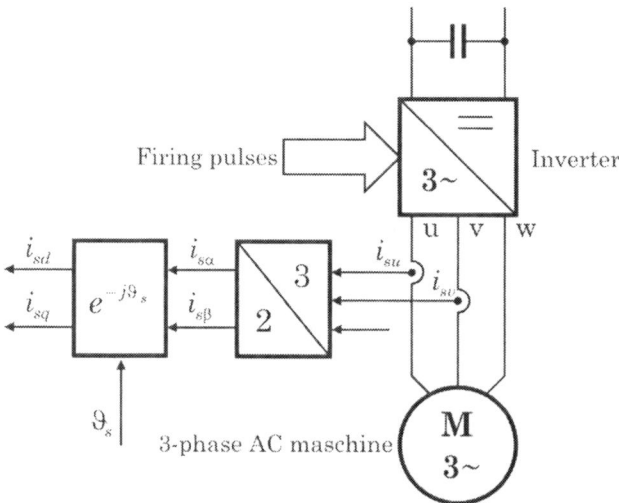

Fig. 1.3 Acquisition of the field synchronous current components

With

$$\begin{cases} i_{sd} = i_{s\alpha}\cos\vartheta_s + i_{s\beta}\sin\vartheta_s \\ i_{sq} = -i_{s\alpha}\sin\vartheta_s + i_{s\beta}\cos\vartheta_s \end{cases} \tag{1.7}$$

the stator current vector is obtained as:

$$\mathbf{i}_s^f = \left[i_{s\alpha}\cos\vartheta_s + i_{s\beta}\sin\vartheta_s \right] + j\left[i_{s\beta}\cos\vartheta_s - i_{s\alpha}\sin\vartheta_s \right]$$

$$\mathbf{i}_s^f = \left[i_{s\alpha} + j i_{s\beta} \right]\left[\cos\vartheta_s - j\sin\vartheta_s \right] = \mathbf{i}_s^s e^{-j\vartheta_s}$$

In generalization of that the following general formula results to transform complex vectors between the coordinate systems:

$$\mathbf{v}^s = \mathbf{v}^f e^{j\vartheta_s} \quad \text{or} \quad \mathbf{v}^f = \mathbf{v}^s e^{-j\vartheta_s} \tag{1.8}$$

v: an arbitrary complex vector

The acquisition of the field synchronous current components, using equations (1.6) and (1.7), is illustrated in figure 1.3.

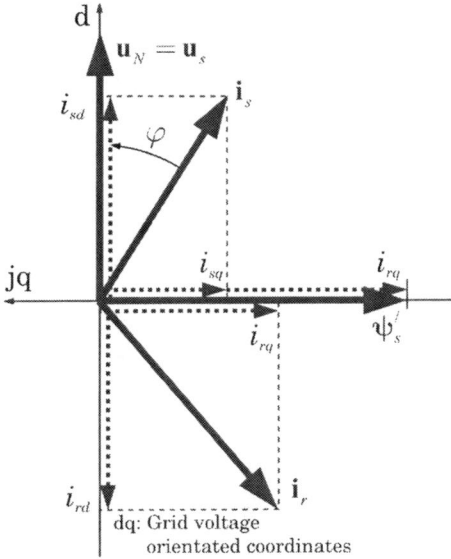

Fig. 1.4 Vectors of the stator and rotor currents of DFIM in grid voltage ($\mathbf{u}_N$) orientated coordinates

In generator systems like wind power plants with the stator connected directly to the grid, the real axis of the grid voltage vector $\mathbf{u}_N$ can be chosen as the d axis (see fig. 1.4). Such systems often use doubly-fed induction machines (DFIM) as generators because of several economic advantages. In Cartesian coordinates orientated to the grid voltage vector, the following relations for the DFIM are obtained.

• The doubly-fed induction machine:

$$\sin\varphi = \frac{|\boldsymbol{\psi}_s|/L_m - i_{rq}}{|\mathbf{i}_s|}; \quad m_G = -\frac{3}{2}z_p\frac{L_m}{L_s}\psi_{sq}i_{rd} \tag{1.9}$$

In equation (1.9), the following symbols are used:

m_G	Generator torque
ψ_{sq}, ψ_s	Stator flux
$\mathbf{i}_s$	Vector of stator current
i_{rd}, i_{rq}	Direct and quadrature components of rotor current
L_m, L_s	Mutual and stator inductance
	with $L_s = L_m + L_{\sigma s}$ ($L_{\sigma s}$: stator leakage inductance)
φ	Angle between vectors of grid voltage and stator current

Because the stator flux ψ_s is determined by the grid voltage and can be viewed as constant, the rotor current component i_{rd} plays the role of a control variable for the generator torque m_G and therefore for the active power P respectively. This fact is illustrated by the second equation in (1.9). The first of both equations (1.9) means that the power factor $\cos\varphi$ or the reactive power Q can be controlled by the control variable i_{rq}.

1.2 Basic structures with field-orientated control for three-phase AC drives

DC machines by their nature allow for a completely decoupled and independent control of the flux-forming field current and the torque-forming armature current. Because of this complete separation, very simple and computing time saving control algorithms were developed, which gave the dc machine preferred use especially in high-performance drive systems within the early years of the computerized feedback control. In contrast to this, the 3-phase AC machine represents a mathematically complicated construct with its multi-phase winding and voltage system, which made it difficult to maintain this important decoupling quality. Thus, the aim of the field orientation can be defined to re-establish the decoupling of the flux and torque forming components of the stator current vector. The field-orientated control scheme is then based on impression the decoupled current components using closed-loop control.

Based on the theoretical statements, briefly outlined in chapter 1.1, the classical structure (see fig. 1.5) of a 3-phase AC drive system with field-orientated control shall now be looked at in some more detail. If block 8 remains outside our scope at first, the structure, similar as for the case of a system with DC motor, contains in the outer loop two controllers: one for the flux (block 1) and one for the speed (block 9). The inner loop is formed of two separate current controllers (blocks 2) with PI behaviour for the field-forming component i_{sd} (comparable with the field current of the DC motor) and the torque-forming component i_{sq} (comparable with the armature current of the DC motor). Using the rotor flux ψ_{rd} and the speed ω, the decoupling network (DN: block 3) calculates the stator voltage components u_{sd} and u_{sq} from the output quantities y_d and y_q of the current controllers R_1. If the field angle ϑ_s between the axis d or the rotor flux axis and the stator-fixed reference axis (e.g. the axis of the winding u or the axis α) is known, the components u_{sd}, u_{sq} can be transformed, using block 4, from the field coordinates dq into the stator-fixed coordinates $\alpha\beta$. After transformation and processing the well known vector modulation (VM: block 5), the stator voltage is finally applied on the motor terminals with respect to amplitude and phase. The flux model (FM: block 8) helps to estimate the values of the rotor flux ψ_{rd} and the field angle ϑ_s from the vector of the stator current $\mathbf{i}_s$ and from the speed ω, and will be subject of chapter 4.4.

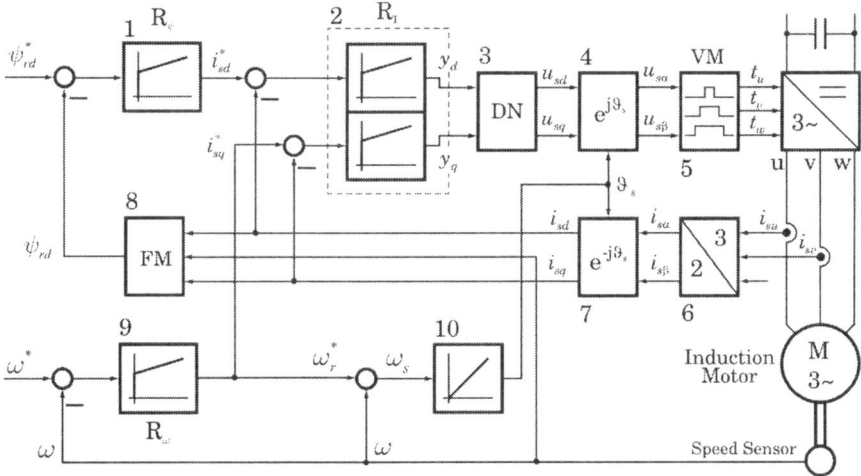

Fig. 1.5 Classical structure of field-orientated control for 3-phase AC drives using IM and voltage source inverter (VSI) with two separate PI current controllers for d and q axes

If the two components i_{sd}, i_{sq} were completely independent of each other, and therefore completely decoupled, the concept would work perfectly with two separate PI current controllers. But the decoupling network DN represents in this structure only an algebraic relation, which performes just the calculation of the voltage components u_{sd}, u_{sq} from the current-like controller output quantities y_d, y_q. The DN with this stationary approach does not show the wished-for decoupling behavior in the control technical sense. This classical structure therefore worked with good results in steady-state, but with less good results in dynamic operation. This becomes particularly clear if the drive is operated in the field weakening range with strong mutual influence between the axes d and q.

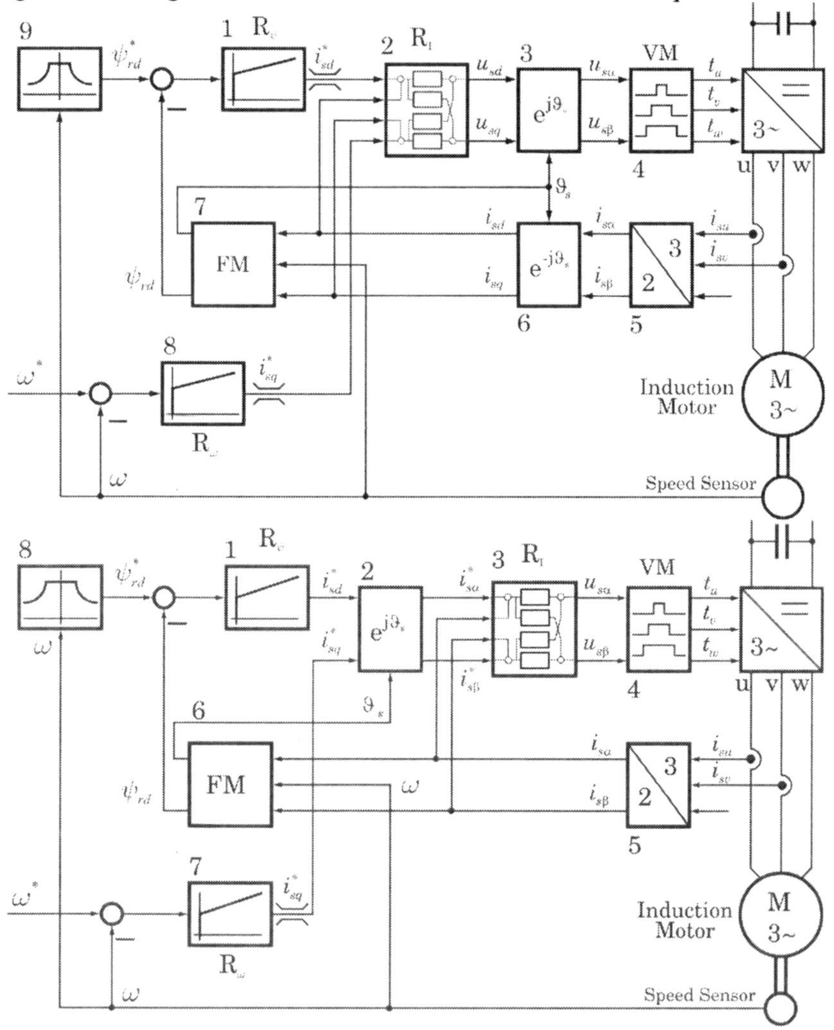

Fig. 1.6 Modern structures with field-orientated control for three-phase AC drives using IM and VSI with current control loop in field coordinates (**top**) and in stator-fixed coordinates (**bottom**)

In contrast to this simple control approach, the 3-phase AC machine, as highlighted above, represents a mathematically complicated structure. The actual internal *dq* current components are dynamically coupled with each other. From the control point of view, the control object „3-phase AC machine" is an object with multi-inputs and multi-outputs (MIMO process), which can only be mastered by a vectorial MIMO feedback controller (see fig. 1.6). Such a control structure generally comprises of decoupling controllers next to main controllers, which provide the actual decoupling.

Figure 1.6 shows the more modern structures of the field-orientated controlled 3-phase AC drive systems with a vectorial multi-variable current controller $\mathbf{R_I}$. The difference between the two approaches only consists in the location of the coordinate transformation before or after the current controller. In the field-synchronous coordinate system, the controller has to process uniform reference and actual values, whereas in the stator-fixed coordinate system the reference and actual values are sinusoidal.

The set point ψ_{rd}^* for the rotor flux or for the magnetization state of the IM for both approaches is provided depending on the speed. In the reality the magnetization state determines the utilization of the machine and the inverter. Thus, several possibilities for optimization (torque or loss optimal) arise from an adequate specification of the set point ψ_{rd}^*. Further functionality like parameter settings for the functional blocks or tracking of the parameters depending on machine states are not represented explicitly in fig. 1.5 and 1.6.

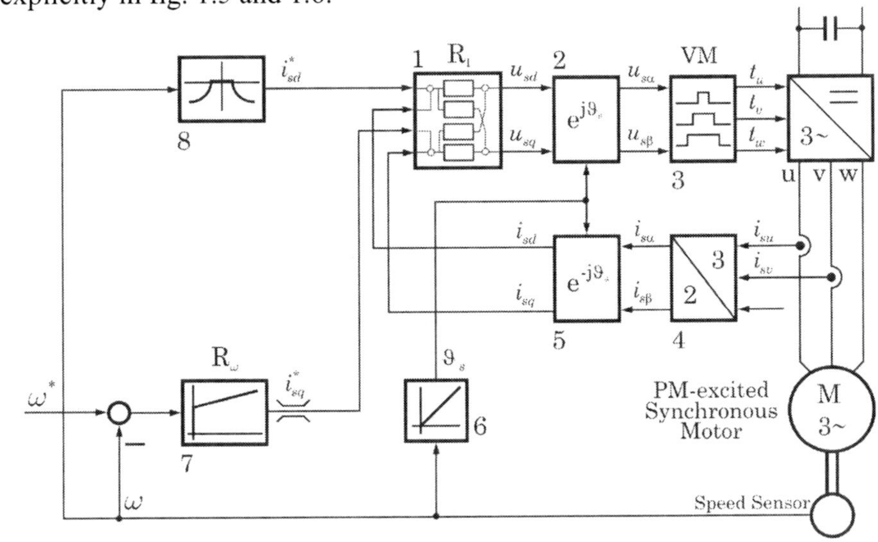

Fig. 1.7 Modern structure with field-orientated control for three-phase AC drives using PMSM and VSI with current control loop in field coordinates

PMSM drive systems with field-orientated control are widely used in practical applications (fig. 1.7). Because of the constant pole flux, the torque in equation (1.5) is directly proportional to the current component i_{sq}. Thus, the stator current does not serve the flux build-up, as in the case of the IM, but only the torque formation and contains only the component i_{sq}. The current vector is located vertically to the vector of the pole flux (fig. 1.8 on the left).

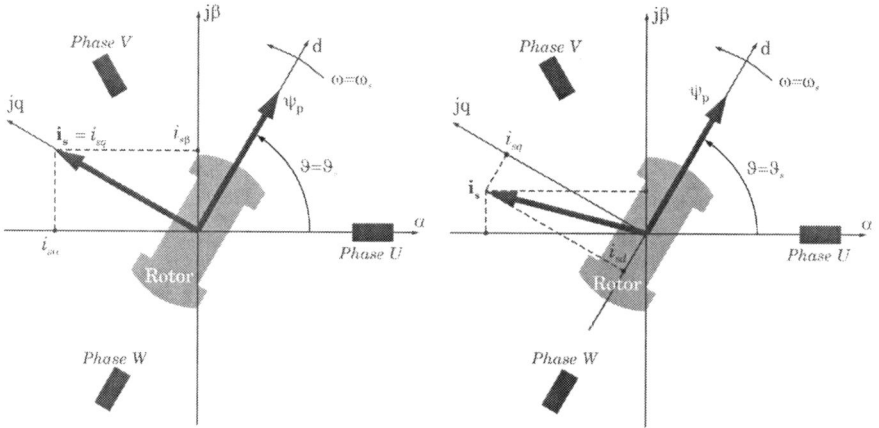

Fig. 1.8 Stator current vector $\mathbf{i}_s$ of the PMSM in the basic speed range (**left**) and in the field-weakening area (**right**)

Using a similar control structure as in the case of the IM, the direct component i_{sd} has the value zero (fig. 1.8 on the left). A superimposed flux controller is not necessary. But a different situation will arise, if the synchronous drive shall be operated in the field-weakening area as well (fig. 1.8 on the right). To achieve this, a negative current will be fed into the d axis depending on the speed (fig. 1.7, block 8). This is primarily possible because the modern magnets are nearly impossible to be demagnetized thanks to state-of-the-art materials. Like for the IM, possibilities for the optimal utilization of the PMSM and the inverter similarly arise by appropriate specification of i_{sd}. The flux angle ϑ_s will be obtained either by the direct measuring – e.g. with a resolver – or by the integration of the measured speed incorporating exact knowledge of the rotor initial position.

1.3 Basic structures of grid voltage orientated control for DFIM generators

One of the main control objectives stated above was the decoupled control of active and reactive current components. This suggests to choose the stator voltage oriented reference frame for the further control design. Let us consider some of the consequences of this choice for other variables of interest.

The stator of the machine is connected to the constant-voltage constant-frequency grid system. Since the stator frequency is always identical to the grid frequency, the voltage drop across the stator resistance can be neglected compared to the voltage drop across the mutual and leakage inductances L_m and $L_{\sigma s}$. Starting point is the stator voltage equation

$$\mathbf{u}_s = R_s \mathbf{i}_s + \frac{d\boldsymbol{\psi}_s}{dt} \quad \Rightarrow \quad \mathbf{u}_s \approx \frac{d\boldsymbol{\psi}_s}{dt} \text{ or } \mathbf{u}_s \approx j\omega_s \boldsymbol{\psi}_s \qquad (1.10)$$

with the stator and rotor flux linkages

$$\begin{cases} \boldsymbol{\psi}_s = \mathbf{i}_s L_s + \mathbf{i}_r L_m \\ \boldsymbol{\psi}_r = \mathbf{i}_s L_m + \mathbf{i}_r L_r \end{cases} \qquad (1.11)$$

Since the stator flux is kept constant by the constant grid voltage (equ. (1.10)) the component i_{rd} in equation (1.9) may be considered as torque producing current.

In the grid voltage orientated reference frame the fundamental power factor, or displacement factor $\cos\varphi$ respectively, with φ being the phase angle between voltage vector $\mathbf{u}_s$ and current vector $\mathbf{i}_s$, is defined according to figure 1.4 as follows:

$$\cos\varphi = \frac{i_{sd}}{|\mathbf{i}_s|} = \frac{i_{sd}}{\sqrt{i_{sd}^2 + i_{sq}^2}} \qquad (1.12)$$

However, it must be considered that according to equation (1.11) for near-constant stator flux any change in $\mathbf{i}_r$ immediately causes a change in $\mathbf{i}_s$ and consequently in $\cos\varphi$. To show this in more detail the stator flux in equation (1.11) can be rewritten in the grid voltage oriented system to:

$$\begin{cases} \psi'_{sd} = \frac{L_s}{L_m} i_{sd} + i_{rd} \approx 0 \\ \psi'_{sq} = \frac{L_s}{L_m} i_{sq} + i_{rq} \approx |\psi'_s| \end{cases} \quad \text{with} \quad \psi'_s = \psi_s / L_m \qquad (1.13)$$

For $L_s / L_m \approx 1$ equation (1.13) may be simplified to:

$$\begin{cases} i_{sd} + i_{rd} \approx 0 \\ i_{sq} + i_{rq} \approx \left| \psi_s' \right| = \psi_{sq}' \end{cases} \tag{1.14}$$

The phasor diagram in figure 1.4 illustrates the context of (1.14). With the torque producing current i_{rd} determined by the torque controller according to (1.13) the stator current i_{sd} is pre-determined as well. To compensate the influence on $\cos\varphi$ according to equation (1.12) an appropriate modification of i_{sq} is necessary. The relation between the stator phase angle φ and i_{sq} is defined by:

$$\sin\varphi = \frac{i_{sq}}{\left| \mathbf{i}_s \right|} = \frac{i_{sq}}{\sqrt{i_{sd}^2 + i_{sq}^2}} \tag{1.15}$$

Equation (1.15) expresses a quasi-linear relation between $\sin\varphi$ and i_{sq}, for small phase angles directly between φ and i_{sq} because of $\sin\varphi \approx \varphi$ in this area. This implies to implement a $\sin\varphi$ control rather than the $\cos\varphi$ control considered initially. Due to the fixed relation between i_{sq} and i_{rq} expressed in the second equation of (1.14) the rotor current component i_{rq} is supposed to serve as $\sin\varphi$- or $\cos\varphi$-producing current component. Another advantage of the $\sin\varphi$ control is the simple distinction of inductive and capacitive reactive power by the sign of $\sin\varphi$.

The DFIM control system consists of two parts: Generator-side control and grid-side control. The generator-side control is responsible for the adjustment of the generator reference values: regenerative torque m_G and power factor $\cos\varphi$. For these values suitable control variables must be found. It was worked out in the previous section, that in the grid voltage reference system the rotor current component i_{rd} may be considered as torque producing quantity, refer to equation (1.9). Therefore, if the generator-side control is working with a current controller to inject the desired current into the rotor winding, the reference value for i_{rd} may be determined by an outer torque control loop.

With this context in mind the generator-side control structure may be assembled now like depicted in figure 1.9. Assuming a fast and accurate rotor current vector control this control structure enables a very good decoupling between torque and power factor in both steady state and dynamic operation. With a fast inner current control loop, torque and power factor might be impressed almost delay-free; the controlled systems for both values have proportional behaviour.

However, in practical implementation measurement noise and current harmonics might cause instability due to the strong correlation of the signals in both control loops. Feedback smoothing low-pass filters are necessary to avoid such effects (fig. 1.9). These feedback filters then form the actual process model and the control dynamics has to be slowed down.

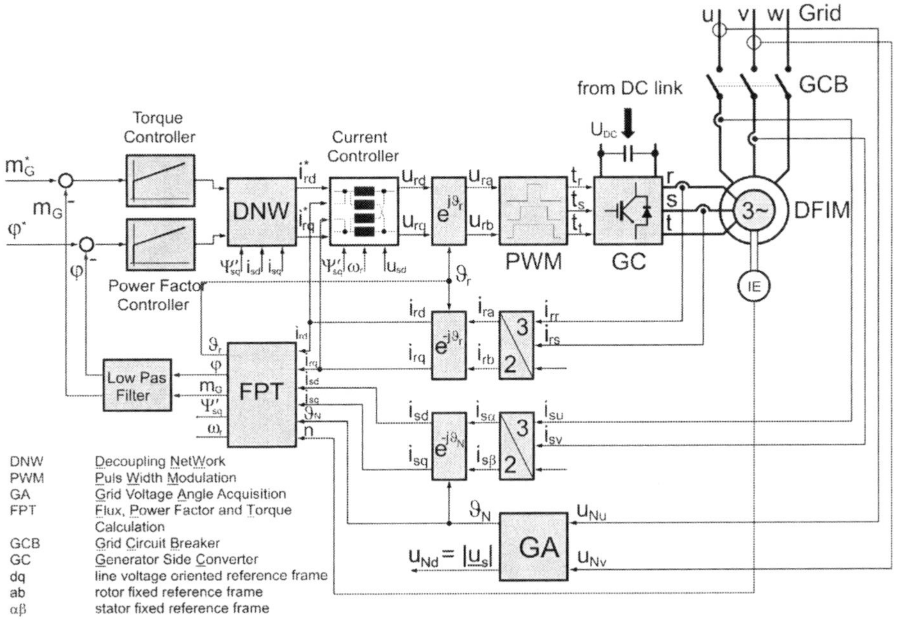

Fig. 1.9 Modern structure with grid voltage orientated control for generator systems using DFIM and VSI with current control loop in grid voltage coordinates

The DFIM is often used in wind power plants thanks to the fundamentally smaller power demand for the power electronic components compared to systems with IM or SM. The demand for improved short-circuit capabilities (ride-through of the wind turbine during grid faults) seems to be invincible for DFIM, because the stator of the generator is directly connected to the grid. Practical solutions require additional power electronics equipment and interrupt the normal system function. Thanks to the power electronic control equipment between the stator and the grid, this problem does not exist for IM or SM systems.

Figure 1.10 presents a nonlinear control structure, which results from the idea of the exact linearization and contains a direct decoupling between active and reactive power. However, the most important advantage of this concept consists of the improvement of the system performance at grid faults, which allows to maintain system operation up to higher fault levels.

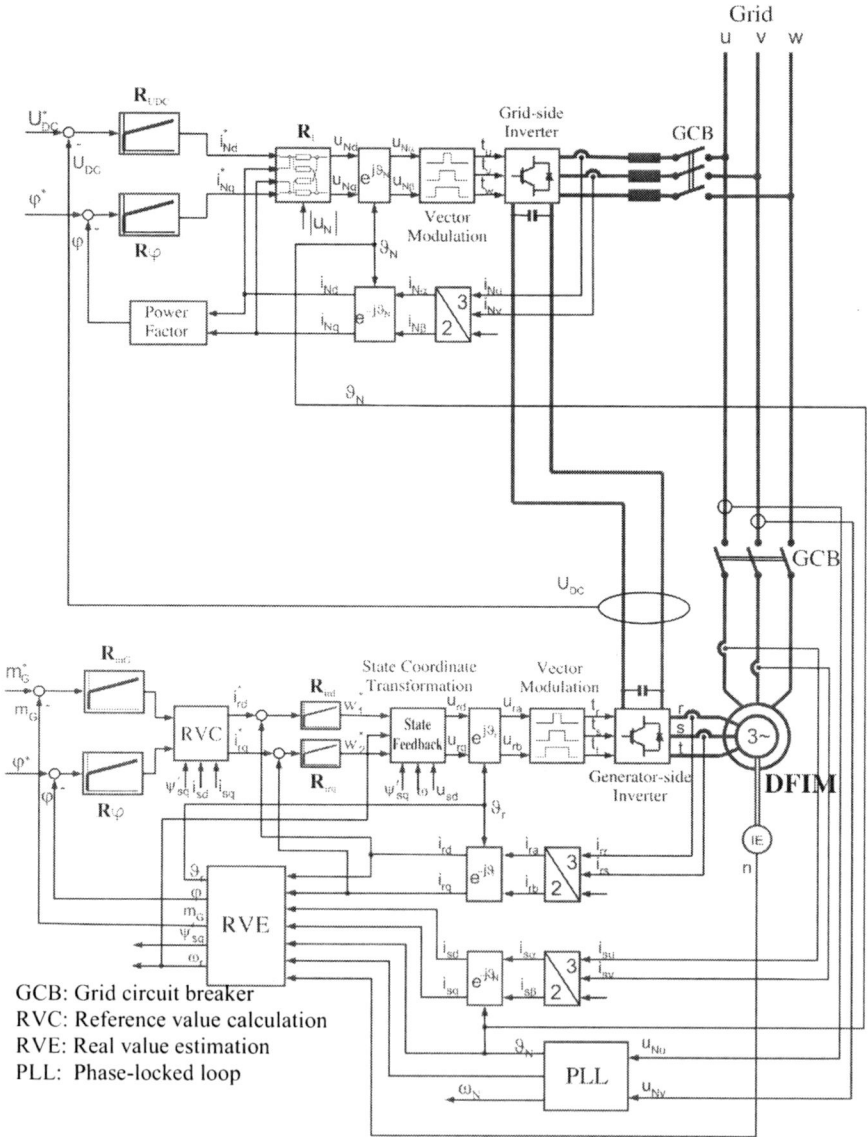

Fig. 1.10 Complete structure of wind power plant with grid voltage orientated control using a nonlinear control loop in grid voltage coordinates

1.4 References

Blaschke F (1972) Das Verfahren der Feldorientierung zur Regelung der Asynchronmaschine. Siemens Forschungs- und Entwicklungsberichte. Bd.1, Nr.1/1972

Hasse K (1969) Zur Dynamik drehzahlgeregelter Antriebe mit stromrichter-gespeisten Asynchron-Kurzschlußläufermaschinen. Dissertation, TH Darmstadt

Leonhard W (1996) Control of Electrical Drives. Springer Verlag, Berlin Heidelberg New York Tokyo

Quang NP, Dittrich A (1999) Praxis der feldorientierten Drehstromantriebs-regelungen. 2. erweiterte Auflage, expert Verlag

Quang NP, Dittrich A, Lan PN (2005) Doubly-Fed Induction Machine as Generator in Wind Power Plant: Nonlinear Control Algorithms with Direct Decoupling. CD Proc. of 11th European Conference on Power Electronics and Applications, 11-14 September, EPE Dresden 2005

Quang NP, Dittrich A, Thieme A (1997) Doubly-fed induction machine as generator: control algorithms with decoupling of torque and power factor. Electrical Engineering / Archiv für Elektrotechnik, 10.1997, pp. 325-335

Schönfeld R (1990) Digitale Regelung elektrischer Antriebe. Hüthig Verlag, Heidelberg

2 Inverter control with space vector modulation

The figure 2.1a shows the principle circuit of an inverter fed 3-phase AC motor with three phase windings u, v and w. The three phase voltages are applied by three pairs of semiconductor switches v_{u+}/v_{u-}, v_{v+}/v_{v-} and v_{w+}/v_{w-} with amplitude, frequency and phase angle defined by microcontroller calculated pulse patterns. The inverter is fed by the DC link voltage U_{DC}. In our example, a transistor inverter is used, which is today realized preferably with IGBTs.

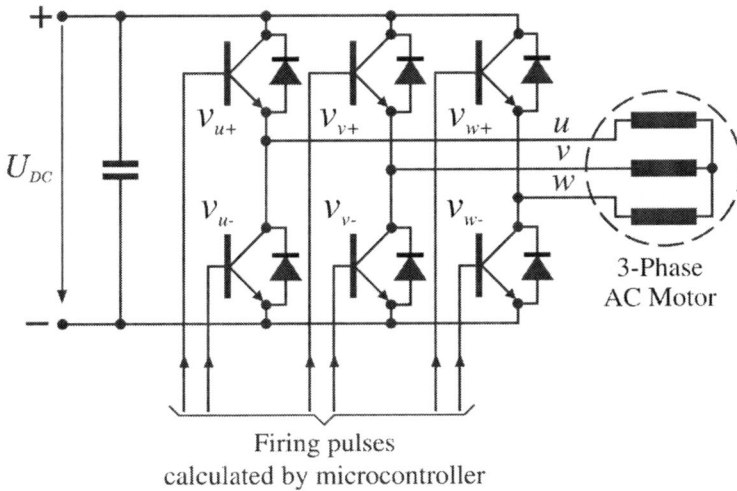

Fig. 2.1a Principle circuit of a VSI inverter-fed 3-phase AC motor

Figure 2.1b shows the spacial assignment of the stator-fixed $\alpha\beta$ coordinate system, which is discussed in the chapter 1, to the three windings u, v and w. The logical position of the three windings is defined as:

     0, if the winding is connected to the negative potential, or as
     1, if the winding is connected to the positive potential

of the DC link voltage. Because of the three windings eight possible logical states and accordingly eight standard voltage vectors $\mathbf{u}_0$, $\mathbf{u}_1$ ... $\mathbf{u}_7$ are obtained, of which the two vectors $\mathbf{u}_0$ - *all windings are on the negative potential* - and $\mathbf{u}_7$ - *all windings are on the positive potential* - are the so called zero vectors.

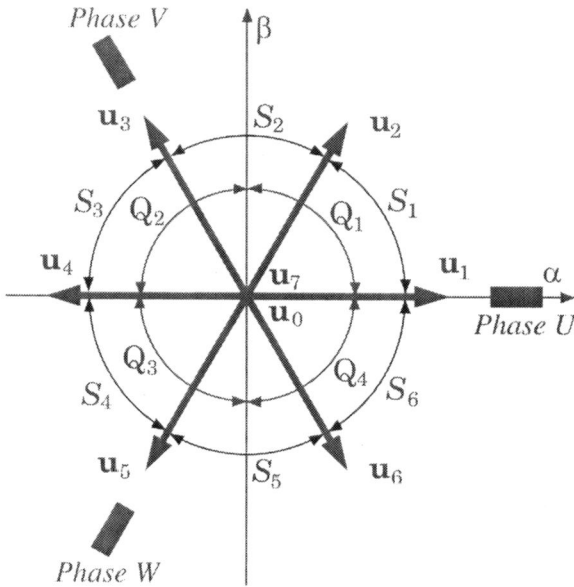

Fig. 2.1b The standard voltage vectors $\mathbf{u}_0$, $\mathbf{u}_1$ ... $\mathbf{u}_7$ formed by the three transistor pairs (Q_1 ... Q_4: quadrants, S_1 ... S_6: sectors)

The spacial positions of the standard voltage vectors in stator-fixed $\alpha\beta$ coordinates in relation to the three windings u, v and w are illustrated in figure 2.1b as well. The vectors divide the vector space into six sectors S_1 ... S_6 and respectively into four quadrants Q_1 ... Q_4. The table 2.1 shows the logical switching states of the three transistor pairs.

Table 2.1 The standard voltage vectors and the logic states

	$\mathbf{u}_0$	$\mathbf{u}_1$	$\mathbf{u}_2$	$\mathbf{u}_3$	$\mathbf{u}_4$	$\mathbf{u}_5$	$\mathbf{u}_6$	$\mathbf{u}_7$
u	0	1	1	0	0	0	1	1
v	0	0	1	1	1	0	0	1
w	0	0	0	0	1	1	1	1

2.1 Principle of vector modulation

The following example will show how an arbitrary stator voltage vector can be produced from the eight standard vectors.

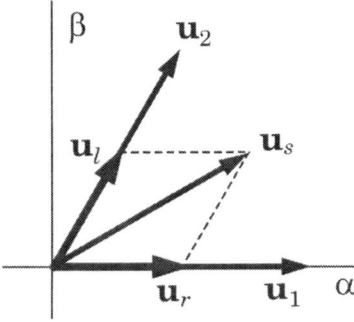

Fig. 2.2 Realization of an arbitrary voltage vector from two boundary vectors

Let us assume that the vector to be realized, $\mathbf{u}_s$ is located in the sector S_1, the area between the standard vectors $\mathbf{u}_1$ and $\mathbf{u}_2$ (fig. 2.2). $\mathbf{u}_s$ can be obtained from the vectorial addition of the two boundary vectors $\mathbf{u}_r$ and $\mathbf{u}_l$ in the directions of $\mathbf{u}_1$ and $\mathbf{u}_2$, respectively. In figure 2.2 mean:

Subscript r, l: boundary vector on the right, left

Supposed the complete pulse period T_p^* is available for the realization of a vector with the maximum modulus (amplitude), which corresponds to the value $2U_{DC}/3$ of a standard vector, the following relation is valid:

$$|\mathbf{u}_s|_{max} = |\mathbf{u}_1| = \ldots = |\mathbf{u}_6| = \frac{2}{3}U_{DC} \qquad (2.1)$$

From this, following consequences result:

1. $\mathbf{u}_s$ is obtained from the addition of $\mathbf{u}_r + \mathbf{u}_l$
2. $\mathbf{u}_r$ and $\mathbf{u}_l$ are realized by the logical states of the vectors $\mathbf{u}_1$ and $\mathbf{u}_2$ within the time span:

$$T_r = T_p^* \frac{|\mathbf{u}_r|}{|\mathbf{u}_s|_{max}}; T_l = T_p^* \frac{|\mathbf{u}_l|}{|\mathbf{u}_s|_{max}} \qquad (2.2)$$

$\mathbf{u}_1$ and $\mathbf{u}_2$ are given by the pulse pattern in table 2.1. Only the switching times T_r, T_l must be calculated. From equation (2.2) the following conclusion can be drawn:

To be able to determine T_r and T_l, the amplitudes of $\mathbf{u}_r$ and $\mathbf{u}_l$ must be known.

It is prerequisite that the stator voltage vector $\mathbf{u}_s$ must be provided by the current controller with respect to modulus and phase. The calculation of the switching times T_r, T_l will be discussed in detail in section 2.2. For now, two questions remain open:

1. What happens in the rest of the pulse period $T_p^* - (T_r + T_l)$?

2. In which sequence the vectors $\mathbf{u}_1$ and $\mathbf{u}_2$, and respectively $\mathbf{u}_r$ and $\mathbf{u}_l$ are realized?

In the rest of the pulse period $T_p^* - (T_r + T_l)$ one of the two zero vectors $\mathbf{u}_0$ or $\mathbf{u}_7$ will be issued to finally fulfil the following equation.

$$\mathbf{u}_s = \mathbf{u}_r + \mathbf{u}_l + \mathbf{u}_0 \text{ (or } \mathbf{u}_7)$$

$$= \frac{T_r}{T_p^*}\mathbf{u}_1 + \frac{T_l}{T_p^*}\mathbf{u}_2 + \frac{T_p^* - (T_r + T_l)}{T_p^*}\mathbf{u}_0 \text{ (or } \mathbf{u}_7) \qquad (2.3)$$

The resulting question is, in which sequence the now three vectors - two boundary vectors and one zero vector - must be issued. Table 2.2 shows the necessary switching states in the sector S_1.

Table 2.2 The switching states in the sector S_1

	$\mathbf{u}_0$	$\mathbf{u}_1$	$\mathbf{u}_2$	$\mathbf{u}_7$
u	0	1	1	1
v	0	0	1	1
w	0	0	0	1

It can be recognized that with respect to transistor switching losses the most favourable sequence is to switch every transistor pair only once within a pulse period.

If the last switching state was $\mathbf{u}_0$, *this would be the sequence*
$$\mathbf{u}_0 \Rightarrow \mathbf{u}_1 \Rightarrow \mathbf{u}_2 \Rightarrow \mathbf{u}_7$$
but if the last switching state was $\mathbf{u}_7$, *this would be*
$$\mathbf{u}_7 \Rightarrow \mathbf{u}_2 \Rightarrow \mathbf{u}_1 \Rightarrow \mathbf{u}_0$$

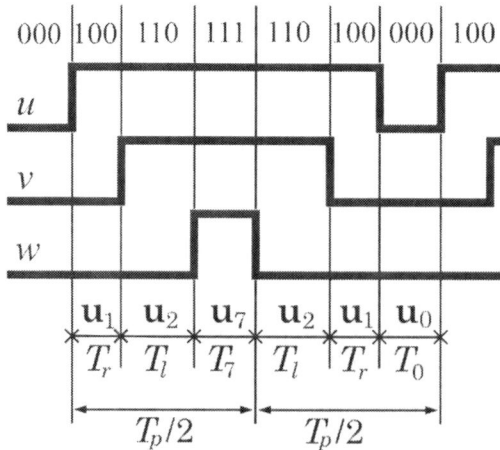

Fig. 2.3 Pulse pattern of voltage vectors in sector S_1

With this strategy the switching losses of the inverter become minimal. Different strategies will arise if other criteria come into play (refer to sections 2.5.1, 2.5.3). If the switching states of two pulse periods succeeding one another are plotted exemplarily a well-known picture from the pulse width modulation technique arises (fig. 2.3).

Figure 2.3 clarifies that the time period T_p^* for the realization of a voltage vector is only one half of the real pulse period T_p. Actually, in the real pulse period T_p two vectors are realized. These two vectors may be the same or different, depending only on the concrete implementation of the modulation.

Until now the process of the voltage vector realization was explained for the sector S_1 independent of the vector position within the sector. With the other sectors S_2 - S_6 the procedure will be much alike: splitting the voltage vector into its boundary components which are orientated in the directions of the two neighbouring standard vectors, every vector of any arbitrary position can be developed within the whole vector space. This statement is valid considering the restrictions which will be discussed in section 2.3. The following pictures give a summary of switching pattern samples in the remaining sectors S_2 ... S_6 of the vector space.

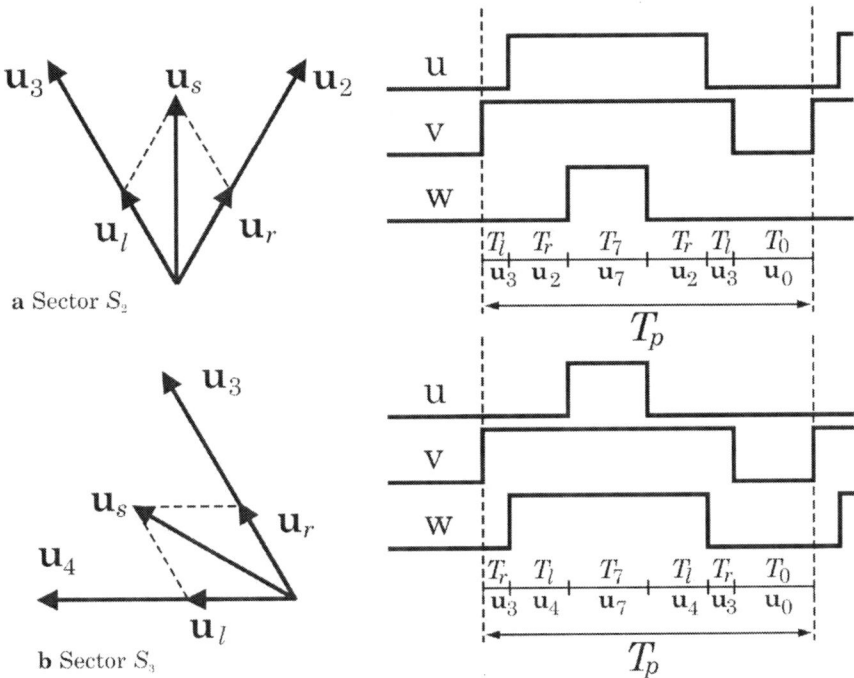

a Sector S_2

b Sector S_3

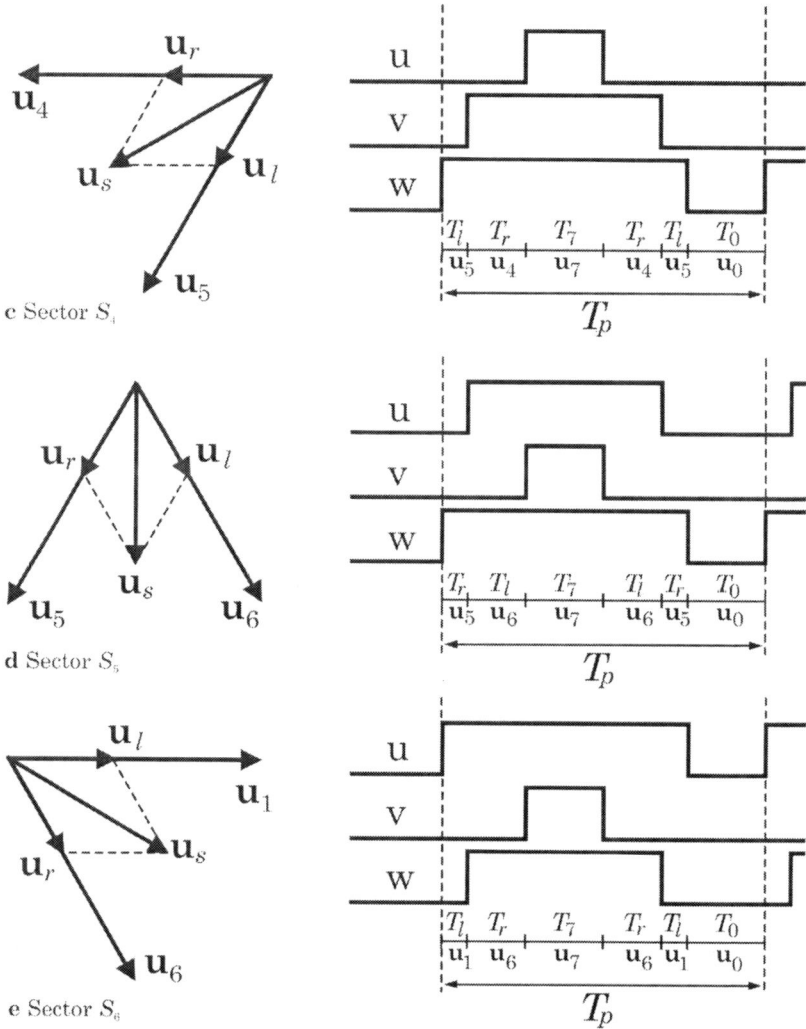

Fig. 2.4 Pulse pattern of the voltage vectors in the sectors $S_2 \dots S_6$

From the fact, that:

1. the current controller delivers the reference value of a new voltage vector $\mathbf{u}_s$ to the modulation after every sampling period T, and
2. every (modulation and) pulse period T_p contains the realization of two voltage vectors,

the relation between the pulse frequency $f_p = 1/T_p$ and the sampling frequency $1/T$ is obtained. The theoretical statement from figure 2.3 is that two sampling periods T correspond to one pulse period T_p. However this relationship is rarely used in practice. In principle it holds *that the new voltage vector $\mathbf{u}_s$ provided by the current controller is realized within at least one or several pulse periods T_p.*

Thereby it is possible to find a suitable ratio of pulse frequency to sampling frequency, which makes a sufficiently high pulse frequency possible at simultaneously sufficiently big sampling period (necessary because of a restricted computing power of the microcontroller). In most systems f_p is normally chosen in the range 2,5...20kHz. The figure 2.5 illustrates the influence of different pulse frequencies on the shape of voltages and currents.

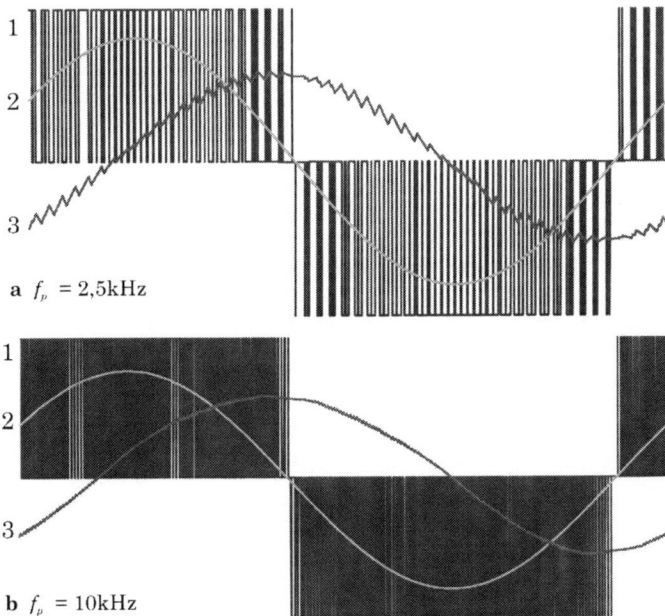

a f_p = 2,5kHz

b f_p = 10kHz

Fig. 2.5 Pulse frequency f_p and the influence on the stator voltage as well as the stator current. 1: Pulsed phase-to-phase voltage; 2: Fundamental wave of the voltage; 3: Current

2.2 Calculation and output of the switching times

After the principle of the space vector modulation has been introduced, the realization of that principle shall be discussed now. Eventually the inverter must be informed on „how" and respectively „how long" it shall switch its transistor pairs, after the voltage vector to be realized is given with respect to modulus and phase angle.

Thanks to the information about phase angle and position (quadrant, sector) of the voltage vector the question „how" can be answered immediately. From the former section the switching samples for all sectors as well as their optimal output sequences with respect to the switching losses are already arranged.

The question „how long" is subject of this section. From equations (2.2), (2.3) it becomes obvious, that the calculation of the switching times T_r, T_l depends only on the information about the moduli of the two boundary vectors $\mathbf{u}_r$, $\mathbf{u}_l$. The vector $\mathbf{u}_s$ (fig. 2.6) is predefined by:

1. Either the DC components u_{sd}, u_{sq} in dq coordinates. From these, the total phase angle is obtained from the addition of the current angular position ϑ_s of the coordinate system (refer to fig. 1.2) and the phase angle of $\mathbf{u}_s$ within the coordinate system.

$$\vartheta_u = \vartheta_s + \arctan\left(\frac{u_{sq}}{u_{sd}}\right) \tag{2.4}$$

2. Or the sinusoidal components $u_{s\alpha}$, $u_{s\beta}$ in $\alpha\beta$ coordinates. This representation does not contain explicitly the information about the phase angle, but includes it implicitly in the components.

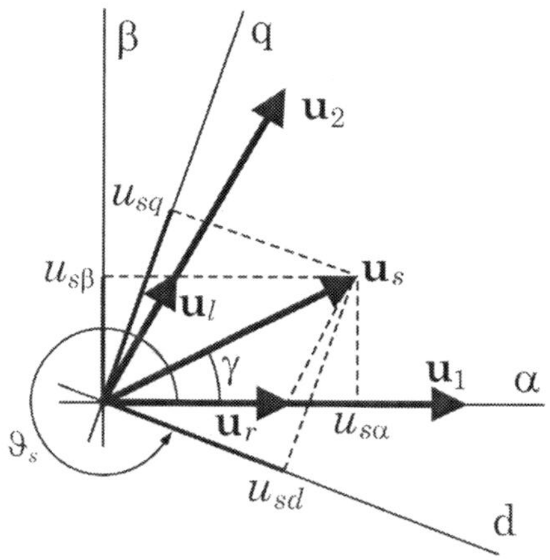

Fig. 2.6 Possibilities for the specification of the voltage vector u_s

Therefore two strategies for calculation of the boundary components exist.

1. *Strategy 1:* At first, the phase angle ϑ_u is found by use of the equation (2.4), and after that the angle γ according to figure 2.6 is calculated, where γ represents the angle ϑ_u reduced to sector 1. Then the calculation of the boundary components can be performed by use of the following formulae, which is valid for the whole vector space:

$$|\mathbf{u}_r| = \frac{2}{\sqrt{3}}|\mathbf{u}_s|\sin\left(60^0 - \gamma\right); |\mathbf{u}_l| = \frac{2}{\sqrt{3}}|\mathbf{u}_s|\sin\left(\gamma\right) \qquad (2.5)$$

With:

$$|\mathbf{u}_s| = \sqrt{u_{sd}^2 + u_{sq}^2} \qquad (2.6)$$

2. *Strategy 2:* After the coordinate transformation, the stator-fixed components $u_{s\alpha}$, $u_{s\beta}$ are obtained from u_{sd}, u_{sq}. For the single sectors, $\mathbf{u}_r$ and $\mathbf{u}_l$ can be calculated using the formulae in table 2.3.

Table 2.3 Moduli of the boundary components $\mathbf{u}_r$, $\mathbf{u}_l$ dependent on the positions of the voltage vectors

		$|\mathbf{u}_r|$	$|\mathbf{u}_l|$
S_1	Q_1	$\left|u_{s\alpha}\right| - \frac{1}{\sqrt{3}}\left|u_{s\beta}\right|$	$\frac{2}{\sqrt{3}}\left|u_{s\beta}\right|$
S_2	Q_1	$\left|u_{s\alpha}\right| + \frac{1}{\sqrt{3}}\left|u_{s\beta}\right|$	$-\left|u_{s\alpha}\right| + \frac{1}{\sqrt{3}}\left|u_{s\beta}\right|$
	Q_2	$-\left|u_{s\alpha}\right| + \frac{1}{\sqrt{3}}\left|u_{s\beta}\right|$	$\left|u_{s\alpha}\right| + \frac{1}{\sqrt{3}}\left|u_{s\beta}\right|$
S_3	Q_2	$\frac{2}{\sqrt{3}}\left|u_{s\beta}\right|$	$\left|u_{s\alpha}\right| - \frac{1}{\sqrt{3}}\left|u_{s\beta}\right|$
S_4	Q_3	$\left|u_{s\alpha}\right| - \frac{1}{\sqrt{3}}\left|u_{s\beta}\right|$	$\frac{2}{\sqrt{3}}\left|u_{s\beta}\right|$
S_5	Q_3	$\left|u_{s\alpha}\right| + \frac{1}{\sqrt{3}}\left|u_{s\beta}\right|$	$-\left|u_{s\alpha}\right| + \frac{1}{\sqrt{3}}\left|u_{s\beta}\right|$
	Q_4	$-\left|u_{s\alpha}\right| + \frac{1}{\sqrt{3}}\left|u_{s\beta}\right|$	$\left|u_{s\alpha}\right| + \frac{1}{\sqrt{3}}\left|u_{s\beta}\right|$
S_6	Q_4	$\frac{2}{\sqrt{3}}\left|u_{s\beta}\right|$	$\left|u_{s\alpha}\right| - \frac{1}{\sqrt{3}}\left|u_{s\beta}\right|$

The proposed strategies for the calculation of the switching times T_r, T_l are equivalent. The output of the switching times itself depends on the hardware configuration of the used microcontroller. The respective procedures will be explained in detail in the section 2.4.

The application of the 2nd strategy seems to be more complicated in the first place because of the many formulae in table 2.3. But at closer look it will become obvious that essentially only three terms exist.

$$\mathbf{a} = \left|u_{s\alpha}\right| + \frac{1}{\sqrt{3}}\left|u_{s\beta}\right|; \mathbf{b} = \left|u_{s\alpha}\right| - \frac{1}{\sqrt{3}}\left|u_{s\beta}\right|; \mathbf{c} = \frac{2}{\sqrt{3}}\left|u_{s\beta}\right| \qquad (2.7)$$

With the help of the following considerations the phase angle of $\mathbf{u}_s$ can be easily calculated.

1. By the signs of $u_{s\alpha}$, $u_{s\beta}$ one finds out in which of the four quadrants the voltage vector is located.

2. Because the moduli of $\mathbf{u}_r$ and $\mathbf{u}_l$ are always positive, and because the term $\mathbf{b}$ changes its sign at every sector transition, $\mathbf{b}$ can be tested on its sign to determine to which sector of the thus found quadrant the voltage vector belongs.

2.3 Restrictions of the procedure

For practical application to inverter control, the vector modulation algorithm (VM) has certain restrictions and special properties which implicitly must be taken into account for implementation of the algorithm as well as for hardware design.

2.3.1 Actually utilizable vector space

The geometry of figure 2.2 may lead to the misleading assumption that arbitrary vectors can be realized in the entire vector space which is limited by the outer circle in fig. 2.7b, i.e. every vector $\mathbf{u}_s$ with $|\mathbf{u}_s| \leq 2U_{DC}/3$ would be practicable. The following consideration disproves this assumption: It is known that the vectorial addition of $\mathbf{u}_r$ and $\mathbf{u}_l$ is not identical with the scalar addition of the switching times T_r and T_l. To simplify the explanation, the constant half pulse period which, according to fig. 2.3, is available for the realization of a vector is replaced by $T_{p/2} = T_p/2$. After some rearrangements of equation (2.2) by use of (2.5) the following formula is obtained.

$$T_\Sigma = T_r + T_l = \sqrt{3}\,T_{p/2}\frac{|\mathbf{u}_s|}{U_{DC}}\cos\left(30^0 - \gamma\right) \tag{2.8}$$

In case of voltage vector limitation, that is $|\mathbf{u}_s| = 2U_{DC}/3$, it follows from (2.8):

$$T_{\Sigma\max} = T_{p/2}\frac{2}{\sqrt{3}}\cos\left(30^0 - \gamma\right) \quad \text{with} \quad 0^0 \leq \gamma \leq 60^0 \tag{2.9}$$

The diagram in figure 2.7a shows the fictitious characteristic of $T_{\Sigma\max}$ with excess of the half pulse period $T_{p/2}$. By limitation of T_Σ to $T_{p/2}$ the actual feasible area is enclosed by the hexagon in fig. 2.7b.

In some cases in the practice - e.g. for reduction of harmonics in the output voltage - the hexagon area is not used completely. Only the area of the inner, the hexagon touching circle will be used. The usable maximum voltage is then:

$$\left|\mathbf{u}_s\right|_{max} = \frac{1}{\sqrt{3}} U_{DC} \qquad (2.10)$$

Thus the area between the hexagon and the inner circle remains unused. Utilization of this remaining area is possible if the voltage modulus is limited by means of a *time limitation* from T_Σ to $T_{p/2}$. To achieve this, the zero vector time is dispensed with, and only one transistor pair is involved in the modulation in each sector (refer to fig. 2.14d, right). A *direct modulus limitation* will be discussed later in connection with the current controller design.

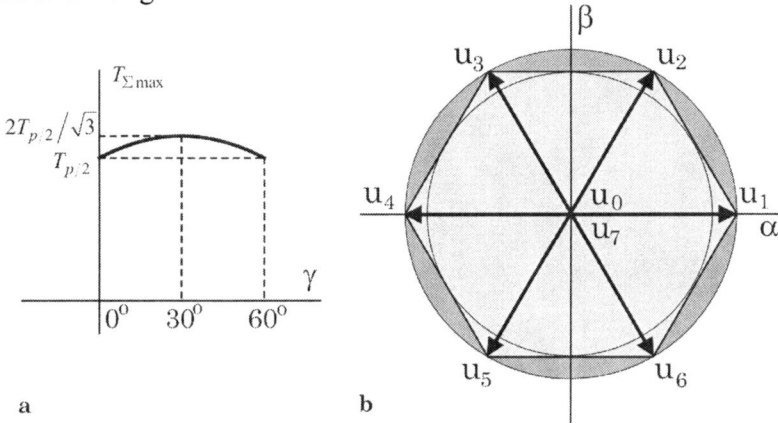

Fig. 2.7 Temporal (**a**) and spacial (**b**) representation of the utilizable area for the voltage vector $\mathbf{u}_s$

An important characteristic for the application of the VM is the voltage resolution Δu, which for the case of limitation to the inner circle or at use of equation (2.10) can be calculated as follows:

$$\Delta u = \frac{2}{\sqrt{3}} \frac{\Delta t}{T_p} U_{DC} \; [\mathrm{V}] \qquad (2.11)$$

At deeper analysis, restricted on the hexagon only, it turns out that the zero vector times become very small or even zero if the voltage vector approaches its maximum amplitude. This is equivalent to an (immediate) switch on or off of the concerned transistor pair after it has been switched off or on. For this reason the voltage vector moduli have to be limited to make sure the zero vector times T_0 and T_7 never fall below the switching times of the transistors. For IGBT's the switching times are approx. <1...4µs, so that this contraction of the voltage vector for usual switching frequencies of 1...5kHz can be considered insignificant. However, the situation becomes more critical for higher switching frequencies or if slow-switching semiconductors, such as thyristors, are used.

The values either of T_r or of T_l become very small in the boundary zone between the sectors or near one of the standard vectors $\mathbf{u}_1$ ... $\mathbf{u}_6$. For some commonly used digital signal processing structures (refer to the application example with TMS 320C20/C25 in section 2.4) the PWM synchronization is directly coupled to the interrupt evaluation of the timer counters for T_r and T_l. For these structures the values of T_r and T_l must never fall below the interrupt reaction times causing another limitation of the utilizable area. The arising forbidden zones are shown in figure 2.8.

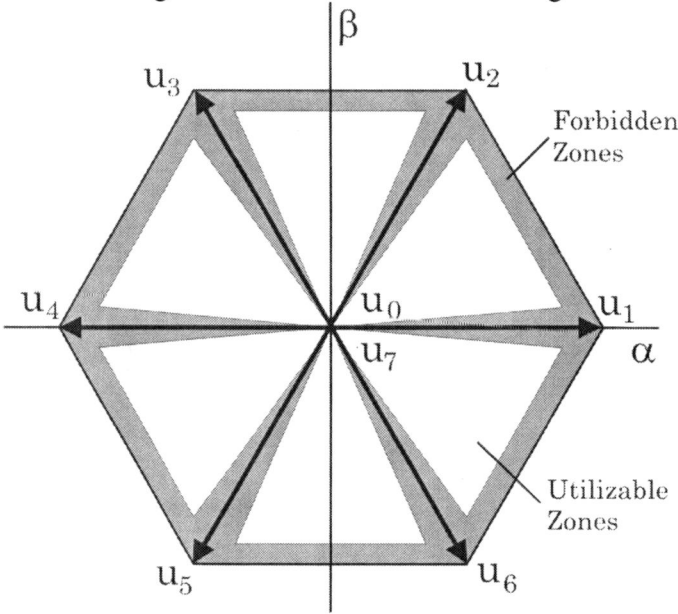

Fig. 2.8 Forbidden zones in the vector space

2.3.2 Synchronization between modulation and signal processing

According to theory (refer to fig. 2.3) the modulated voltage in the context of control or digital signal processing looks like in figure 2.9 for the samplings periods $(k-1)$, (k) and $(k+1)$. The voltage output sequence in period (k)

$$T_r(\mathbf{u}_r) \Rightarrow T_l(\mathbf{u}_l) \Rightarrow T_7(\mathbf{u}_7) / T_l(\mathbf{u}_l) \Rightarrow T_r(\mathbf{u}_r) \Rightarrow T_0(\mathbf{u}_0)$$

leads to the following time relation:

$$T_{synch} = T_p - \frac{T_0(k)}{2} + \frac{T_0(k-1)}{2}$$

For a dynamic process with $\mathbf{u}_s(k-1) \neq \mathbf{u}_s(k)$ is also $T_0(k-1)/2 \neq T_0(k)/2$. That means, that T_{synch} would be not constant (fig. 2.9b), making the use of up/down counters – like usually done in PWM units – impossible. Therefore, a different sequence shall be used for voltage output:

$$\frac{T_0}{2}(\mathbf{u}_0) \Rightarrow T_r(\mathbf{u}_r) \Rightarrow T_l(\mathbf{u}_l) \Rightarrow T_7(\mathbf{u}_7)/T_l(\mathbf{u}_l) \Rightarrow T_r(\mathbf{u}_r) \Rightarrow \frac{T_0}{2}(\mathbf{u}_0)$$

Figure 2.9b shows this alternative sequence. It is obvious from the figure that this sequence is absolutely stable and therefore the use of up/down counters is supported. This means also *a strict synchronization between control and pulse periods* which must be considered already in the design phase of the signal processing hardware.

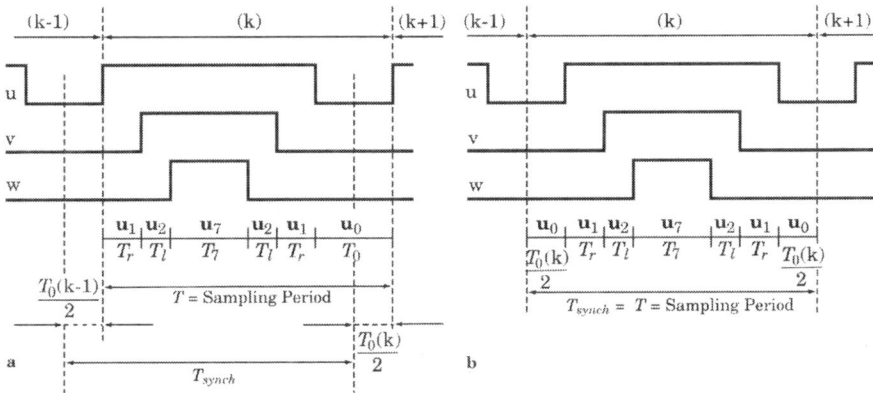

Fig. 2.9 For ensuring synchronization between modulation and control: the theoretical sequence (**a**) must be modified (**b**)

2.3.3 Consequences of the protection time and its compensation

So far, the semiconductors had been regarded as ideal switches with un-delayed turn-on and turn-off characteristics. However, the IGBT's physically reach their safe switched-on or switched-off state only after a certain turn-on or turn-off period t_{on}, t_{off}. To avoid inverter short circuit, the switch-on edge of the control signal must be delayed for a time t_D which is greater than the turn-off time t_{off}. This time is called protection time or blanking time (fig. 2.10). In practice t_D is chosen so that t_{off} will be nearly 70 ... 80% of t_D.

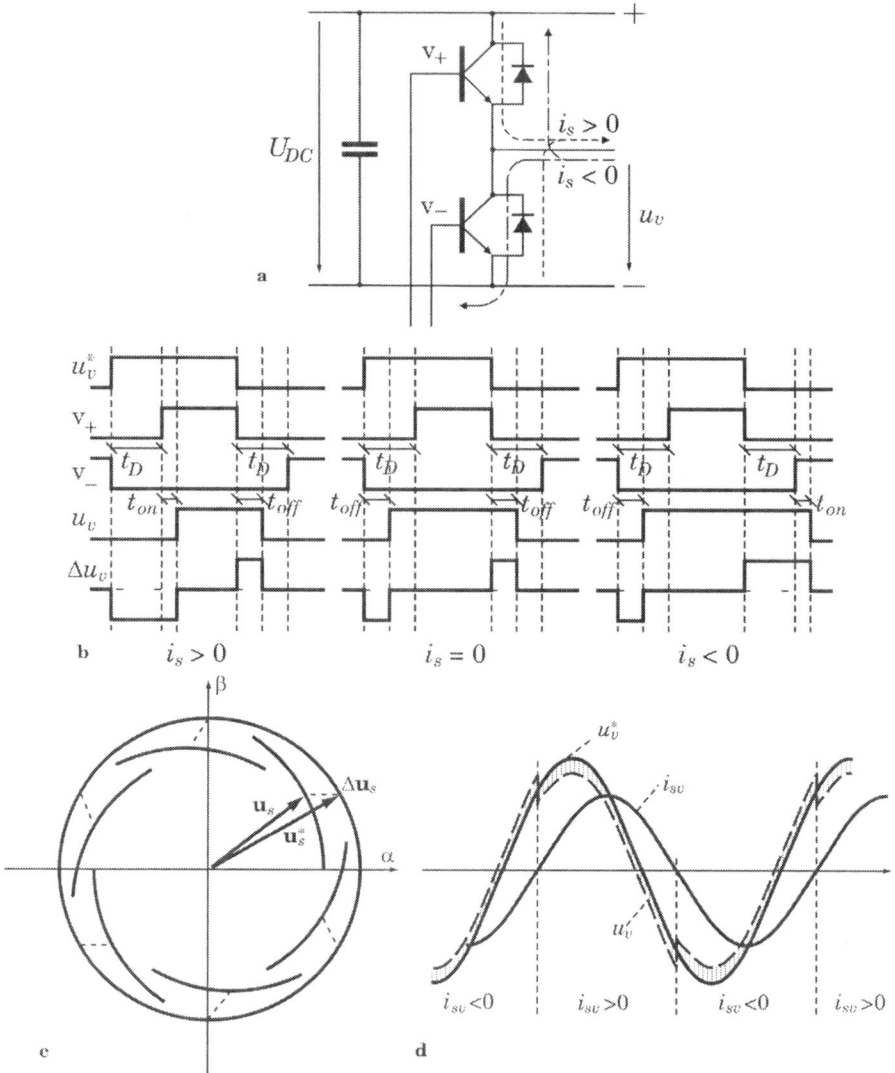

Fig. 2.10 Origin of the protection time t_D and its influence on the output voltage

Figure 2.10b shows in turn: **1.** The reference voltage u_v^* for the phase v. **2.** The actual IGBT control signals v_+ and v_-, modified by the protection time t_D. **3.** The actual voltage u_v of phase v. **4.** The voltage errors Δu_v. The influence of t_D on the trajectory of the stator voltage vector $\mathbf{u}_s$ as well as on the fundamental wave of the phase voltage are illustrated in figures 2.10c,d.

The voltage error Δu_v, caused by t_D and shown in figure 2.10d, can be calculated as follows:

$$\Delta u_v = u_v^* - u_v = \begin{cases} -\dfrac{t_D}{T_p}\left(\dfrac{2}{3}U_{DC}\right) & \text{for } i_{sv} > 0 \\[3mm] \dfrac{t_D}{T_p}\left(\dfrac{2}{3}U_{DC}\right) & \text{for } i_{sv} < 0 \end{cases} \tag{2.12}$$

The voltage error depends on the sign of the phase current and may be effectively compensated with respect to the voltage mean average value. This compensation can be realized either in hardware or in software. *Software compensation* is more widely used today. Preferably, the compensation is done without using the actual current feedbacks which could be critical because of the pulsed current as well as the measuring noise at zero crossings. This is possible if the current controller works without delay. In chapter 5 it will be shown that this condition is largely fulfilled for the control algorithms to be introduced there.

In this case the reference value can be used to capture the sign instead of the actual value. The reference values i_{su}^*, i_{sv}^* and i_{sw}^* of the phase currents can be calculated from i_{sd}^*, i_{sq}^* by use of a coordinate transformation. With that the error components in $\alpha\beta$-coordinates are obtained as follows:

$$\begin{cases} \Delta u_{s\alpha} = \left[-sign\left(i_{su}^*\right) + \dfrac{1}{2}sign\left(i_{sv}^*\right) + \dfrac{1}{2}sign\left(i_{sw}^*\right)\right]\dfrac{t_D}{T_p}\left(\dfrac{2}{3}U_{DC}\right) \\[3mm] \Delta u_{s\beta} = \left[-sign\left(i_{sv}^*\right) + sign\left(i_{sw}^*\right)\right]\dfrac{t_D}{T_p}\left(\dfrac{2}{3}U_{DC}\right)\dfrac{\sqrt{3}}{2} \end{cases} \tag{2.13}$$

The error components according to (2.13) are added to the stator-fixed voltage components $u_{s\alpha}$, $u_{s\beta}$ before they are forwarded to the modulation.

2.4 Realization examples

The realization of the space vector modulation requires a suitable periphery, which has to be added to the processor hardware when normal microprocessors (µP) or digital signal processors (DSP) are used. However, a number of microprocessors with this periphery on chip, so called micro controllers (µC) are available on the market today, allowing implementation of advanced modulation algorithms without additional hardware.

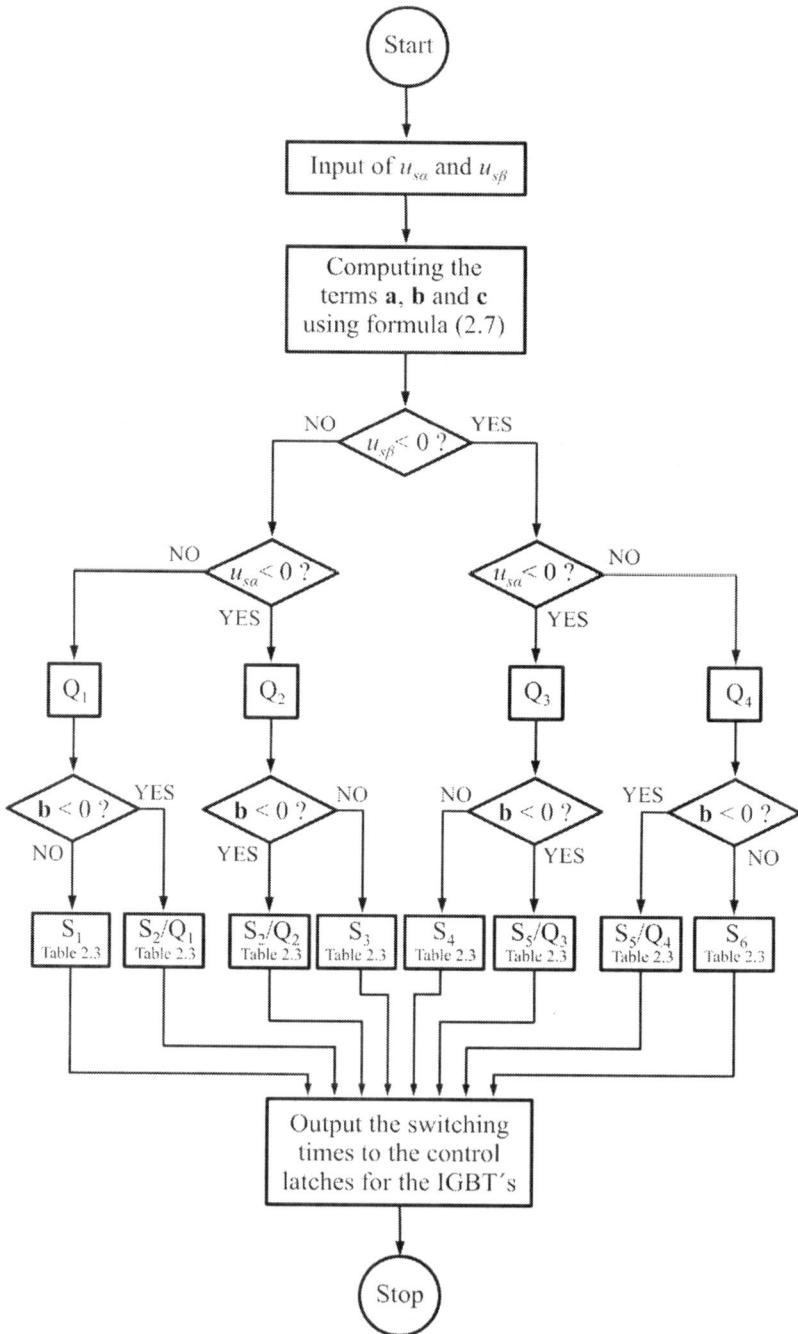

Fig. 2.11 Flow chart for the computing of the switching times according to the space vector modulation

Microcontrollers, which are utilizable for 3-phase AC machine systems due to their internal PWM units as well as other on-chip periphery units, are e.g.

1. SAB 80C166, SAB C167 (Siemens, Infineon): The time resolution Δt of C166 is 400 ns, of C167 50 ns. The upper and the lower transistor of a phase leg are not controllable separately using the C167-PWM unit[1], which would be necessary for an efficient, software based generation of the protection time. A 32 bit single chip microcontroller of the TC116x series can be used very advantageously today for a high-quality drive.

2. TMS 320C240/F240 (Texas Instruments): $\Delta t = 50ns$. The µC supports the direct generation of the protection time t_D, and the transistors of a pair are controllable independently. Also, chips of the family TMS 320 F281x are used very widely today.

In many systems a double processor configuration is used due to the strong price collapse of the processors in the last years. For such applications, the digital signal processors from Texas Instruments TMS 320C25 (16 Bit, fixed-point arithmetic) or TMS 320C32 (32 Bit, floating-point arithmetic) can be recommended particularly.

The application of the modulation algorithm, described in sections 2.1 and 2.2, shall be illustrated now in detail on 4 examples, orientated essentially on the Siemens microcontrollers SAB 80C166, SAB C167 and the Texas Instruments DSP TMS 320C20/C25. The calculation of the switching times is carried out according to the 2nd strategy of section 2.2, i.e. by means of the $\alpha\beta$ voltage components.

In principle, the concrete formulae for the computing of the switching times in all sectors shall be worked out first using table 2.3. These formulae will then be used on-line. The computing and output will be independent of the hardware following the flow chart in the figure 2.11. The flow chart clarifies the steps to determine the space vector area in which the voltage vector to be realized is located. After that, the computing, dependent on the respective hardware, will follow.

2.4.1 Modulation with microcontroller SAB 80C166

The microcontroller SAB 80C166 is a special high-performance microprocessor with an extensive periphery on the chip. Particularly the Capture/Compare register unit supports the space vector modulation for

---

[1] This is possible, however, if the modulation is not realized with PWM units but with CAPCOM registers

3-phase AC machines. The double register compare mode is used in the following example.

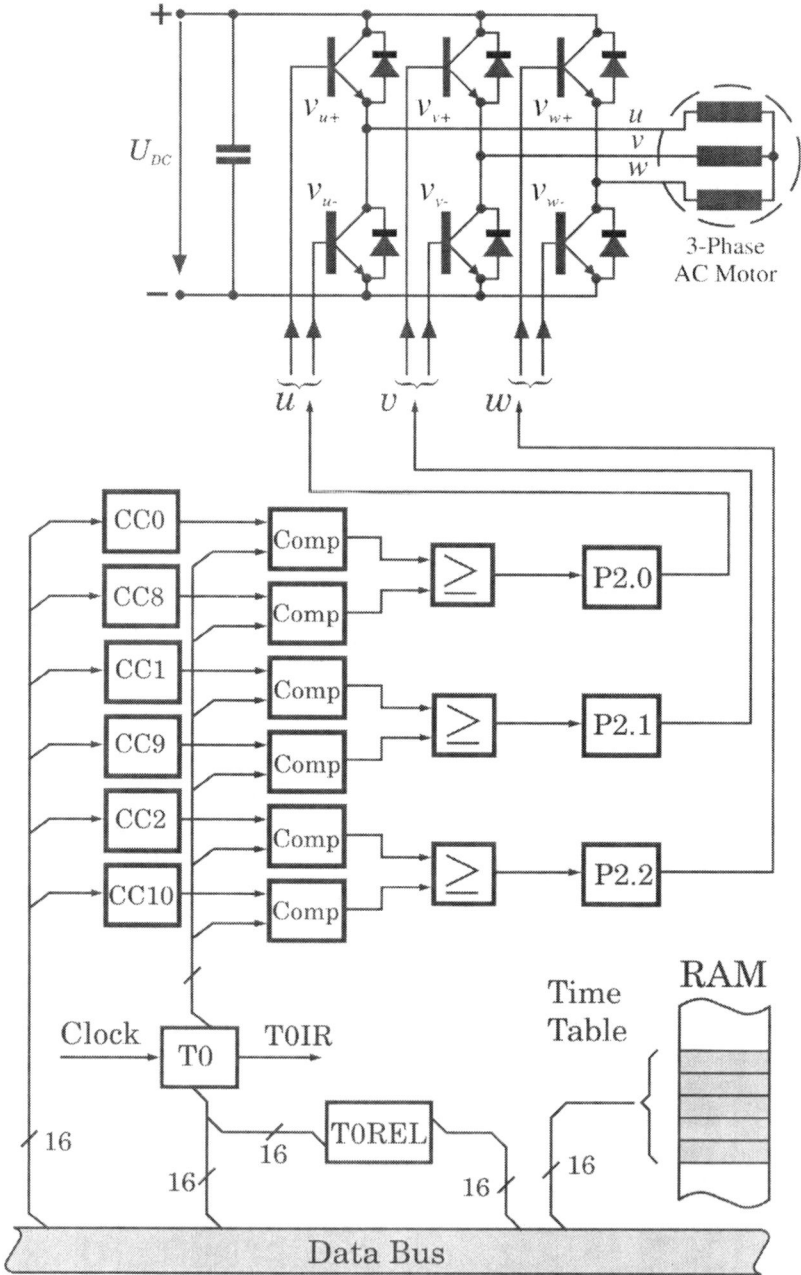

Fig. 2.12a Hardware configuration for the space vector modulation using the microcontroller SAB 80C166 in double register compare mode

In double register compare mode the 16 CapCom registers CC0-CC15 are configured in two register banks and assigned in pairs to one of the two timers T0 or T1 respectively. E.g. the three pairs CC0/CC8, CC1/CC9 and CC2/CC10 with the inputs/outputs CC0IO/P2.0, CC1IO/P2.1 and CC2IO/P2.2, which are configured as outputs here, shall be used. The simplified hardware structure to control the inverter is shown in the figure 2.12a. The assignment of the register pairs to the inverter phase legs is represented in the figure 2.12b.

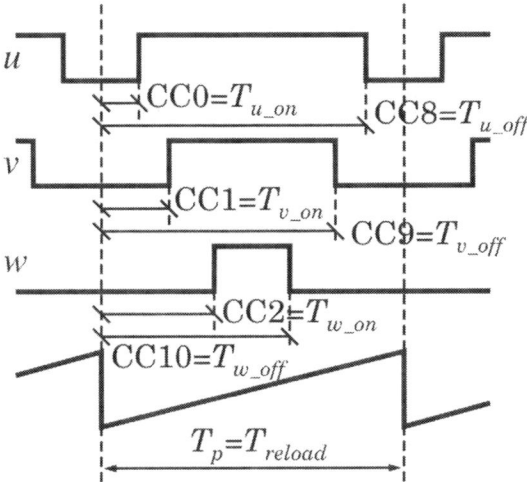

Fig. 2.12b Assignment of the register pairs to the switching times of the inverter legs

The modulation works in a fixed time frame with the pulse period T_p, which represents at the same time the reload value T_{reload} for the timer T0. This stable time frame supports the synchronization between the hardware hierarchies as well as between digital control, modulation and current measurement, which shall be discussed later. Thus, the reload register T0REL must be loaded with T_p only once at processor initialization. In the current sampling period (k) the turn-on/turn-off times T_{u_on}, T_{u_off}, T_{v_on}, T_{v_off}, T_{w_on} and T_{w_off} of the inverter legs will be calculated and stored intermediately in a RAM table. An interrupt signal T0IR is triggered at overflow of the timer T0 which causes the transfer of the reload value from the register T0REL into timer T0. The interrupt signal T0IR at the same time activates an interrupt service routine to load the new switching times from the RAM table into the register pair for the following sampling period. In the next sampling period ($k+1$) and while the timer T0 is counting up, the compare matches between:

T0 and CC0, CC1 and CC2 as well as

T0 and CC8, CC9 and CC10

cause the switchover of the phases u, v and w to

the positive or respectively

the negative potential

of the DC link voltage U_{DC}. The voltage components $u_{s\alpha}$ and $u_{s\beta}$ are normalized to the maximum value $2U_{DC}/3$ in the following calculations, so that an extra index is neglected following the definition of the times introduced in figure 2.12b, and using equations (2.1), (2.2), (2.7) and table 2.3 the following formulae are obtained for the different sectors:

1. Sector 1:

$$T_{u\_on} = \frac{T_p}{2}(1-\mathbf{a}); \ T_{v\_on} = \frac{T_p}{2}(1+\mathbf{b}-\mathbf{c}); \ T_{w\_on} = \frac{T_p}{2}(1+\mathbf{a})$$

$$T_{u\_off} = \frac{T_p}{2}(3+\mathbf{a}); \ T_{v\_off} = \frac{T_p}{2}(3-\mathbf{b}+\mathbf{c}); \ T_{w\_off} = \frac{T_p}{2}(3-\mathbf{a})$$

$$(2.14)$$

2. Sector 2:

$$\text{Quadrant 1:} T_{u\_on} = \frac{T_p}{2}(1-\mathbf{a}-\mathbf{b}); \ T_{u\_off} = \frac{T_p}{2}(3+\mathbf{a}+\mathbf{b})$$

$$\text{Quadrant 2:} T_{u\_on} = \frac{T_p}{2}(1+\mathbf{a}+\mathbf{b}); \ T_{u\_off} = \frac{T_p}{2}(3-\mathbf{a}-\mathbf{b})$$

$$(2.15)$$

$$T_{v\_on} = \frac{T_p}{2}(1-\mathbf{c}); \ T_{w\_on} = \frac{T_p}{2}(1+\mathbf{c})$$

$$T_{v\_off} = \frac{T_p}{2}(3+\mathbf{c}); \ T_{w\_off} = \frac{T_p}{2}(3-\mathbf{c})$$

3. Sector 3:

$$T_{u\_on} = \frac{T_p}{2}(1+\mathbf{a}); \ T_{v\_on} = \frac{T_p}{2}(1-\mathbf{a}); \ T_{w\_on} = \frac{T_p}{2}(1-\mathbf{b}+\mathbf{c})$$

$$T_{u\_off} = \frac{T_p}{2}(3-\mathbf{a}); \ T_{v\_off} = \frac{T_p}{2}(3+\mathbf{a}); \ T_{w\_off} = \frac{T_p}{2}(3+\mathbf{b}-\mathbf{c})$$

$$(2.16)$$

4. Sector 4:

$$T_{u\_on} = \frac{T_p}{2}(1+\mathbf{a}); \ T_{v\_on} = \frac{T_p}{2}(1-\mathbf{b}+\mathbf{c}); \ T_{w\_on} = \frac{T_p}{2}(1-\mathbf{a})$$

$$T_{u\_off} = \frac{T_p}{2}(3-\mathbf{a}); \ T_{v\_off} = \frac{T_p}{2}(3+\mathbf{b}-\mathbf{c}); \ T_{w\_off} = \frac{T_p}{2}(3+\mathbf{a})$$

$$(2.17)$$

5. Sector 5:

$$\text{Quadrant 3: } T_{u\_on} = \frac{T_p}{2}(1 + \mathbf{a} + \mathbf{b}); \ T_{u\_off} = \frac{T_p}{2}(3 - \mathbf{a} - \mathbf{b})$$

$$\text{Quadrant 4: } T_{u\_on} = \frac{T_p}{2}(1 - \mathbf{a} - \mathbf{b}); \ T_{u\_off} = \frac{T_p}{2}(3 + \mathbf{a} + \mathbf{b})$$

$$(2.18)$$

$$T_{v\_on} = \frac{T_p}{2}(1 + \mathbf{c}); \ T_{w\_on} = \frac{T_p}{2}(1 - \mathbf{c})$$

$$T_{v\_off} = \frac{T_p}{2}(3 - \mathbf{c}); \ T_{w\_off} = \frac{T_p}{2}(3 + \mathbf{c})$$

6. Sector 6:

$$T_{u\_on} = \frac{T_p}{2}(1 - \mathbf{a}); \ T_{v\_on} = \frac{T_p}{2}(1 + \mathbf{a}); \ T_{w\_on} = \frac{T_p}{2}(1 + \mathbf{b} - \mathbf{c})$$

$$T_{u\_off} = \frac{T_p}{2}(3 + \mathbf{a}); \ T_{v\_off} = \frac{T_p}{2}(3 - \mathbf{a}); \ T_{w\_off} = \frac{T_p}{2}(3 - \mathbf{b} + \mathbf{c})$$

$$(2.19)$$

Using these equations provides an easily comprehensible realization of the space vector modulation following the control flow of the structogram of fig. 2.11. The afore mentioned limitation to the maximum voltage vector should be already carried out in the current controller because of the necessary feedback correction discussed later. The normalization of the voltage components to $2U_{DC}/3$ permits the calculation of the switching times independent of the motor nominal voltage.

2.4.2 Modulation with digital signal processor TMS 320C20/C25

Unlike the Siemens microcontroller the digital signal processor is not equipped with the intelligent Capture/Compare register unit, but provides a superior computing power instead. In principle, there are two possibilities for the realization of the space vector modulation.
1. Using additional hardware: The processor is extended by a latch-counter-unit providing the process interface to the inverter (refer to fig. 2.13a).
2. Without additional hardware: The internal processor timer is used to generate the pulse pattern.

Since the 2nd variant causes certain disadvantages, such as an inaccurate voltage realization, particularly at the sector boundaries as well as in the area of small stator voltage (important for the low speed region), the 1st variant (following the realization with microcontroller) is discussed first.

The figure 2.13a illustrates the hardware configuration. The figure 2.13b shows, representative of the complete vector space, the definition and respectively the assignment of the turn-on / turn-off times to the inverter legs.

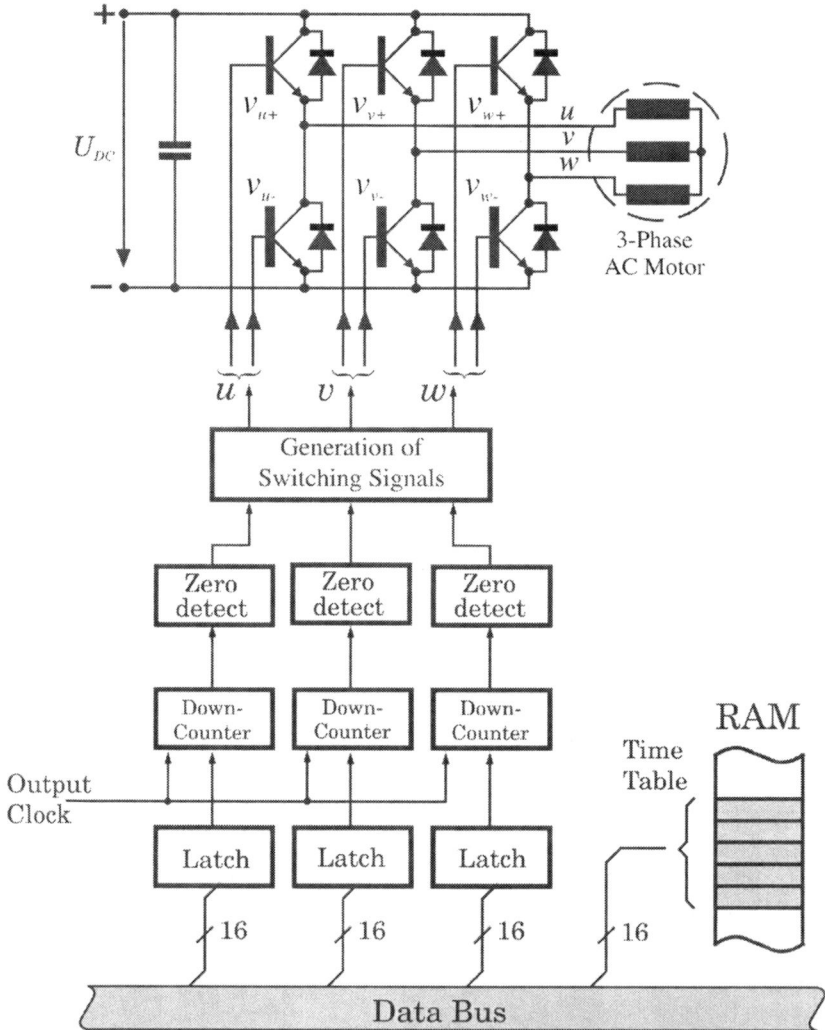

Fig. 2.13a Hardware configuration using DSP with external down-counter

According to the definition the switching times of the sampling period (k) can be calculated and stored in a RAM table. In the next period ($k+1$) they are output half-pulse wise. Computing and output of the switching times are processed in a time frame with the fixed period $T_p/2$, which is provided by either the internal timer of the signal processor or possibly also by the master-processor in the case of a multi-processor system. At the synchronization instants the switching times for the actual half pulse are automatically transferred from the latches to the down-counters, giving way to write the switching times for the next half pulse from the RAM table into the latch. Thus, output of the switching times independent of the interrupt reaction time is achieved, which results in a very precise voltage realization particularly in the area of small voltage values. After having been loaded with the switching times the counter starts to count backwards. Once the counter reading is zero, a zero detector will generate turn-on / turn-off pulses to control the inverter.

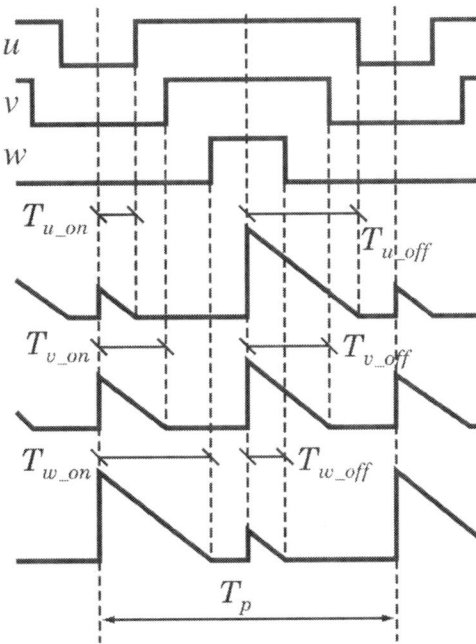

Fig. 2.13b Definition of the turn-on and turn-off times in the version DSP with additional hardware

According to the definition in the figure 2.13b, equation (2.2) and table 2.3 the switching times can be calculated as follows.
1. Sector 1:

$$T_{u\_on} = T_{w\_off} = \frac{T_p}{2}(1-\mathbf{a}); \; T_{u\_off} = T_{w\_on} = \frac{T_p}{2}(1+\mathbf{a})$$

$$T_{v\_on} = \frac{T_p}{2}(1+\mathbf{b}-\mathbf{c}); \; T_{v\_off} = \frac{T_p}{2}(1-\mathbf{b}+\mathbf{c})$$

(2.20)

2. Sector 2:

$$\text{Quadrant 1:} T_{u\_on} = \frac{T_p}{2}(1 - \mathbf{a} - \mathbf{b}); \ T_{u\_off} = \frac{T_p}{2}(1 + \mathbf{a} + \mathbf{b})$$

$$\text{Quadrant 2:} T_{u\_on} = \frac{T_p}{2}(1 + \mathbf{a} + \mathbf{b}); \ T_{u\_off} = \frac{T_p}{2}(1 - \mathbf{a} - \mathbf{b}) \quad (2.21)$$

$$T_{v\_on} = T_{w\_off} = \frac{T_p}{2}(1 - \mathbf{c}); \ T_{v\_off} = T_{w\_on} = \frac{T_p}{2}(1 + \mathbf{c})$$

3. Sector 3:

$$T_{u\_on} = T_{v\_off} = \frac{T_p}{2}(1 + \mathbf{a}); \ T_{u\_off} = T_{v\_on} = \frac{T_p}{2}(1 - \mathbf{a})$$

$$T_{w\_on} = \frac{T_p}{2}(1 - \mathbf{b} + \mathbf{c}); \ T_{w\_off} = \frac{T_p}{2}(1 + \mathbf{b} - \mathbf{c}) \quad (2.22)$$

4. Sector 4:

$$T_{u\_on} = T_{w\_off} = \frac{T_p}{2}(1 + \mathbf{a}); \ T_{u\_off} = T_{w\_on} = \frac{T_p}{2}(1 - \mathbf{a})$$

$$T_{v\_on} = \frac{T_p}{2}(1 - \mathbf{b} + \mathbf{c}); \ T_{v\_off} = \frac{T_p}{2}(1 + \mathbf{b} - \mathbf{c}) \quad (2.23)$$

5. Sector 5:

$$\text{Quadrant 3:} T_{u\_on} = \frac{T_p}{2}(1 + \mathbf{a} + \mathbf{b}); \ T_{u\_off} = \frac{T_p}{2}(1 - \mathbf{a} - \mathbf{b})$$

$$\text{Quadrant 4:} T_{u\_on} = \frac{T_p}{2}(1 - \mathbf{a} - \mathbf{b}); \ T_{u\_off} = \frac{T_p}{2}(1 + \mathbf{a} + \mathbf{b}) \quad (2.24)$$

$$T_{v\_on} = T_{w\_off} = \frac{T_p}{2}(1 + \mathbf{c}); \ T_{v\_off} = T_{w\_on} = \frac{T_p}{2}(1 - \mathbf{c})$$

6. Sector 6:

$$T_{u\_on} = T_{v\_off} = \frac{T_p}{2}(1 - \mathbf{a}); \ T_{u\_off} = T_{v\_on} = \frac{T_p}{2}(1 + \mathbf{a})$$

$$T_{w\_on} = \frac{T_p}{2}(1 + \mathbf{b} - \mathbf{c}); \ T_{w\_off} = \frac{T_p}{2}(1 - \mathbf{b} + \mathbf{c}) \quad (2.25)$$

    The shown variant with additional hardware fulfils highest requirements regarding the precision of the voltage realization. The additional hardware costs are faced by a time resolution, which is practically limited only by the word length of the three counters and their clock frequency.

    With regard to a very exact and dynamic feedback control this solution has to be preferred to the one with microcontroller if one considers that the controller has a maximum time resolution of only 400 ns (with hardware expansion also 200 ns possible). This time resolution permits a voltage resolution of only 7 bits at a pulse frequency of 10 kHz (approx. 4 V/time increment) and a resolution of 8 bits at 5 kHz (approx. 2 V/time increment). This is a rather coarse resolution. In contrast to this, a time resolution of 50 ns corresponding to a voltage resolution of 10 bits (approx. 0.5 V/time increment) can easily be achieved, which requires just the use of counters with 10 bit word length and 10 MHz clock frequency. Another drawback of the microcontroller solution is due to the fact that the CAP/COM registers of the SAB 80C166 cannot be switched simultaneously because they are subject to a skew of 50 ns from register to register. This necessitates a hardware-based compensation to attain a high precision of the voltage realization. Such a compensation is particularly important at the sector boundaries as well as in the area of small voltages or small speeds.

For the DSP solution, the version without additional hardware offers itself as an alternative possibility. The switching times are generated using the only internal timer. The figure 2.14a shows the used hardware. The figure 2.14b shows the time frame, in which the switching time calculation as well as their output are processed. As familiar, the switching times are calculated and stored into a RAM table already in the period (k) for the following period ($k+1$). The difference, compared with the two previous solutions, consists in the switching times not being output to the inverter separately for every phase in the form of T_{on} and T_{off}, but in original form as T_r, T_l or $T_{0,7}$ together with the needed switching state. The respective switching state is sent as a 3 bit data block to a buffer latch ahead of the inverter which holds it for the complete period.

From figure 2.14b it becomes evident, that two information are relevant about the modulation: the switching time and the switching state. These information are determined using table 2.3 and the flow chart in the figure 2.11, depending on the sector the voltage vector is located in. The hold time of the switching state was fetched from the RAM table and loaded into the period register PRD before. The timer counts backwards and when reaching zero activates the automatic loading of the new time constant from the PRD into its own counter register. At the same time, it triggers an interrupt request Tint, which activates an interrupt routine for handing over the following switching state (pulse pattern) into the latch as well as reloading the PRD.

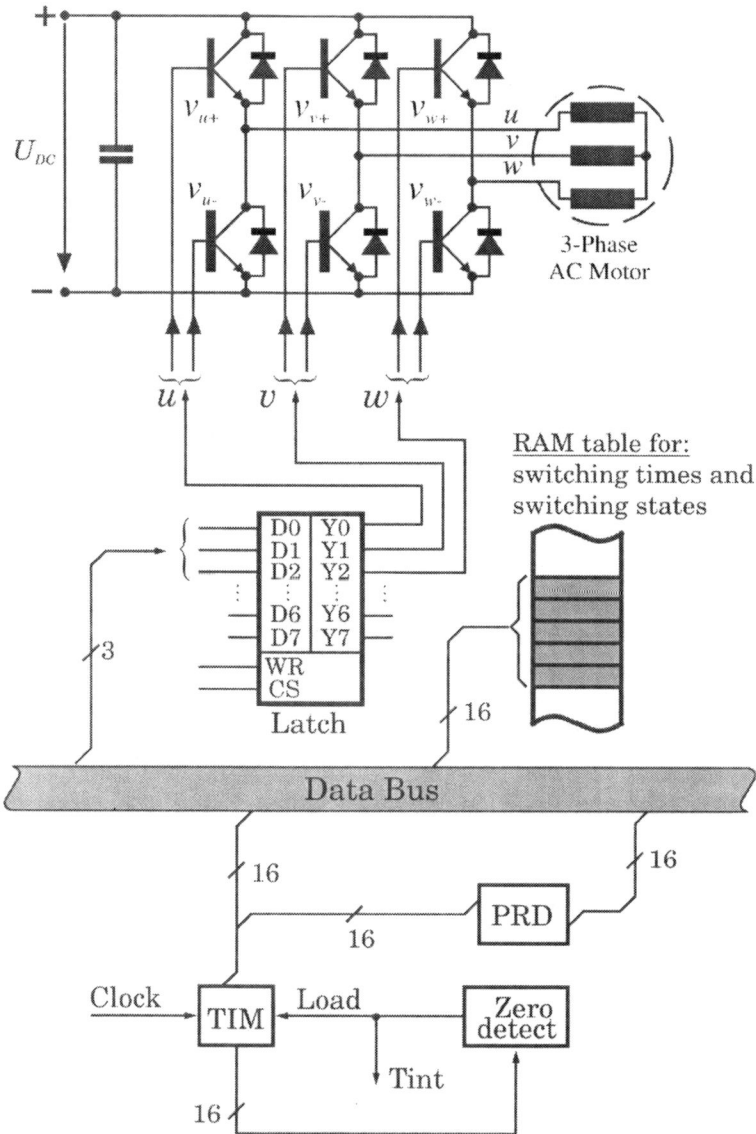

Fig. 2.14a Hardware configuration using DSP with its internal timer TIM

To output the switching states the following simple algorithm can be used. If again the figure 2.14b for the sector S_1 is viewed as an example the following assignment table can be composed.

Switching times	T_r	T_l	T_7	T_l	T_r	T_0
Switching states	100	110	111	110	100	000

The phases u, v and w are assigned to the data bits D0, D1 and D2. If the switching states of the above table are now written in reversed order

$$000 / 001 / 011 / 111 / 011 / 001,$$

a so-called control word (CW) results with CW=17D9h as a hexadecimal number. The control words for all six sectors can be summarized like in table 2.4.

Table 2.4 Control words of all sectors

Sectors	S_1	S_2	S_3	S_4	S_5	S_6
Control words	17D9h	27DAh	2DF2h	4DF4h	4BECh	1BE9h

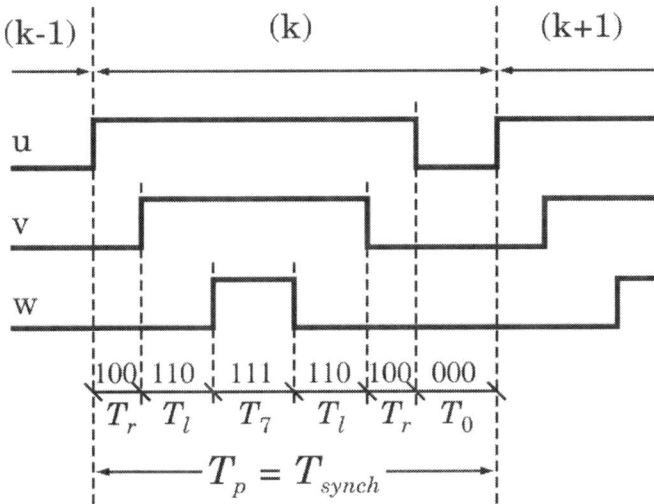

Fig. 2.14b Modulation time frame of the solution without additional hardware

The control word, corresponding to the determined sector is loaded by the interrupt routine from the memory into the accumulator, submitted to the latch, shifted three times to the right (to remove the switching state), and then stored back into the RAM. Every time after the control word has arrived in the accumulator, a zero test is carried out. The value zero indicates a new control word for the next sampling period. The described handling of the control words is illustrated again by the flow chart in figure 2.14c.

Some disadvantages of this method shall be mentioned now. The figure 2.14a is redrawn for two extreme cases:

1. the areas of small voltages at the sector boundaries, and
2. the area of voltage limitation (refer to fig. 2.14d).

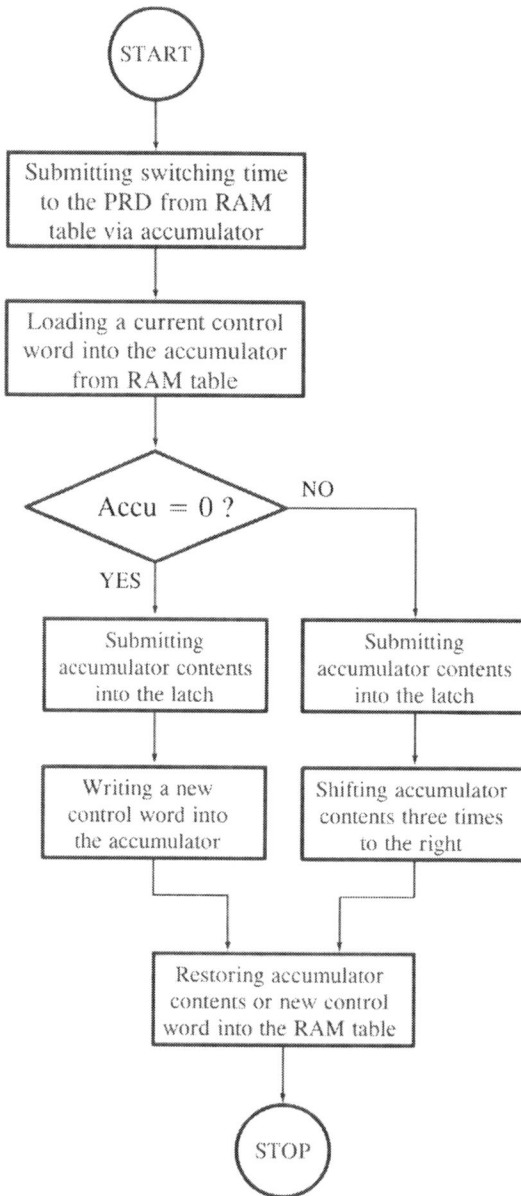

Fig. 2.14c Flow chart of the interrupt routine to output switching times and switching states

Changing the switching states by means of an interrupt routine reacting to T_{int} implies that the interval between two changes must be longer than the interrupt reaction time and respectively the run time of the interrupt routine itself. The figure 2.14d (refer to fig. 2.6) shows for sector S_1 that:

1. near the sector boundaries one of the two times T_r or T_l, and
2. in case of small voltages both times T_r and T_l

may fall below the reaction time of the interrupt routine. In the 1st case the boundary vector with the smaller switching time must be suppressed, and the second one will be realized for the whole period instead. This, of course, causes an inaccuracy of the voltage realization. In the 2nd case the voltage amplitude in the vicinity of zero is limited on the lower end, which has a negative effect on the speed control at small speeds.

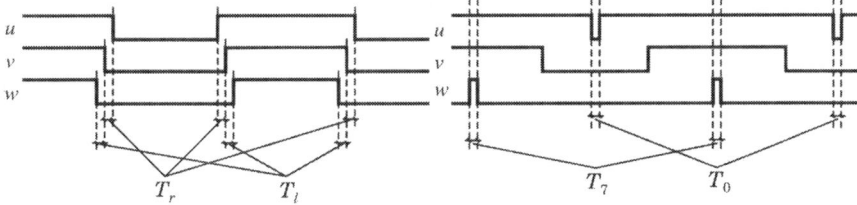

Fig. 2.14d Switching times at sector boundaries, in the area of small voltage (**left**) or of voltage at upper limits (**right**)

At large voltage amplitudes or during transients (magnetization, field-weakening, speed-up, speed reversal) the zero times T_0 and T_7 can become very small, and also fall below the reaction time of the interrupt routine (fig. 2.14d right). This means a limitation of the voltage amplitude on its upper end.

2.4.3 Modulation with double processor configuration

In this section a double processor system is introduced combining harmonically the strength of the digital signal processor TMS 320C25 – with respect to computing power – with the strength of the microcontroller SAB C167 – with respect to peripheries.

In this configuration the DSP is responsible for the processing of the near-motor control functions, and the µC has to process the tasks of the superposed control loops. The DSP allows to calculate the extensive real time algorithms, part of which is also the space vector modulation within a small sampling time of 100...200 µs. In every sampling period the DSP stores the newly calculated switching times into its own RAM, they are read from the µC memory driver using HOLD/HOLDA signals and submitted to the µC-internal PWM units. That means, with respect to the modulation the µC is only responsible for the output of the switching times and for the control of the transistor legs.

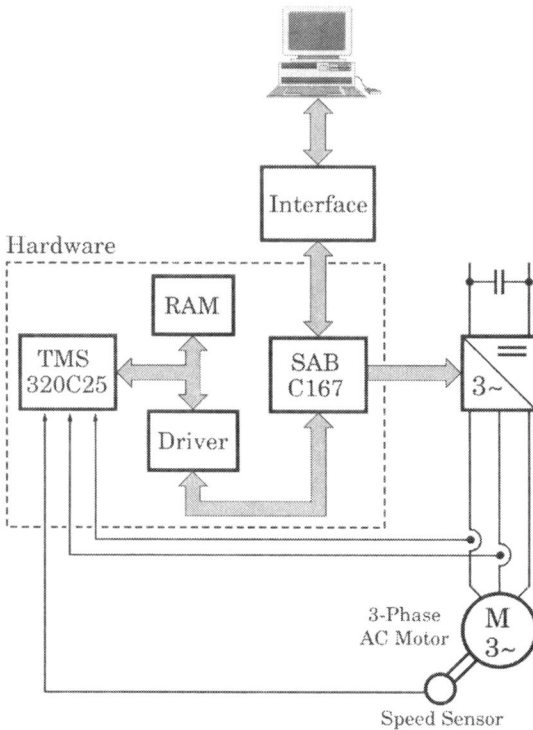

Fig. 2.15a Overview of the double processor configuration DSP - μC

The microcontroller SAB C167 contains, different to the earlier SAB 80C166, four timers PT0...PT3. In the symmetrical modulation mode these timers work as up/down counters. After every forward and the following backward counting process, when the counter content has reached the value zero, the timer/counter automatically receives the new maximum counter content from one of the four period registers PP0...PP3 for the new counting period. For the modulation only three registers of each category are needed (fig. 2.15b). It can be easily recognized that the three registers PP0, PP1 and PP2 have to get the same value simultaneously to realize the same counting or modulation periods. Furthermore it can be easily recognized that these three registers have to be initialized only once with the value $T_p=1/f_p$ because of the constant pulse frequency f_p.

In comparison with the SAB 80C166 the registers PT0, PT1, PT2 play the role of T0, and the registers PP0, PP1, PP2 the role of T0REL (refer to fig. 2.12a). The registers PW0, PW1 and PW2 generate the pulse widths. The assignment of the registers to the transistor legs is shown in the figure 2.15b.

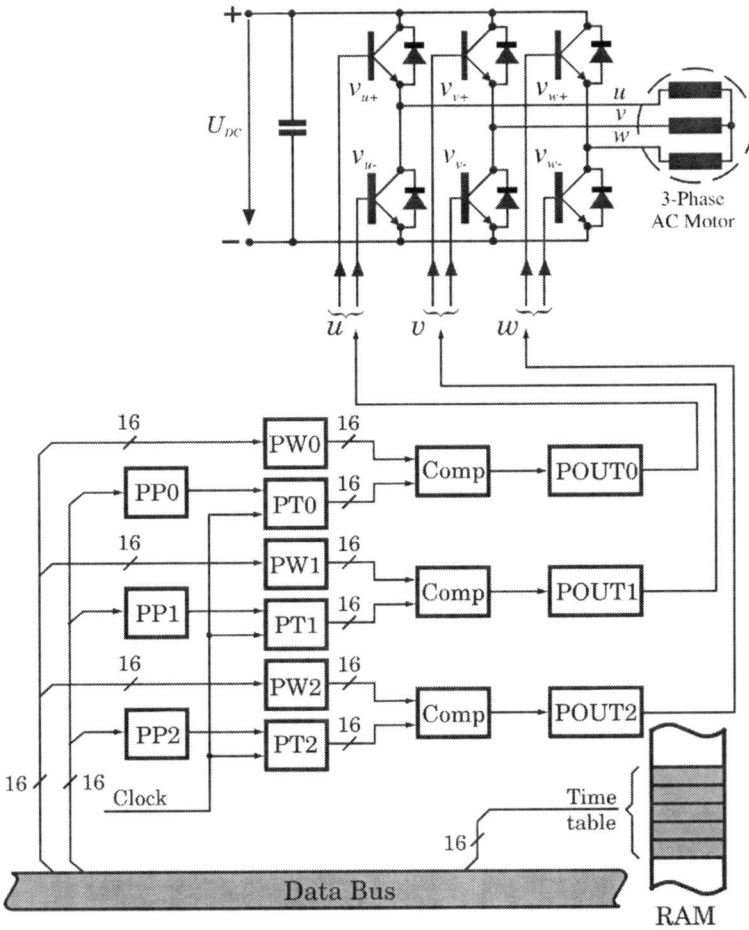

Fig. 2.15b Simplified structure of the modulation registers of the SAB C167

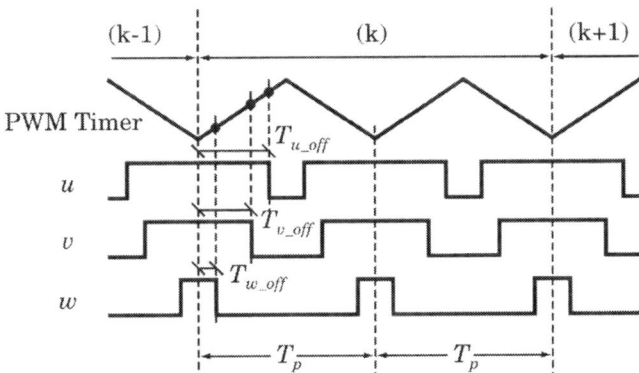

Fig. 2.15c Definition of the switching times for the structure in the figure 2.15b

While PT0, PT1 and PT2, which are represented as PWM timers in figure 2.15c, are counting forwards and backwards, their values are permanently compared with the contents of the corresponding pulse width registers PW0, PW1 and PW2. Respective compare-match events cause the output ports POUT0, POUT1, POUT2 to toggle and in due course the switchover of the corresponding inverter legs.

The figure 2.15c shows that the pulse width registers PW0, PW1 and PW2 have to be reloaded with new switching times, for turn-on and turn-off, only once per modulation period. The switching times can be calculated according to the definition in figure 2.15c as follows.

1. Sector 1:

$$T_u = \frac{T_p}{2}(1+\mathbf{a}); \ T_v = \frac{T_p}{2}(1-\mathbf{a}+2\mathbf{c}); \ T_w = \frac{T_p}{2}(1-\mathbf{a}) \tag{2.26}$$

2. Sector 2:

$$\text{Quadrant } 1: T_u = \frac{T_p}{2}(1+2\mathbf{b}+\mathbf{c}); \ \text{Quadrant } 2: T_u = \frac{T_p}{2}(1-2\mathbf{a}+\mathbf{c})$$

$$T_v = \frac{T_p}{2}(1+\mathbf{c}); \ T_w = \frac{T_p}{2}(1-\mathbf{c})$$

$$\tag{2.27}$$

3. Sector 3:

$$T_u = \frac{T_p}{2}(1-\mathbf{a}); \ T_v = \frac{T_p}{2}(1+\mathbf{a}); \ T_w = \frac{T_p}{2}(1+\mathbf{a}-2\mathbf{c}) \tag{2.28}$$

4. Sector 4:

$$T_u = \frac{T_p}{2}(1-\mathbf{a}); \ T_v = \frac{T_p}{2}(1+\mathbf{a}-2\mathbf{c}); \ T_w = \frac{T_p}{2}(1+\mathbf{a}) \tag{2.29}$$

5. Sector 5:

$$\text{Quadrant } 3: T_u = \frac{T_p}{2}(1-2\mathbf{a}+\mathbf{c}); \ \text{Quadrant } 4: T_u = \frac{T_p}{2}(1+2\mathbf{b}+\mathbf{c})$$

$$T_v = \frac{T_p}{2}(1-\mathbf{c}); \ T_w = \frac{T_p}{2}(1+\mathbf{c})$$

$$\tag{2.30}$$

6. Sector 6:

$$T_u = \frac{T_p}{2}(1+\mathbf{a}); \ T_v = \frac{T_p}{2}(1-\mathbf{a}); \ T_w = \frac{T_p}{2}(1-\mathbf{a}+2\mathbf{c}) \tag{2.31}$$

To complete the chapter of the realization examples the figure 2.16 shows the switching time plots, produced by the structure in figure 2.15b, at great voltages. This may be easily recognized by the fact that the switching times also show values near zero.

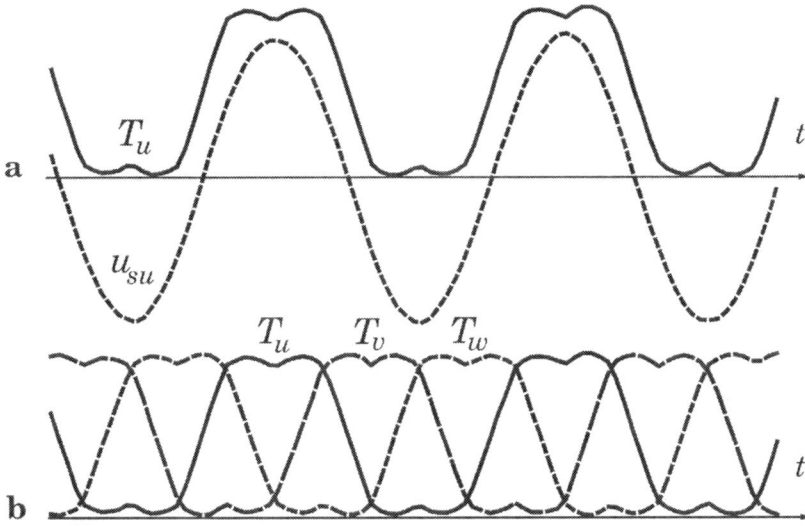

Fig. 2.16 The phase voltage u_{su} with the corresponding switching time T_u (**top**), and the switching times T_u, T_v, T_w (**bottom**)

2.5 Special modulation procedures

2.5.1 Modulation with two legs

Starting point for this section is figure 2.9, which represents the standard modulation in a stable time frame. The standard modulation realizes the same voltage vector, which is determined by the lengths of its boundary vector times T_r, T_l, twice per pulse period. For the purpose of comparison it is represented again for the sector S_1 in figure 2.17a.

We will try now to combine the zero times T_0, T_7 such that their sum is output either equally distributed at the ends (fig. 2.17b) or concentrated in the center (fig. 2.17c) of the pulse period. The times T_r, T_l or the voltage vector to be realized remain unchanged. With respect to the mean average value the two new sequences realize the same vector as in figure 2.17a.

It is obvious in the newly arisen sequences, that only two inverter legs are actually switched over. If this method, which will be called *modulation with two legs* from now on, is used consistently for the whole vector space, then *the switching losses automatically go down to approx. 2/3 of the original value.*

From the figures 2.17b,c it becomes obvious that either *the phase with the smallest pulse width* (for S_1: phase w) or *the phase with the smallest pause time* (for S_1: phase u) would be clamped to *negative potential* (the lower transistor of a phase leg is conducting) or to *positive potential* (the upper transistor is conducting). The formulae for the calculation of the switching times depend on the hardware and can be derived according to the definition from section 2.4.

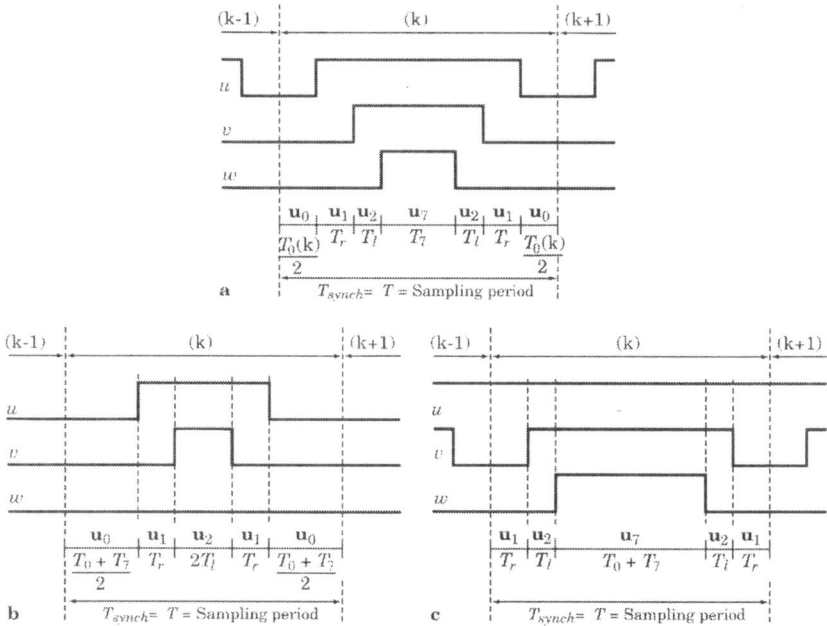

Fig. 2.17 Modulation with two legs for sector S_1

With the help of the firing pulse patterns in figures 2.4a...e, the suitable clamping phases or transistor legs can be found for all sectors and are summarized in the table in figure 2.18. For each sector two phases are available alternatively.

To obtain the same switching losses for all transistors, the upper transistor of one leg (corresponding phase on $+$) and then the lower transistor of the next leg (corresponding phase on $-$) are alternately switched on permanently for an angular range of 60^0. To switch-over the clamping to the next phase,

    1. either the sector boundaries (figure 2.18b),

    2. or the middle points of the sectors (figure 2.18c)

can be used. For all variants every transistor of the inverter conducts only for 60^0 per rotation of the voltage vector. With regard to the switching time calculation, which already requires a sector selection (refer to table 2.3), the version shown in figure 2.18b, seems to be more suitable for the practical implementation compared to the one in figure 2.18c.

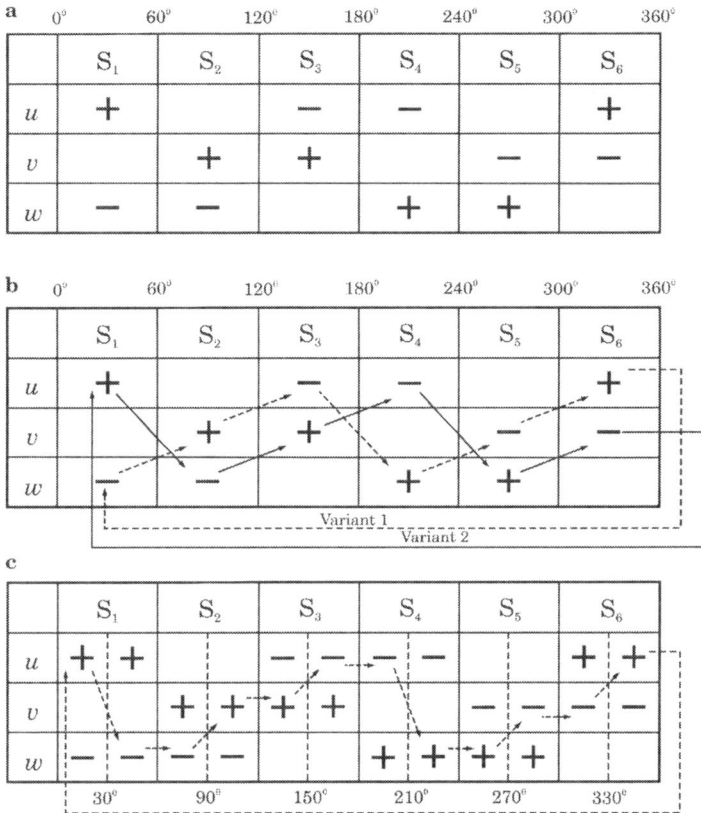

a

$0°$	$60°$	$120°$	$180°$	$240°$	$300°$	$360°$
	S_1	S_2	S_3	S_4	S_5	S_6
u	$+$		$-$	$-$		$+$
v		$+$	$+$		$-$	$-$
w	$-$	$-$		$+$	$+$	

b

$0°$	$60°$	$120°$	$180°$	$240°$	$300°$	$360°$
	S_1	S_2	S_3	S_4	S_5	S_6
u	$+$		$-$	$-$		$+$
v		$+$	$+$		$-$	$-$
w	$-$	$-$		$+$	$+$	

Variant 1
Variant 2

c

	S_1	S_2	S_3	S_4	S_5	S_6
u	$+$ $+$		$-$ $-$ $-$ $-$			$+$ $+$
v		$+$ $+$ $+$ $+$		$-$ $-$ $-$ $-$		
w	$-$ $-$	$-$ $-$		$+$ $+$ $+$ $+$		

$30°$	$90°$	$150°$	$210°$	$270°$	$330°$

Fig. 2.18 Possibilities of modulation with two legs: Possible phases for clamping per sector (**a**), and ways to switchover the clamped phases (**b, c**)

The advantage of the lower switching losses, however, is faced by considerably higher current harmonics, about twice the ripple amplitude has to be expected compared to the standard PWM algorithm.

2.5.2 Synchronous modulation

For the modulation algorithms discussed so far, it was always assumed that the pulse period T_p or the pulse frequency $f_p = 1/T_p$ is kept constant. However, since the fundamental frequency or the stator frequency f_s of the driven motor depends on the speed as well as on the load and is therefore variable, the relationship f_p/f_s is not constant. In this case one speaks of *asynchronous modulation*. The pulse period and the fundamental voltage period are not in any fixed relation.

This asynchronous characteristic causes subharmonics and losses as well as torque oscillations, which do not play an important role, as long as the relationship f_p/f_s is sufficiently large. The negative influence of the asynchronous characteristic may become a significant problem for high-speed drives (centrifuges, vacuum pumps etc.) in the speed range of 30000...60000 rpm. This problem can be avoided by keeping f_p and f_s in a fixed relationship.

$$N = \frac{f_p}{f_s} = const$$

$$T_p = \frac{1}{f_p} = \frac{1}{N f_s} = const$$

(2.32)

N is the number of the pulse periods per fundamental wave and may assume – because of the three-phase symmetry of the machines – only values, which fit the following relationship.

$$N = 9 + 6n \qquad n = 0, 1, 2, 3, ...$$
$$N = 9, 15, 21, 27, ...$$

(2.33)

In principle the modulation is processed in the same way as for the asynchronous algorithm, only, that the length of the pulse period T_p – depending on the working frequency f_s – must be recalculated permanently. It has to be taken into account for the practical implementation, that the value of the period register cannot be changed during the current pulse period, although the new value is already available after the recalculation is finished. This requires a *double buffering of the period register*. However, not every microcontroller will have the ability of double buffering. Regarding this feature the SAB C167 is very recommendable because the registers PP0, PP1, PP2 and PP3 are doubly buffered[1] by the so-called „*shadow register*".

The following problems must be taken into account for the application of the method:

1. Switching over of the pulse number N is carried out depending on the working (fundamental) frequency, and a *hysteresis – to prevent continuous to- and from-switching* – must be installed.

---

[1] Note: This ability is a further development of the SAB C167 in newer versions. The SAB C167 in the first version does not have double buffering for period registers.

2. Switching over of the pulse number N as well as switching over between asynchronous and synchronous modulation *must – to reduce transient effects – take place at the sector boundaries* where one of the phase voltages u_{su}, u_{sv} and u_{sw} reaches its peak value. At the sector boundaries the current harmonics pass through their zero crossings.

2.5.3 Stochastic modulation

In this chapter we shall take a closer look at the switching frequency harmonics produced by the modulation and discuss certain ways to take influence on their appearance. Typical spectra of inverter voltage and current for the standard modulation with fixed pulse width are shown in figure 2.19. Their shape depends on the modulation ratio $m = |\mathbf{u}_s|/u_{max}$ and in case of the current on the load characteristic.

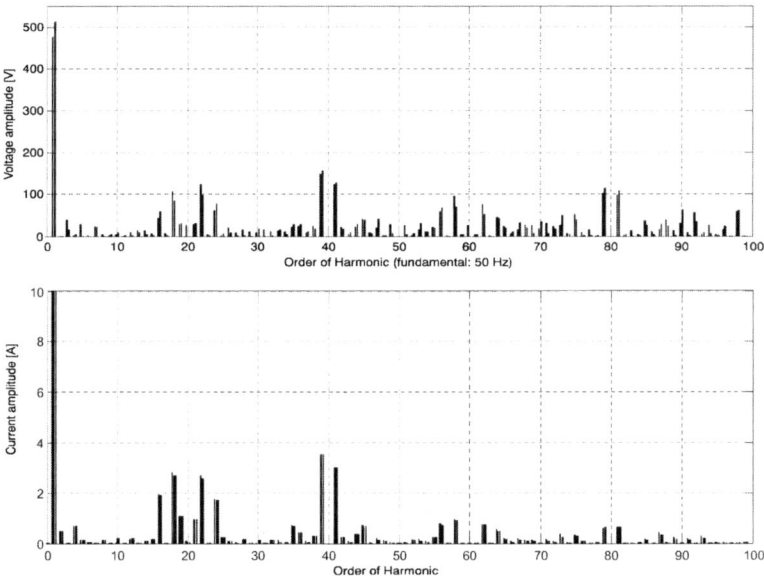

Fig. 2.19 Voltage (**top**) and current (**bottom**) spectra for standard modulation with m = 0.4 and pulse frequency = 1.0 kHz; current fundamental is truncated!

The spectra show pronounced maxima at the pulse frequency and its multiples with the overall maximum at the 2nd harmonic. Because of the low-pass characteristic of the load (R – L) harmonics beyond the 4th are suppressed in the current. Depending on the application and performance requirements, both positive and negative effects arise from this kind of spectrum:

- Below the switching frequency and its sidebands appear only low harmonic amplitudes and consequently their effect on ripple control frequencies in grid applications (active front-end converters) is negligible.
- The maximum harmonic current amplitudes are concentrated around two specific frequencies (1st and 2nd order), which facilitates filtering.
- Especially for grid applications, the maxima at 1st and 2nd switching frequency harmonic may exceed the limits specified in the applicable grid codes, which requires additional filtering for their suppression.
- The pronounced single-frequency harmonics produce noise which may be unwanted and experienced as disturbing in many environments.

To overcome the mentioned negative effects, it would in the first place be necessary to get rid of the pronounced 1st and 2nd harmonics and to obtain a more uniformly distributed spectrum. A straightforward solution could be to elude to control strategies with variable pulse period, such as bang-bang control, predictive control or direct torque/flux control. This is however outside the scope of this book, since we want to rely on the current control procedures to be discussed in the later chapters. So the question is how we can achieve a distributed spectrum while keeping a constant pulse period at the same time.

To derive respective procedures, it is first necessary to take a closer look on how the harmonic frequencies originate. Figure 2.20 shall help to do this. In both phase voltage and inverter control signals two repeating patterns may be identified:

1. The first pattern is formed by the ever repeating sequence of zero and active vectors … 0-R-L-7-7-L-R-0 … which appears with switching frequency and multiples of it.
2. The second one is formed by regular blocks of the active vectors R and L which are interrupted by zero vectors 0/7 with symmetric distribution within one period. This pattern is responsible for the especially strong 2nd harmonic in the spectrum and its multiples.

To shape a distributed spectrum, the regularity of these patterns has to be overcome, we have to "break the symmetries". Two methods shall be discussed to achieve this task.

The first approach, "sequence randomizing", breaks the first symmetry pattern by randomly changing the start vector of the pulse period between 0 and 7. This implies to add an additional simultaneous switchover of all three phase legs at the beginning of the pulse period. The start vector for each period is determined by a pseudo-random binary sequence (PRBS) which can easily be generated in a microcontroller. The resulting pulse patterns and spectra are shown in figures 2.21 and 2.22.

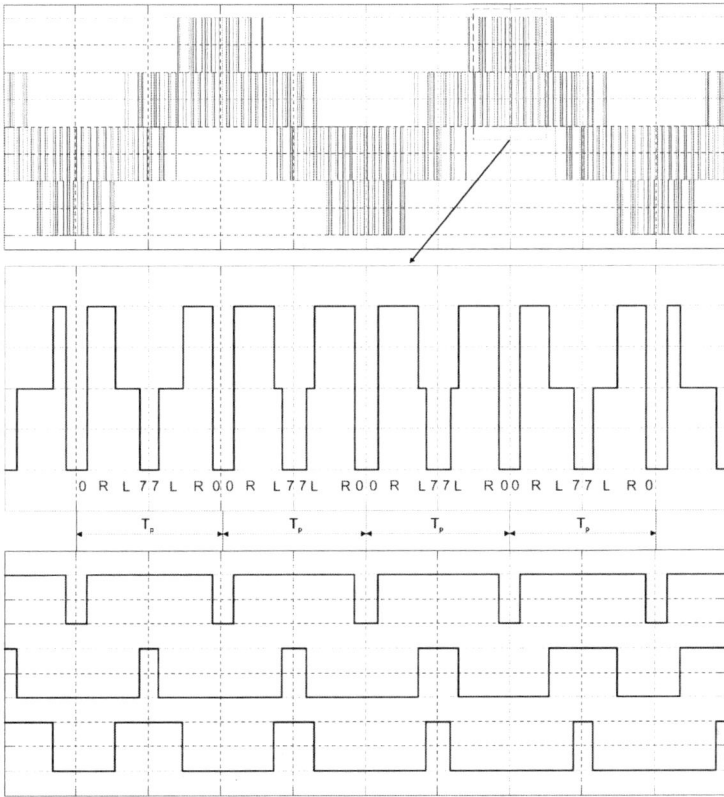

Fig. 2.20 Switching pattern of phase voltage (**top**, center) and phase control signals *u*/*v*/*w* (**bottom**) for standard modulation

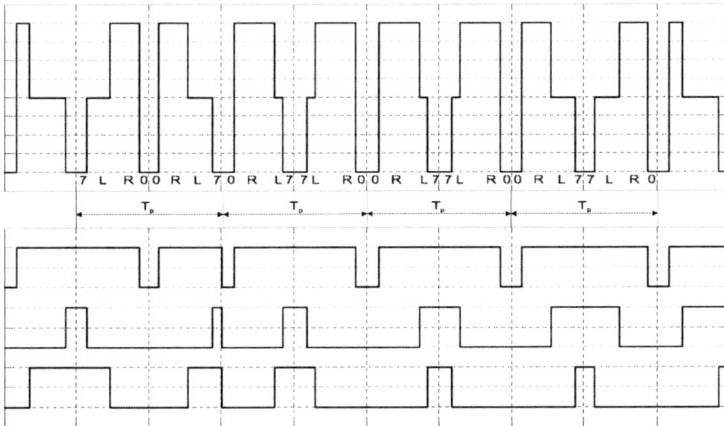

Fig. 2.21 Switching pattern of phase voltage (**top**) and phase control signals *u*/*v*/*w* (**bottom**) for modulation with sequence randomizing

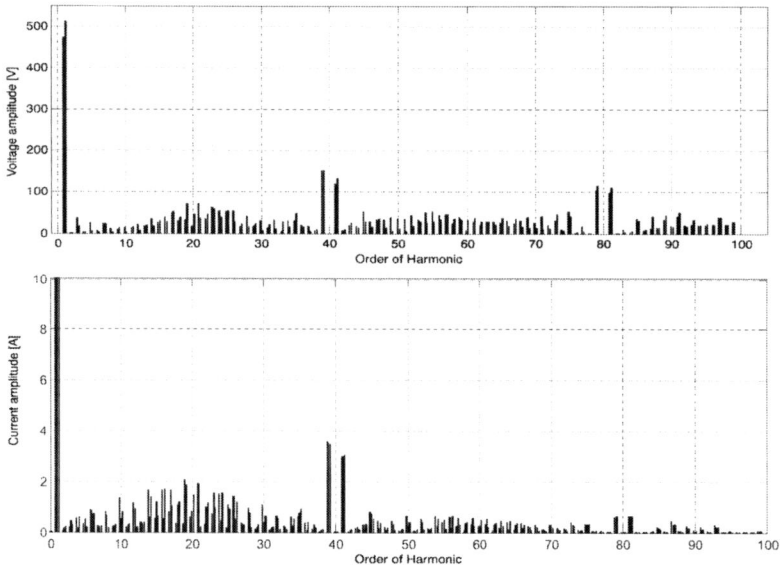

Fig. 2.22 Voltage (**top**) and current (**bottom**) spectra for modulation with sequence randomizing

The described change of the vector sequence occurs in the example between first and second pulse period in fig. 2.21. The peak value of the first harmonic is clearly reduced but, since nothing is changed on the zero vector lengths, the second symmetry pattern and therefore the second harmonic remain largely unaffected.

The 2nd harmonic is addressed with a different approach, which we will call "zero vector randomizing". The symmetrical distribution of u_0 and u_7 inside one period in the standard modulation scheme is dropped in favor of a randomly chosen ratio between both vectors, while keeping their symmetry with regard to the center of the pulse period. The latter is an important condition to maintain the coincidence between sampling instant of the phase current and the current fundamental (refer to chapter 4.1). With an uniformly distributed random number $r(k)$ where $0 \leq r(k) \leq 1$ and the original zero vector time T_{00}, the resulting zero vector times can be calculated from:

$$T_0 = r(k)T_{00}$$
$$T_7 = (1 - r(k))T_{00}$$

(2.34)

As it turns out in the practical implementation, the results become more impressive when the extremes of the $r(k)$ interval {0; 1} are stronger emphasized, i.e. $r(k)$ is calculated by:

$$r(k) = k_r(r_1(k) - 0.5) + 0.5$$

$$k_r = 2,3,...,8$$

$$0 \le r(k) \le 1$$

(2.35)

$0 \le r_1(k) \le 1$ an uniformly distributed random number

It must be mentioned, that the effectiveness of zero vector randomizing of course depends on the modulation ratio $m = |\mathbf{u}_s|/u_{max}$, since m determines the available space for the zero vector variation. Near the maximum voltage vector the effect will be minimal.

It must also be noted, that the total harmonic current, and therefore the total harmonic distortion (THD) value cannot essentially be changed by modifying the modulation scheme. Thus, reducing harmonics in one area of the spectrum inevitably will shift them to and increase them in another area.

Figures 2.23 and 2.24 again show resulting sample pulse patterns and spectra, both figures for combined sequence randomizing and zero vector randomizing and at the same operating point as in the figures above.

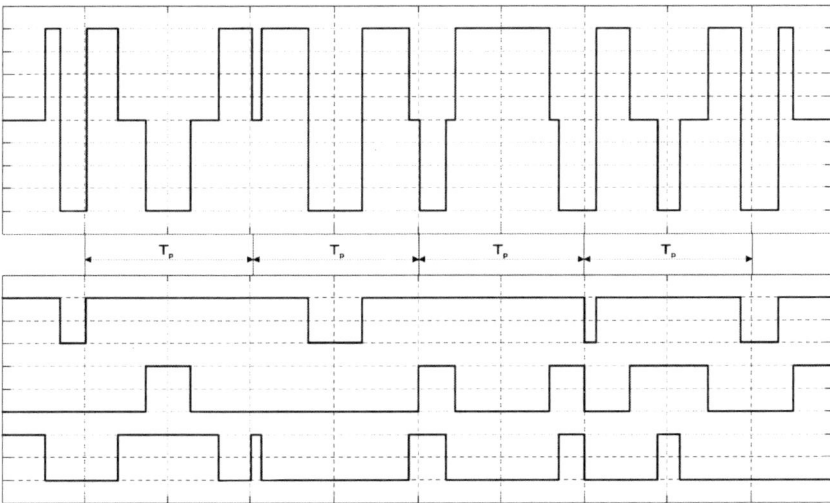

Fig. 2.23 Switching pattern of phase voltage (**top**) and phase control signals $u/v/w$ (**bottom**) for modulation with combined sequence randomizing and zero vector randomizing

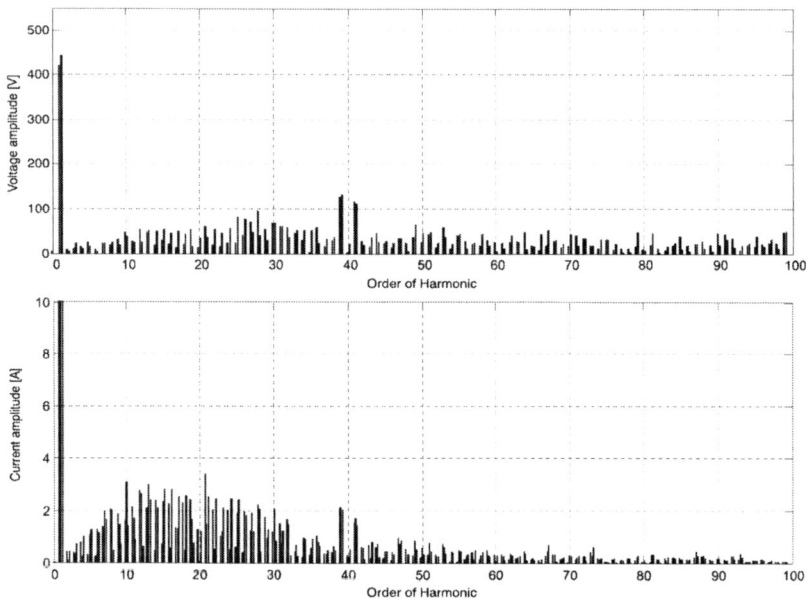

Fig. 2.24 Voltage (**top**) and current (**bottom**) spectra for modulation with combined sequence randomizing and zero vector randomizing

2.6 References

Choi JW, Yong SI, Sul SK (1994) Inverter output voltage synthesis using novel dead time compensation. Proc. IEEE Applied Power and Electronics Conf., pp. 100-106

Kiel E (1994) Anwendungsspezifische Schaltkreise in der Drehstrom-Antriebstechnik. Dissertation, TU Carolo-Wilhemina zu Braunschweig

Lipp A (1989) Beitrag zur Erarbeitung und Untersuchung von online Algorithmen für die digitale Ansteuerung von Pulsspannungswechselrichtern zur Drehzahlstellung von Drehstromasynchronmaschinen. Dissertation, TU Ilmenau

Mohan N, Undeland TM, Robbins WP (1995) Power electronics: Converters, Applications, and Design. 2nd Edition, John Wiley & Sons, Inc.

Pollmann A (1984) Ein Beitrag zur digitalen Pulsbreitenmodulation bei pulswechselrichtergespeisten Asynchronmaschinen. Dissertation, TU Carolo-Willhelmina zu Braunschweig

Quang NP (1991) Schnelle Drehmomenteinprägung in Drehstromstellantrieben. Dissertation, TU Dresden

Quang NP, Wirfs R (1995) Mehrgrößenregler löst PI-Regler ab: Ein Umrichterkonzept für Drehstromantriebe. Elektronik, H.7, S. 106-110

Rashid MH (2001) Power Electronics Handbook. Academic Press

Schröder D (1998) Elektrische Antriebe: Leistungselektronische Schaltungen. Springer Verlag, Berlin Heidelberg Paris London New York Tokyo

Siemens AG (1990) Microcomputer Components SAB 80C166 / 83C166. User's Manual

Siemens AG (2003) C167CR Derivatives 16-Bit Single-Chip Microcontrollers. User'Manual, V3.2

Siemens AG (2005) TC116x Series 32-Bit Single-Chip Microcontrollers. User'Manual, V1.2

Texas Instruments (1990) Digital Signal Processing Products TMS 320C2x. User's Guide

Texas Instruments (2004) TMS 320F2812 Digital Signal Processors. Data Manual

Yen-Shin L (1999) Sensorless Vector-Controlled IM Drives using Random Switching Technique, Proc. EPE '99 Lausanne

3 Machine models as a prerequisite to design the controllers and observers

3.1 General issues of state space representation

The mathematical modelling of the physical relations in 3-phase machines generally leads to differential equations of higher order and to state models with mutual coupling of the state variables respectively. For such systems the state space representation provides a very clear notation and a suitable starting point for the design of controllers, process models or observers.

Consistently, the equations to be derived in the following chapters will be predominantly based on the state space representation, making it sensible to introduce this chapter with some basic ideas. There the main focus will be on some important topics of the modelling of 3-phase machines such as time variance of the parameters and nonlinearity of the system equations, and their consequences for the discretization of the state equations.

3.1.1 Continuous state space representation

A time-continuous dynamic system can generally be represented in the following form:

$$\dot{\mathbf{x}}(t) = \mathbf{f}\big(\mathbf{x}(t), \mathbf{u}(t)\big); \quad \mathbf{x} \in \mathbf{R}^n; \ \mathbf{u} \in \mathbf{R}^m; \ \mathbf{x}_0 = \mathbf{x}(t_0)$$

$$\mathbf{y}(t) = \mathbf{h}\big(\mathbf{x}(t), \mathbf{u}(t)\big); \quad \mathbf{y} \in \mathbf{R}^p$$

(3.1)

In equation (3.1) $\mathbf{f}$ and $\mathbf{h}$ are general analytical vector functions of the state vector $\mathbf{x}$ and the input vector $\mathbf{u}$. The equation (3.1) describes a system of differential equations of first order, in which the system order n is equal to the number of contained independent energy storages. The system has n state, m input and p output quantities.

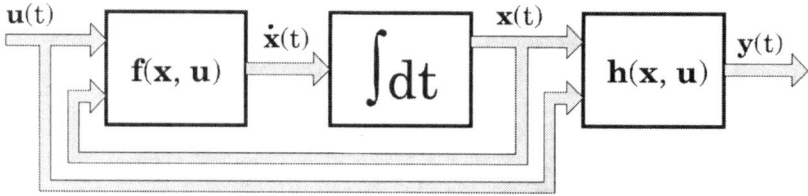

Fig. 3.1   Block circuit diagram of a system in state space representation

In many cases, a system description will not be required in the general form of equation (3.1), or for analysis and controller design a model must be found, which represents an adequately exact approximation of the physical conditions and is more accessible to the further processing. The usual way to achieve this is *the linearization of* (3.1) *along a (quasi-stationary) trajectory* (**X**(*t*), **U**(*t*)) *or around a stationary operating point* (**X**₀, **U**₀). After TAYLOR series expansion and truncation after the linear term, the following system is obtained:

$$\dot{\mathbf{x}}(t) = \mathbf{f}_{\mathbf{x}}\big(\mathbf{X}(t), \mathbf{U}(t)\big)\,\mathbf{x}(t) + \mathbf{f}_{\mathbf{u}}\big(\mathbf{X}(t), \mathbf{U}(t)\big)\,\mathbf{u}(t)$$
$$\mathbf{y}(t) = \mathbf{h}_{\mathbf{x}}\big(\mathbf{X}(t), \mathbf{U}(t)\big)\,\mathbf{x}(t) \tag{3.2}$$

Depending on the choice of the trajectory (**X**(*t*), **U**(*t*)), the operating point (**X**₀, **U**₀) or the degree of the linearization respectively the following special cases can be distinguished.

1. *Linear system with time-variant parameters*

The linearization is performed along the trajectory of a slowly variable quantity. Nonlinear combinations of state quantities are interpreted as products of a state quantity and a time variable parameter. Such a representation proffers itself primarily if products of state quantities with appropriately big differences of their eigendynamics appear. The equation system takes on the following form:

$$\dot{\mathbf{x}}(t) = \mathbf{A}(t)\,\mathbf{x}(t) + \mathbf{B}(t)\,\mathbf{u}(t); \quad \mathbf{x}_0 = \mathbf{x}(t_0); t \ge t_0$$
$$\mathbf{y}(t) = \mathbf{C}(t)\,\mathbf{x}(t) \tag{3.3}$$

In (3.3) **A** is the system matrix, **B** the input matrix and **C** the output matrix. Because no direct feed-through of the input to the output vector **y** exists in electrical drive control systems (*no step-change capability*), we will abstain from explicitly including this dependency in the following.

2. *Bilinear system*

If the transfer matrices are constant in time, and if a nonlinearity only exists regarding the control input, and not regarding the state vector, we speak of a bilinear system.

$$\dot{\mathbf{x}}(t) = \mathbf{A}\,\mathbf{x}(t) + \sum_{i=1}^{m}\mathbf{N}_i\,u_i(t)\mathbf{x}(t) + \mathbf{B}\,\mathbf{u}(t); \quad \mathbf{x}_0 = \mathbf{x}(t_0)$$

$$\mathbf{y}(t) = \mathbf{C}\,\mathbf{x}(t)$$

(3.4)

The multiplicative couplings between input and state quantities are summarized in the matrices $\mathbf{N}_i$.

3. *Linear system with constant parameters*

The class of the linear time-invariant systems finally represents the most simple case. The system equations are:

$$\dot{\mathbf{x}}(t) = \mathbf{A}\,\mathbf{x}(t) + \mathbf{B}\,\mathbf{u}(t); \quad \mathbf{x}_0 = \mathbf{x}(t_0) = \mathbf{x}(0)$$

$$\mathbf{y}(t) = \mathbf{C}\,\mathbf{x}(t)$$

(3.5)

3.1.2 Discontinuous state space representation

Control algorithms and models are processed in micro computers, and therefore in discrete time. The computer receives the system output quantity $\mathbf{y}(t)$ at definite equidistant points of time – e.g. after sampling and A/D conversion or U/f conversion and integration – as discrete quantity $\mathbf{y}(k)$. The calculated control variables are realized discontinuously as voltages by a PWM inverter. The complete control system represents a sampling system (fig. 3.2).

Because of the sampling operation of the computer, as a rule, a discrete design of the control system will be preferred. This is motivated firstly, because special phenomena caused by the sampling can specifically be considered in the design. Secondly, the application of special design methods particularly adopted to sampling operation, such as the dead beat design, will be possible. It is prerequisite that an equivalent time-discrete description can be found for the continuous system, which exactly reflects the dynamic behaviour of the continuous system at the sampling instants.

Unfortunately, for time-variant or nonlinear systems it will only in some rare cases be possible to find such an equivalent time-discrete system representation. The reasons will become clear at the derivation of the discontinuous state equations in later sections. Therefore, to design a discontinuous control system we can principally choose between the following two alternatives:

1. Controller design for the continuous system and then discrete implementation (*quasi-continuous design*).
2. Derivation of an approximated time-discrete process model and then *discrete controller design*.

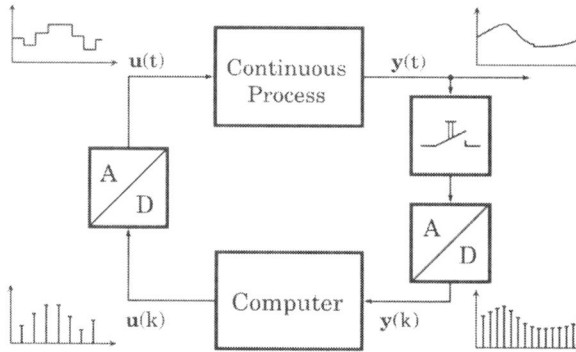

Fig. 3.2   Overview of a sampling system

In many cases it will not be required to take into account all or single nonlinearities of the system because certain approximations for the time-variant parameters and nonlinearities are possible and acceptable depending on the concrete application. In addition, because also in the linear case the discretization of the process model raises some important problems, the second way will be discussed in more detail in the following. The description of the continuous system shall be idealized as far as possible, to enable its discretization like for the linear time-invariant case.

For this purpose it is assumed that *the time-variant and state dependent parameters in equation* (3.3) are *constant within a sampling period,* therefore the sampling period has to be chosen sufficiently small. Thus equation (3.3) can be regarded piecewise linear and time-invariant for each sampling period, and the discretization of the continuous model is possible in a conventional way like for linear time-invariant systems. The discretization starts from the system equation (3.5) with the sampling period T presumed constant.

The discrete-time state model arises from the solution of the continuous state equation, yielding for the time-variant system (3.3) with continuous matrices $\mathbf{A}(t)$ and $\mathbf{B}(t)$:

$$\boldsymbol{\varphi}(t,\mathbf{x_0},t_0) = \boldsymbol{\Phi}(t,t_0)\mathbf{x_0} + \int_{t_0}^{t} \boldsymbol{\Phi}(t,\tau)\,\mathbf{B}(\tau)\,\mathbf{u}(\tau)\,d\tau, \quad t \geq t_0 \quad (3.6)$$

The matrix $\boldsymbol{\Phi}(t, t_0)$ describes the transition of the system from the state $\mathbf{x}(t_0) = \mathbf{x_0}$ to the state $\mathbf{x}(t)$ on the trajectory $\boldsymbol{\varphi}$, and is therefore called the fundamental matrix or transition matrix. The matrix $\boldsymbol{\Phi}$ fulfils the following matrix differential equation with the system matrix $\mathbf{A}(t)$:

$$d\Phi(t,t_0)/dt = \mathbf{A}(t)\Phi(t,t_0); \quad \Phi(t,t_0) = \mathbf{I} \tag{3.7}$$

For a constant system matrix $\mathbf{A}$ the fundamental matrix from equation (3.7) can be calculated analytically and represented as a matrix exponential function:

$$\Phi(t,t_0) = e^{\mathbf{A}(t-t_0)} \tag{3.8}$$

For the derivation of the discrete state equation the transient response between two sampling instants is of interest. That means equation (3.6) must be integrated over a sampling period T. With (3.8) and $t_0 = 0$ the following result is obtained:

$$x\big((k+1)T\big) = e^{\mathbf{A}T}\, x(kT) + \int_0^T e^{\mathbf{A}(T-\tau)}\, \mathbf{B}\,u(kT+\tau)\,d\tau \tag{3.9}$$

To comply with (3.8) and (3.9), for the discretization of a time-variant system, the system matrix $\mathbf{A}$ must be presumed constant over one sampling period, as already indicated above. The transition matrix $\Phi((k+1)T,\ kT)$ becomes the discrete system matrix $\Phi(k)$ and has to be recalculated online for every sampling period. Thus a time-variant discrete system is obtained. If one assumes further that the input vector $\mathbf{u}(t)$ is sampled by a zero-order hold function and therefore is also constant over one sampling period, $\mathbf{u}(t)$ may be extracted from the integral, and the complete system of state equations can be rewritten into the following matrix form:

$$x(k+1) = \Phi(k)\,x(k) + \mathbf{H}(k)\,\mathbf{u}(k); \quad x_0 = x(0); \quad k \geq 0$$
$$y(k) = \mathbf{C}(k)\,x(k) \tag{3.10}$$

The system matrix $\Phi(k)$ is defined by:

$$\Phi(k) = e^{\mathbf{A}(kT)T} \tag{3.11}$$

Because the input matrix $\mathbf{B}$ is constant, $\mathbf{B}$ can also be written outside the integral in (3.9). After substitution of the integration variable τ, the discrete input matrix $\mathbf{H}$ can be written as follows:

$$\mathbf{H} = \int_0^T e^{\mathbf{A}(kT)\tau}\, d\tau\, \mathbf{B} = \int_0^T \Phi(k)\big|_{T=\tau}\, d\tau\, \mathbf{B} \tag{3.12}$$

With regular $\mathbf{A}$, (3.12) can be solved further to:

$$\mathbf{H} = \mathbf{A}(kT)^{-1}\left[e^{\mathbf{A}(kT)T} - \mathbf{I}\right]\mathbf{B} \tag{3.13}$$

The output matrix $\mathbf{C}$ is identical to the continuous system. The system matrix $\Phi(k)$ is the decisive component of the discretization procedure. It determines the dynamics and stability of the discrete system. For its evaluation different methods are known, characterized by more or less calculation effort and higher or lower degree of approximation. Some of them, which are suitable for real time applications, shall be discussed in the following in more detail.

1. Series expansion

With this method, (3.11) is expanded directly into a power series.

$$\mathbf{\Phi} = e^{\mathbf{A}T} = \mathbf{I} + \mathbf{A}T + \frac{(\mathbf{A}T)^2}{2} + ... = \sum_{\nu=0}^{\infty} \frac{(\mathbf{A}T)^{\nu}}{\nu!} \tag{3.14}$$

After truncating the series expansion after the linear term, we obtain the solution for the Euler or RK1 procedure. This quite simple and easily comprehensible solution already suffices for many electrical drive applications at usual sampling times in the range 0.1 ... 1ms with respect to stability and precision. Because of possible numerical stability problems of the Euler procedure a more exact analysis is, however, appropriate.

The stability range of the Euler procedure in the continuous state plane is a circle with the radius $1/T$ and center at $-1/T$ on the real axis. Therefore all eigenvalues λ_i of the continuous system must hold to the following inequality:

$$\left| \lambda_i + \frac{1}{T} \right| < \frac{1}{T} \tag{3.15}$$

Particularly for complex frequency dependent eigenvalues of the system matrix $\mathbf{A}$ an exact check of this stability condition is required. Discretization-induced instabilities may be avoided by:

- Increasing the order of the series expansion of (3.14).
- Eluding to an integration method of higher order, e.g. RK4, which however, probably will be less feasible for real time applications.
- Avoiding complex eigenvalues of the system matrix $\mathbf{A}$ or its partial matrices.

For the latter variant the discretization of the state equations has to be first carried out in a coordinate system in which no frequency dependent eigenvalues of $\mathbf{A}$ or partial matrices $\mathbf{A}_{ii}$ appear. After that the discrete state equations are transformed into the final coordinate system (refer to example in the section 12.2). This procedure already yields decisive improvements for the Euler method. The use of suitable coordinate systems for the discretization can at the same time help to avoid errors, which result from the necessarily idealizing assumption of constant parameters of the system matrix $\mathbf{A}$ within a sampling period.

A similar approach would consist in transforming the input quantities of the partial system of interest into the respective natural coordinate system (without frequency dependent eigenvalues for the A_{ii}). In these coordinates all required calculations (model, controller and observer) would be processed, and then the output quantities would be transformed back into the original reference system.

2. Equivalent function

The matrix function $F(A) = e^{AT}$ is recreated by an equivalent polynomial function $R(A)$ with:

$$R(A) = \sum_{i=0}^{n-1} r_i A^i = e^{AT} \tag{3.16}$$

In this function, n is the order of the continuous system. This substitution is based on the Cayley Hamilton theorem, which states that every square matrix satisfies its own characteristic equation. As a consequence, it can be derived that every ($n{\times}n$) matrix function of order $p \geq n$, therefore also $p \to \infty$ like in (3.14), may be represented by a function of not more than $(n-1)^{th}$ order. The equivalent function (3.16) corresponds exactly to this statement.

With known factors r_i the system matrix Φ can be calculated from (3.16) whereby completely avoiding discretization errors, as in the case of truncated series expansion. For the calculation of the factors r_i the already mentioned property of (3.16) is used, that it is satisfied not only by the matrix A but also by the eigenvalues λ_j. This leads to the following linear system of equations:

$$\sum_{i=0}^{n-1} r_i \lambda_j^i = e^{\lambda_j T} \qquad (j = 1, 2 \dots n), \tag{3.17}$$

which holds at first for single eigenvalues. For p-fold eigenvalues ($p{>}1$) equation (3.17) is differentiated ($p-1$) times with respect to λ_j:

$$T e^{\lambda_j T} = \sum_{i=1}^{n-1} r_i \, i \, \lambda_j^{i-1}$$

$$\vdots \tag{3.18}$$

$$T^{p-1} e^{\lambda_j T} = \sum_{i=p-1}^{n-1} r_i \, i(i-1)\dots(i-p+2) \lambda_j^{i-p+1}$$

A second possibility for the calculation of r_i is offered by the *Sylvester-Lagrange equivalent polynomial method*. A minimal polynomial $M(\lambda)$ is defined, which is equal to the characteristic polynomial for the case of exclusively single eigenvalues of $\mathbf{A}$:

$$M(\lambda) = |\lambda \mathbf{I} - \mathbf{A}| = \prod_{i=1}^{n}(\lambda - \lambda_i) \tag{3.19}$$

In the case of multiple eigenvalues, $M(\lambda)$ contains only the eigenvalues different from each other with number $m<n$. Furthermore the following auxiliary functions are defined.

$$M_i(\lambda) = \frac{M(\lambda)}{\lambda - \lambda_i}; \quad i = 1, 2, ..., n \tag{3.20}$$

$$m_i = M_i(\lambda)\big|_{\lambda = \lambda_i}; \quad i = 1, 2, ..., n \tag{3.21}$$

With these, the substitute function $R(\lambda) = \mathbf{R}(\mathbf{A})\big|_{\mathbf{A}=\lambda}$

$$R(\lambda) = \sum_{i=1}^{n} \frac{e^{\lambda_i T}}{m_i} M_i(\lambda) \tag{3.22}$$

is finally calculated, from which the factors r_i are obtained by organizing after powers of λ. In the case of multiple eigenvalues, n is to be replaced by m in equations (3.19) to (3.22).

The previous explanations for the state space representation shall promote the understanding of the procedure for the later controller and observer design. The example in the section 12.2 (appendices) shall clarify the theoretical explanations.

As opposed to linear systems, no representation of an equivalent time-discrete system can be given for general nonlinear and time-variant systems. The bilinear systems (3.4) are an exception up to a certain point. The system

$$\dot{\mathbf{x}}(t) = \left[\mathbf{A} + \sum_{i=1}^{m} \mathbf{N}_i \, u_i(t) \right] \mathbf{x}(t) + \mathbf{B}\,\mathbf{u}(t); \quad \mathbf{x}_0 = \mathbf{x}(t_0)$$

$$\mathbf{y}(t) = \mathbf{C}\,\mathbf{x}(t) \tag{3.23}$$

can be integrated over T like an ordinary linear system under the prerequisite of the constancy of the control vector $\mathbf{u}$ during a sampling period. For the system matrices of the equivalent discrete system the following results are obtained:

$$\mathbf{\Phi}(k) = e^{\left(\mathbf{A} + \sum_{i=1}^{m} \mathbf{N}_i \, u_i(k)\right)T}; \quad \mathbf{H}(k) = \int_{0}^{T} e^{\left(\mathbf{A} + \sum_{i=1}^{m} \mathbf{N}_i \, u_i(k)\right)\tau} d\tau \, \mathbf{B} \tag{3.24}$$

However, this derivation also will have practical meaning only in special cases.

3.2 Induction machine with squirrel-cage rotor (IM)

As indicated in the previous section, the 3-phase AC machine can be described by a complicated system of higher order differential equations. To derive a machine or a system model, which allows a convenient handling from the control point of view, a series of simplifying assumptions must be met regarding the reproduction accuracy of constructive and electrical details (refer to chapter 6).

The reference axis for the field angle is the axis of the phase winding u and therefore the α axis of the stator-fixed coordinate system. The coordinate transformations (vector rotations for voltage output and current measurement) are assumed as well-known methods. The same applies to the inverter control by means of space vector modulation. These transfer blocks are regarded as error-free with respect to phase and amplitude, and will be considered negligible for the benefit of a clear control structure representation.

In this book, the three-phase machines will be represented using their state space models. In the classical, computer-based control structures the controller designs almost always were based on continuous state models. This approach does not suffice any more today. Therefore, in the first step the continuous state space models of the 3-phase AC machines shall be worked out in this section. Then the equivalent discrete state models will be derived to support the design of the discrete controllers.

The electrical quantities are represented as vectors with real components. As a reminder the important indices to be used shall be listed here.

a) *Superscript*:

	f	field synchronous (or field orientated, rotor flux / pole flux orientated) quantities
	s	stator-fixed quantities
	r	rotor-fixed (or rotor orientated) quantities

b) *Subscript*:

1st letter:	s	stator quantities
	r	rotor quantities
2nd letter:	d, q	field synchronous components
	α, β	stator-fixed components
c) *Letters in bold*:		vectors, matrices

3.2.1 Continuous state space models of the IM in stator-fixed and field-synchronous coordinate systems

Starting-point for all derivations are the stator and rotor voltage equations in their natural and easily comprehensible winding systems: The stator-fixed coordinate system, and the rotor-fixed coordinate system.

- *Stator voltage in the stator winding system:*

$$\mathbf{u}_s^s = R_s\,\mathbf{i}_s^s + \frac{d\psi_s^s}{dt} \tag{3.25}$$

R_s: Stator resistance; ψ_s^s : Stator flux vector

- *Rotor voltage in the short-circuited rotor winding system:*

$$\mathbf{u}_r^r = R_r\,\mathbf{i}_r^r + \frac{d\psi_r^r}{dt} = \mathbf{0} \tag{3.26}$$

R_r: Rotor resistance; ψ_r^r : Rotor flux vector, $\mathbf{0}$: Zero vector

- *Stator and rotor flux:*

$$\begin{cases} \psi_s = L_s\,\mathbf{i}_s + L_m\,\mathbf{i}_r \\ \psi_r = L_m\,\mathbf{i}_s + L_r\,\mathbf{i}_r \end{cases} \text{with} \begin{cases} L_s = L_m + L_{\sigma s} \\ L_r = L_m + L_{\sigma r} \end{cases} \tag{3.27}$$

L_m: Mutual inductance; L_s, L_r: Stator and rotor inductances

$L_{\sigma s}$, $L_{\sigma r}$: Leakage inductances on the side of the stator and rotor

Due to the mechanically symmetrical construction the inductances are equal in all Cartesian coordinate systems. Therefore the superscripts are dropped in equation (3.27). The mechanical equations also are part of the machine description.

- *Torque equation:*

$$m_M = \frac{3}{2}z_p\left(\psi_s \times \mathbf{i}_s\right) = -\frac{3}{2}z_p\left(\psi_r \times \mathbf{i}_r\right)^{[1]} \tag{3.28}$$

$$m_M = \frac{3}{2}z_p\,\mathrm{Im}\left\{\psi_s^*\,\mathbf{i}_s\right\} = -\frac{3}{2}z_p\,\mathrm{Im}\left\{\psi_s\,\mathbf{i}_s^*\right\}^{[2]} \tag{3.29}$$

- *Equation of motion:*

$$m_M = m_W + \frac{J}{z_p}\frac{d\omega}{dt} \tag{3.30}$$

m_M, m_W: Motor and load torque, z_p: Number of pole pair

J: Torque of inertia, ω: Mechanical angular velocity

---

[1] $\times$ cross product of vectors
[2] Im{ } Imaginary part of the term in brackets; * conjugated complex value

Now a coordinate system is introduced which rotates with angular frequency ω_k, as shown in section 1.1, and all quantities are transformed from the winding-coupled systems into the rotating one:

1. *Stator voltage equation*

After applying the transformation rules the following results are obtained:

$$\mathbf{u}_s^s = \mathbf{u}_s^k \, e^{j\vartheta_k}, \mathbf{i}_s^s = \mathbf{i}_s^k \, e^{j\vartheta_k}, \psi_s^s = \psi_s^k \, e^{j\vartheta_k}, \frac{d\psi_s^s}{dt} = \frac{d\psi_s^k}{dt} e^{j\vartheta_k} + j\omega_k \, \psi_s^k \, e^{j\vartheta_k}$$

Inserting the transformed quantities into equation (3.25), the equation (3.31) of the stator voltage in the new rotating system is obtained:

$$\mathbf{u}_s^k = R_s \, \mathbf{i}_s^k + \frac{d\psi_s^k}{dt} + j\omega_k \, \psi_s^k \qquad (3.31)$$

However, the voltage equation is not to be represented in an arbitrary system, but for special practically relevant cases: in the stator-fixed or in the field synchronous (field-orientated) systems. These representations are obtained by setting:

- $\omega_k = \omega_s$: Here ω_s is the angular velocity of the stator-side space vectors or the rotating rotor flux vector.

$$\mathbf{u}_s^f = R_s \, \mathbf{i}_s^f + \frac{d\psi_s^f}{dt} + j\omega_s \, \psi_s^f \qquad (3.32)$$

This coordinate system is chosen to lock the real or the *d*-axis of the system to the rotor flux (refer to section 1.2). Thus the cross component of the rotor flux becomes equal to zero. The axes of the system are denoted by *dq coordinates*.

- $\omega_k = 0$: This means, that the system is fixed in space, whereat the real axis or the α-axis of the coordinate system coincides with the axis of the phase winding *u*.

$$\mathbf{u}_s^s = R_s \, \mathbf{i}_s^s + \frac{d\psi_s^s}{dt} \qquad (3.33)$$

The axes of this stator-fixed coordinate system are denoted as $\alpha\beta$-coordinates. For the case $\omega_k = \omega$ (mechanical angular velocity or respectively motor speed) a rotor-orientated equation of the stator voltage can also be derived. However, since there is hardly any advantage to be obtained from this representation we will not follow it further.

2. *Rotor voltage equation*

The transformation rules are applied in similar way to the stator voltage equation.

$$\mathbf{i}_r^r = \mathbf{i}_r^k \, e^{j\vartheta_k} \,, \, \boldsymbol{\psi}_r^r = \boldsymbol{\psi}_r^k \, e^{j\vartheta_k} \,, \, \frac{d\boldsymbol{\psi}_r^r}{dt} = \frac{d\boldsymbol{\psi}_r^k}{dt} e^{j\vartheta_k} + j\omega_k \, \boldsymbol{\psi}_r^k \, e^{j\vartheta_k}$$

After inserting the transformed quantities into equation (3.26) the following result is obtained:

$$0 = R_r \, \mathbf{i}_r^k + \frac{d\boldsymbol{\psi}_r^k}{dt} + j\omega_k \, \boldsymbol{\psi}_r^k \tag{3.34}$$

The equation (3.34) can also be written for the field-orientated and stator-fixed coordinate systems.

- $\omega_k = \omega_s - \omega = \omega_r$: This coordinate system is rotating ahead of the rotor with angular velocity ω_r and coincides with the field synchronous coordinate system. Inserting ω_r into equation (3.34) yields:

$$0 = R_r \, \mathbf{i}_r^f + \frac{d\boldsymbol{\psi}_r^f}{dt} + j\omega_r \, \boldsymbol{\psi}_r^f \tag{3.35}$$

Equation (3.35) represents the rotor voltage in dq-coordinates.

- $\omega_k = -\omega$: Assuming the rotor to rotate with the mechanical angular velocity ω, this coordinate system turns with the same angular velocity in the opposite direction. Therefore, the coordinate system is fixed to the stator and can be chosen to coincide with the $\alpha\beta$-coordinates mentioned above.

$$0 = R_r \, \mathbf{i}_r^s + \frac{d\boldsymbol{\psi}_r^s}{dt} - j\omega \, \boldsymbol{\psi}_r^s \tag{3.36}$$

The equation (3.36) represents the rotor voltage equation in the stator-fixed, $\alpha\beta$-coordinates.

So far the transformation of all voltage equations from their original winding systems into the required dq- or $\alpha\beta$-coordinates is complete. With the equations (3.32), (3.33), (3.35) and (3.36) the starting point to derive the continuous state space models of the IM is reached.

3. *Continuous state space model of the IM in the stator-fixed coordinate system ($\alpha\beta$-coordinates)*

The equations (3.33) and (3.36) are combined into the following equation system:

$$\begin{cases} \mathbf{u}_s^s = R_s\,\mathbf{i}_s^s + \dfrac{d\mathbf{\psi}_s^s}{dt} \\[2mm] \mathbf{u}_r^s = R_r\,\mathbf{i}_r^s + \dfrac{d\mathbf{\psi}_r^s}{dt} - j\omega\,\mathbf{\psi}_r^s = \mathbf{0} \\[2mm] \mathbf{\psi}_s^s = L_s\,\mathbf{i}_s^s + L_m\,\mathbf{i}_r^s \\[2mm] \mathbf{\psi}_r^s = L_m\,\mathbf{i}_s^s + L_r\,\mathbf{i}_r^s \end{cases} \qquad (3.37)$$

Not all electrical quantities in the system (3.37) are actually of interest. These are e.g. the not measurable rotor current $\mathbf{i}_r^s$, or, depending on the viewpoint of the observer, also the stator flux $\mathbf{\psi}_s^s$. Therefore these quantities shall be eliminated from the equation system. From the two flux equations it follows:

$$\mathbf{i}_r^s = \frac{1}{L_r}\left(\mathbf{\psi}_r^s - L_m\,\mathbf{i}_s^s\right); \quad \mathbf{\psi}_s^s = L_s\,\mathbf{i}_s^s + \frac{L_m}{L_r}\left(\mathbf{\psi}_r^s - L_m\,\mathbf{i}_s^s\right)$$

Now $\mathbf{i}_r^s$ and $\mathbf{\psi}_s^s$ are substituted into the voltage equations (3.37) to yield:

$$\begin{cases} \mathbf{u}_s^s = R_s\,\mathbf{i}_s^s + \sigma L_s\dfrac{d\mathbf{i}_s^s}{dt} + \dfrac{L_m}{L_r}\dfrac{d\mathbf{\psi}_r^s}{dt} \\[3mm] \mathbf{0} = -\dfrac{L_m}{T_r}\mathbf{i}_s^s + \left(\dfrac{1}{T_r} - j\omega\right)\mathbf{\psi}_r^s + \dfrac{d\mathbf{\psi}_r^s}{dt} \end{cases} \qquad (3.38)$$

With: $\sigma = 1 - L_m^2/(L_s\,L_r)$ Total leakage factor

$\quad\quad T_s = L_s/R_s\,; T_r = L_r/R_r$ Stator, rotor time constants

After separating the real and imaginary components from (3.38) we finally obtain:

$$\begin{cases} \dfrac{di_{s\alpha}}{dt} = -\left(\dfrac{1}{\sigma T_s} + \dfrac{1-\sigma}{\sigma T_r}\right)i_{s\alpha} + \dfrac{1-\sigma}{\sigma T_r}\psi_{r\alpha}' + \dfrac{1-\sigma}{\sigma}\omega\psi_{r\beta}' + \dfrac{1}{\sigma L_s}u_{s\alpha} \\[3mm] \dfrac{di_{s\beta}}{dt} = -\left(\dfrac{1}{\sigma T_s} + \dfrac{1-\sigma}{\sigma T_r}\right)i_{s\beta} - \dfrac{1-\sigma}{\sigma}\omega\psi_{r\alpha}' + \dfrac{1-\sigma}{\sigma T_r}\psi_{r\beta}' + \dfrac{1}{\sigma L_s}u_{s\beta} \\[3mm] \dfrac{d\psi_{r\alpha}'}{dt} = \dfrac{1}{T_r}i_{s\alpha} - \dfrac{1}{T_r}\psi_{r\alpha}' - \omega\psi_{r\beta}' \\[3mm] \dfrac{d\psi_{r\beta}'}{dt} = \dfrac{1}{T_r}i_{s\beta} + \omega\psi_{r\alpha}' - \dfrac{1}{T_r}\psi_{r\beta}' \end{cases} \qquad (3.39)$$

With: $\psi_r^{s/} = \psi_r^s / L_m$ and $\psi_{r\alpha}^/ = \psi_{r\alpha}/L_m$; $\psi_{r\beta}^/ = \psi_{r\beta}/L_m$

To get the complete model of the IM the $\alpha\beta$-components of flux and current have to be inserted into the torque equation. The vector $\mathbf{i}_r^s$ is extracted from the last equation of the system (3.37) and substituted into equation (3.28).

$$m_M = \frac{3}{2} z_p \frac{L_m^2}{L_r} \left(\psi_{r\alpha}^/ i_{s\beta} - \psi_{r\beta}^/ i_{s\alpha} \right) \tag{3.40}$$

The equations (3.39), (3.40) can now be summarized to a complete continuous model of the IM. The figure 3.3 illustrates the block structure of this model.

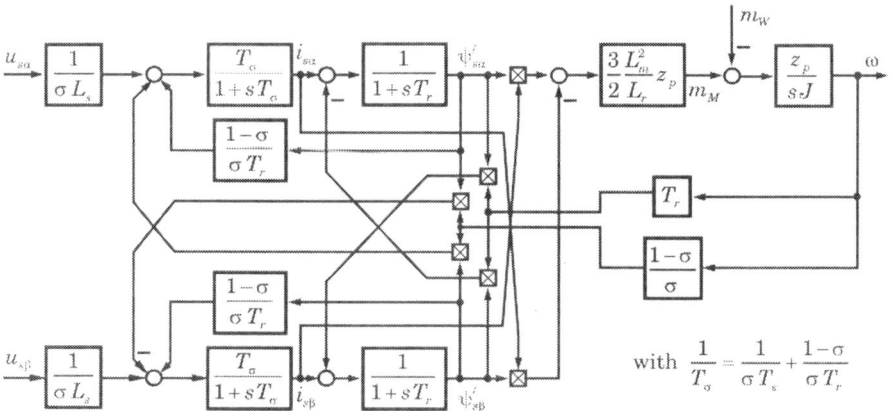

Fig. 3.3   Model of the IM with squirrel-cage rotor in stator-fixed coordinate system

The α- and β-components of stator voltage, stator current and rotor flux may be comprised in the following vectors with real components.

$$\mathbf{x}^{sT} = \left[i_{s\alpha}, i_{s\beta}, \psi_{r\alpha}^/, \psi_{r\beta}^/ \right]; \mathbf{u}_s^{sT} = \left[u_{s\alpha}, u_{s\beta} \right]$$

Superscript index T: Transposed vector

With the newly defined state vector $\mathbf{x}$ the continuous state space model of the IM with squirrel-cage rotor is finally obtained from the equations (3.39).

$$\frac{d\mathbf{x}^s}{dt} = \mathbf{A}^s \, \mathbf{x}^s + \mathbf{B}^s \, \mathbf{u}_s^s \tag{3.41}$$

$\mathbf{A}^s, \mathbf{B}^s$:   System and input matrix

$\mathbf{x}^s$:   State vector in stator-fixed coordinate system

$\mathbf{u}_s^s$:   Input vector in stator-fixed coordinate system

The equations (3.42) show in detail the matrices $\mathbf{A}^s$ and $\mathbf{B}^s$ with the machine parameters.

$$
\mathbf{A}^s = \left[
\begin{array}{cc|cc}
-\left(\dfrac{1}{\sigma T_s}+\dfrac{1-\sigma}{\sigma T_r}\right) & 0 & \dfrac{1-\sigma}{\sigma T_r} & \dfrac{1-\sigma}{\sigma}\omega \\[2ex]
0 & -\left(\dfrac{1}{\sigma T_s}+\dfrac{1-\sigma}{\sigma T_r}\right) & -\dfrac{1-\sigma}{\sigma}\omega & \dfrac{1-\sigma}{\sigma T_r} \\[2ex]
\dfrac{1}{T_r} & 0 & -\dfrac{1}{T_r} & -\omega \\[2ex]
0 & \dfrac{1}{T_r} & \omega & -\dfrac{1}{T_r}
\end{array}
\right]
;\quad
\mathbf{B}^s = \left[
\begin{array}{cc}
\dfrac{1}{\sigma L_s} & 0 \\[2ex]
0 & \dfrac{1}{\sigma L_s} \\[1ex]
0 & 0 \\[1ex]
0 & 0
\end{array}
\right]
$$

$$(3.42)$$

The equation (3.41) introduces a time-variant state system with the rotor speed ω as a measurable time-variant parameter in the system matrix $\mathbf{A}^s$. This continuous state model of the IM (figure 3.4) forms the basis for the design of discrete controllers in the stator-fixed coordinate system in which the components of the state vector $\mathbf{x}^s$ appear as sinusoidal quantities.

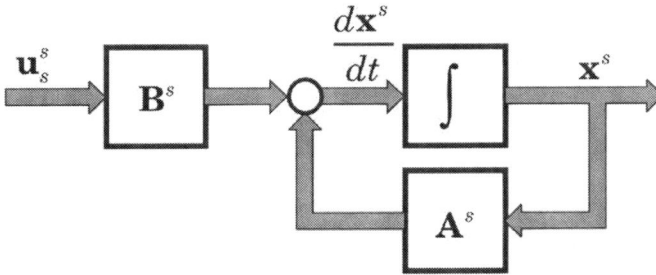

Fig. 3.4   Continuous state model of the IM in stator-fixed $\alpha\beta$-coordinates

4. *Continuous state space model of the IM in the field synchronous or field-orientated coordinate system (dq-coordinates)*:
   The equations (3.32), (3.35) are summarized in the following system.

$$
\left\{
\begin{aligned}
\mathbf{u}_s^f &= R_s\,\mathbf{i}_s^f + \frac{d\psi_s^f}{dt} + j\omega_s\,\psi_s^f \\[1ex]
0 &= R_r\,\mathbf{i}_r^f + \frac{d\psi_r^f}{dt} + j\omega_r\,\psi_r^f \\[1ex]
\psi_s^f &= L_s\,\mathbf{i}_s^f + L_m\,\mathbf{i}_r^f \\[1ex]
\psi_r^f &= L_m\,\mathbf{i}_s^f + L_r\,\mathbf{i}_r^f
\end{aligned}
\right.
\qquad (3.43)
$$

As in the case of the stator-fixed coordinate system the not measurable rotor current as well as the stator flux are eliminated.

$$\left\{ \begin{array}{l} \dfrac{di_{sd}}{dt} = -\left(\dfrac{1}{\sigma T_s} + \dfrac{1-\sigma}{\sigma T_r}\right) i_{sd} + \omega_s\, i_{sq} + \dfrac{1-\sigma}{\sigma T_r}\psi'_{rd} + \dfrac{1-\sigma}{\sigma}\omega\psi'_{rq} + \dfrac{1}{\sigma L_s}u_{sd} \\[3mm] \dfrac{di_{sq}}{dt} = -\omega_s\, i_{sd} - \left(\dfrac{1}{\sigma T_s} + \dfrac{1-\sigma}{\sigma T_r}\right) i_{sq} - \dfrac{1-\sigma}{\sigma}\omega\psi'_{rd} + \dfrac{1-\sigma}{\sigma T_r}\psi'_{rq} + \dfrac{1}{\sigma L_s}u_{sq} \\[3mm] \dfrac{d\psi'_{rd}}{dt} = \dfrac{1}{T_r}i_{sd} - \dfrac{1}{T_r}\psi'_{rd} + (\omega_s - \omega)\psi'_{rq} \\[3mm] \dfrac{d\psi'_{rq}}{dt} = \dfrac{1}{T_r}i_{sq} - (\omega_s - \omega)\psi'_{rd} - \dfrac{1}{T_r}\psi'_{rq} \end{array} \right.$$

$$(3.44)$$

Here are: $\psi'_{rd} = \psi_{rd}/L_m$; $\psi'_{rq} = \psi_{rq}/L_m$; $\omega_s - \omega = \omega_r$

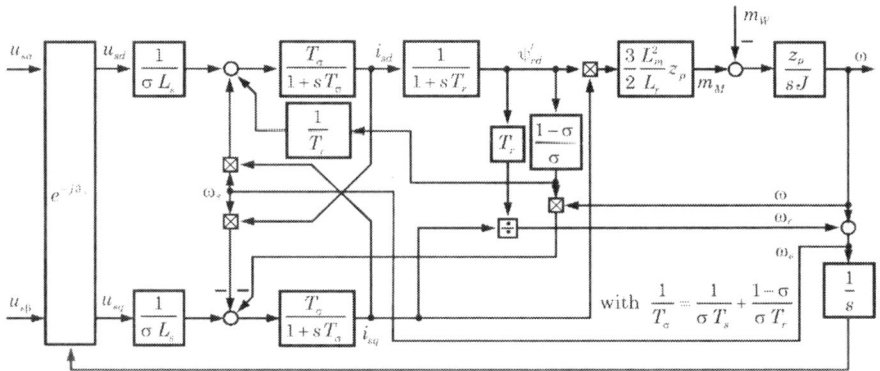

Fig. 3.5 Model of the IM with squirrel-cage rotor in field synchronous coordinate system

After extraction of i_r^f from equation (3.27), substituting into (3.28) or (3.29) and setting ψ_{rq} to zero due to fixing of the rotor flux vector to the real axis of the coordinate system, the equation (3.45) for the torque is arrived at:

$$m_M = \frac{3}{2}z_p \frac{L_m^2}{L_r}\psi'_{rd}\, i_{sq} = \frac{3}{2}z_p(1-\sigma)L_s\,\psi'_{rd}\, i_{sq} \qquad (3.45)$$

The equations (3.44) and (3.45) together form the complete, continuous model of the IM like shown in figure 3.5. The equation system (3.44) can be condensed into the following state space model:

$$\frac{dx^f}{dt} = A^f\, x^f + B^f\, u_s^f + N x^f\, \omega_s \qquad (3.46)$$

with the state vector x_f, the input vector u_s^f :

$$\mathbf{x}^{fT} = \left[i_{sd}, i_{sq}, \psi'_{rd}, \psi'_{rq} \right]; \mathbf{u}_s^{fT} = \left[u_{sd}, u_{sq} \right]$$

the system matrix $\mathbf{A}^f$, the input matrix $\mathbf{B}^f$ and the nonlinear coupling matrix $\mathbf{N}$:

$$\mathbf{A}^f = \begin{bmatrix} -\left(\dfrac{1}{\sigma T_s} + \dfrac{1-\sigma}{\sigma T_r} \right) & 0 & \dfrac{1-\sigma}{\sigma T_r} & \dfrac{1-\sigma}{\sigma}\omega \\[2mm] 0 & -\left(\dfrac{1}{\sigma T_s} + \dfrac{1-\sigma}{\sigma T_r} \right) & -\dfrac{1-\sigma}{\sigma}\omega & \dfrac{1-\sigma}{\sigma T_r} \\[2mm] \dfrac{1}{T_r} & 0 & -\dfrac{1}{T_r} & -\omega \\[2mm] 0 & \dfrac{1}{T_r} & \omega & -\dfrac{1}{T_r} \end{bmatrix}$$
(3.47)

$$\mathbf{B}^f = \begin{bmatrix} \dfrac{1}{\sigma L_s} & 0 \\[2mm] 0 & \dfrac{1}{\sigma L_s} \\[2mm] 0 & 0 \\[2mm] 0 & 0 \end{bmatrix}; \mathbf{N} = \begin{bmatrix} 0 & 1 & 0 & 0 \\ -1 & 0 & 0 & 0 \\ 0 & 0 & 0 & 1 \\ 0 & 0 & -1 & 0 \end{bmatrix}$$

The state equation (3.46) with the matrices (3.47) obviously points to *a bilinear characteristic* (refer to section 3.1.1, equation (3.4)). Here the field synchronous components u_{sd}, u_{sq} of the stator voltage and the angular velocity ω_s of the stator circuit represent the input quantities. The mechanical angular velocity ω in the system matrix $\mathbf{A}^f$ is regarded as a measurable variable system parameter. The only formal difference between the two continuous state models (3.41) and (3.46) is the nonlinear term with the matrix $\mathbf{N}$. The other matrices of both models are identical. The figure 3.6 illustrates the derived state model.

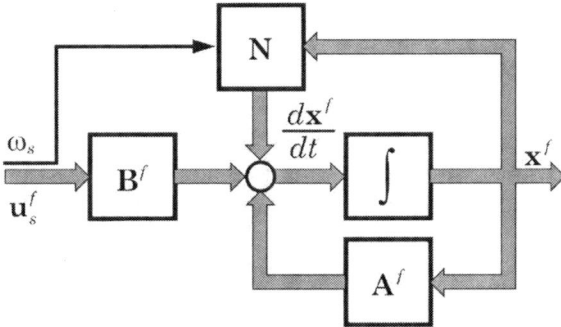

Fig. 3.6   Continuous state model of the IM in field synchronous dq-coordinates

So far the basic prerequisites for the further work are completed. However, for the controller design the continuous models are not particularly suitable. The microcomputer works discretely and processes only the motor quantities measured at discrete instants. A discrete model of the motor corresponding to this reality is therefore necessary for the controller design. The development of the discrete models is subject of the following section. It is useful to derive the models in the field synchronous as well as in the stator-fixed coordinate system because in practice control methods are developed in both coordinate systems.

3.2.2 Discrete state space models of the IM

Depending on the choice of the control coordinate system, the starting point for the derivation of a discrete state model for the IM is given by one of the two continuous state models (3.41) or (3.46).

In principle the discretization of the continuous model is relatively simple for linear and time-invariant systems. This presumption is fulfilled to a large degree if the IM model in the stator-fixed coordinate system is used and one assumes that the electrical transient processes settle essentially faster than the mechanical ones. Thus the system matrix $\mathbf{A}^s$ or the stator-fixed system (3.41) can be considered as virtually time-invariant within one sampling period of the current control. The mechanical angular velocity ω of the rotor can be regarded as a slowly variable parameter and is measured by a resolver or an incremental encoder.

This condition however is no longer fulfilled if the system is processed in the field synchronous coordinate system. The system model (3.46) indicates a *bilinear characteristic* additionally, the stator frequency ω_s consisting of the mechanical speed ω and the load ω_r leads to a time-variant system, further complicating the derivation of the required model. But under the prerequisite that the input quantities u_{sd}, u_{sq} and ω_s are constant within one sampling period T the discretization of this *bilinear and time-variant system* becomes feasible. The result is *a time-variant however linear system* which allows the application of a similar design methodology as for linear systems, like in stator-fixed coordinates. The demanded prerequisite is largely fulfilled for modern drive systems with sampling periods below 500µs. The pulsed stator voltage is processed as mean average over one period, and therefore also regarded constant in T.

1. *Discrete state model in the stator-fixed coordinate system*
    After integrating the equation (3.41) (refer to equations (3.10) to (3.14)) the following equivalent discrete state model of the IM is obtained.

$$\mathbf{x}^s\left(k+1\right)=\mathbf{\Phi}^s\,\mathbf{x}^s\left(k\right)+\mathbf{H}^s\,\mathbf{u}_s^s\left(k\right) \tag{3.48}$$

$$\mathbf{\Phi}^s = e^{\mathbf{A}^s T} = \sum_{\nu=0}^{\infty}\left(\mathbf{A}^s\right)^{\nu}\frac{T^{\nu}}{\nu!};\;\mathbf{H}^s = \int_{kT}^{(k+1)T} e^{\mathbf{A}^s \tau}\,d\tau\,\mathbf{B}^s = \sum_{\nu=1}^{\infty}\left(\mathbf{A}^s\right)^{\nu-1}\frac{T^{\nu}}{\nu!}\mathbf{B}^s$$

$$(3.49)$$

The input vector $\mathbf{u}_s^s$ (k) is given by the microcontroller and therefore has step-shaped components. The transition matrix $\mathbf{\Phi}_s$ and the input matrix $\mathbf{H}_s$ depend on the sampling period T and the mechanical angular velocity ω. The two matrices can be derived from the matrix exponential function e^{AT}, which may be developed into a series expansion like in (3.49). But for the practical application a further simplification would be very helpful and wished for. Here the consideration may help that the discrete model to be developed is not intended for mathematical simulation of the IM, but to serve the design of the discrete controller. For this purpose the series expansion may be truncated at an early stage if the inaccuracy hereby produced is compensated by appropriate control means, e.g. by an implicit integral part in the control algorithms.

The practical experience shows that an approximation of first order for $\mathbf{\Phi}_s$ and $\mathbf{H}_s$ suffices completely for small sampling times (under 500μs). An approximation of higher order would increase the needed computation power unnecessarily. A special issue is the investigation of the stability of such discrete systems. It shall only be mentioned at this place that the stability very strongly depends on the sampling time T. The smaller the sampling time T, the larger becomes the stability area and thus also the utilizable speed range. Therefore a compromise must be found between decreasing the sampling time and increasing the stability area as well as the speed range, and the acceptable computation power or the computing time. The following formulae (3.50) show the approximation of first order for transition and input matrix.

The representation of the discrete state models with partial matrices (figure 3.7) gives a good insight into the inner physical structure of the IM.

$$\mathbf{\Phi}^s = \left[\begin{array}{cc|cc} 1-\dfrac{T}{\sigma}\left(\dfrac{1}{T_s}+\dfrac{1-\sigma}{T_r}\right) & 0 & \dfrac{1-\sigma}{\sigma}\dfrac{T}{T_r} & \dfrac{1-\sigma}{\sigma}\omega T \\[2ex] 0 & 1-\dfrac{T}{\sigma}\left(\dfrac{1}{T_s}+\dfrac{1-\sigma}{T_r}\right) & -\dfrac{1-\sigma}{\sigma}\omega T & \dfrac{1-\sigma}{\sigma}\dfrac{T}{T_r} \\[2ex] \hline \dfrac{T}{T_r} & 0 & 1-\dfrac{T}{T_r} & -\omega T \\[2ex] 0 & \dfrac{T}{T_r} & \omega T & 1-\dfrac{T}{T_r} \end{array}\right] = \left[\begin{array}{c|c} \mathbf{\Phi}_{11}^s & \mathbf{\Phi}_{12}^s \\ \hline \mathbf{\Phi}_{21}^s & \mathbf{\Phi}_{22}^s \end{array}\right]$$

$$\mathbf{H}^s = \left[\begin{array}{cc} \dfrac{T}{\sigma L_s} & 0 \\[2ex] 0 & \dfrac{T}{\sigma L_s} \\[1ex] 0 & 0 \\ 0 & 0 \end{array}\right] = \left[\begin{array}{c} \mathbf{H}_1^s \\ \hline \mathbf{H}_2^s \end{array}\right]$$

(3.50)

The equation (3.48) can be written in detail as follows:

$$\begin{cases} \mathbf{i}_s^s(k+1) = \mathbf{\Phi}_{11}^s\,\mathbf{i}_s^s(k) + \mathbf{\Phi}_{12}^s\,\mathbf{\psi}_r^{s/}(k) + \mathbf{H}_1^s\,\mathbf{u}_s^s(k) \\[1ex] \mathbf{\psi}_r^{s/}(k+1) = \mathbf{\Phi}_{21}^s\,\mathbf{i}_s^s(k) + \mathbf{\Phi}_{22}^s\,\mathbf{\psi}_r^{s/}(k) \end{cases}$$

(3.51)

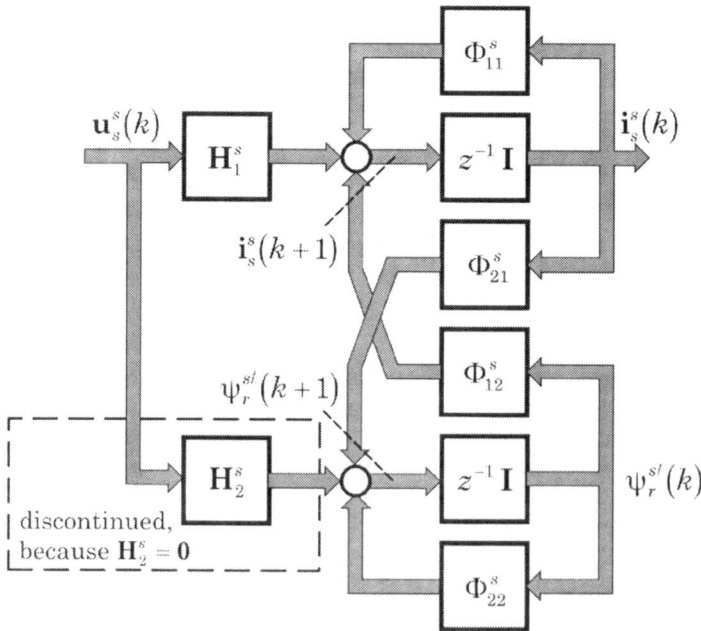

Fig. 3.7   Block structure of the state model of the IM in stator-fixed coordinates represented with partial matrices

With the separation of the complete model (3.48) into two submodels (3.51) a favourable starting point arises for the practical controller design. The first equation of (3.51) represents the current process model of the IM. The system has two input vectors: The stator voltage $\mathbf{u}_s^s(k)$ and the slowly variable rotor flux $\boldsymbol{\psi}_r^{s\prime}(k)$ (figure 3.8a).

In the chapter 5 it will be worked out, that the slowly variable rotor flux can be understood as a disturbance variable and therefore can be eliminated separately by a disturbance feed-forward compensation. The rotor flux is not measurable, it must be estimated. For this purpose the second equation of (3.51) may be used and is for this reason designated as i-ω flux model (figure 3.8b). From the measured currents and speed the rotor flux can be calculated using this model.

The special issue of the flux estimation has been treated in some detail in earlier works. Besides this simple flux model, different flux observers have been proposed for flux estimation (refer to section 4.4). The rotor flux estimates are used:

• to calculate the slip frequency ω_r or the field angle ϑ_s for the field orientation, and
• as actual flux values for the flux controller.

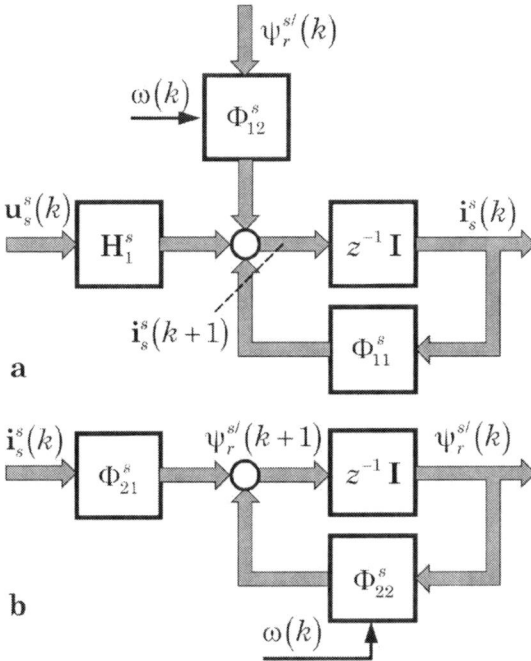

Fig. 3.8 Structure of the current process model (**a**) and the i-ω flux model (**b**) of the IM in stator-fixed coordinates

The structures in figure 3.8 have been derived by splitting-up the structure in the figure 3.7 with $\mathbf{H}_2^s = \mathbf{0}$.

2. Discrete state model in the field synchronous coordinate system

The derivation of a discrete state model or the discretization of the continuous bilinear state model (3.46) is carried out under the prerequisite that the input components u_{sd}, u_{sq} and ω_s are constant within a sampling period T. It was already indicated in the introduction of this section that this demand can be looked-at as largely fulfilled for modern three-phase AC drives with PWM inverters due to their high sampling and pulse frequencies.

After iterative integration of the equation (3.46) the following equivalent discrete state model of the IM is obtained.

$$\mathbf{x}^f\left(k+1\right) = \boldsymbol{\Phi}^f \, \mathbf{x}^f\left(k\right) + \mathbf{H}^f \, \mathbf{u}_s^f\left(k\right) \tag{3.52}$$

$$\boldsymbol{\Phi}^f = e^{\left[\mathbf{A}^f + \mathbf{N}\omega_s(k)\right]T} = \sum_{\nu=0}^{\infty}\left[\mathbf{A}^f + \mathbf{N}\omega_s\left(k\right)\right]^{\nu} \frac{T^{\nu}}{\nu!}$$

$$\mathbf{H}^f = \int_{kT}^{(k+1)T} e^{\left[\mathbf{A}^f + \mathbf{N}\omega_s(k)\right]\tau} \, d\tau \, \mathbf{B}^f = \sum_{\nu=1}^{\infty}\left[\mathbf{A}^f + \mathbf{N}\omega_s\left(k\right)\right]^{\nu-1} \frac{T^{\nu}}{\nu!} \mathbf{B}^f \tag{3.53}$$

The discrete model (3.52) is a time-variant, however linear model, unlike the continuous one. The elements of the transition matrix $\boldsymbol{\Phi}^f(k)$ and the input matrix $\mathbf{H}^f(k)$ are calculated on-line. Like for the discrete state model in the stator-fixed system, useable formulae are obtained by first-order approximation of the series expansions of the exponential functions in (3.53).

$$\boldsymbol{\Phi}^f = \begin{vmatrix} 1 - \dfrac{T}{\sigma}\left(\dfrac{1}{T_s} + \dfrac{1-\sigma}{T_r}\right) & \omega_s T & \dfrac{1-\sigma}{\sigma}\dfrac{T}{T_r} & \dfrac{1-\sigma}{\sigma}\omega T \\ -\omega_s T & 1 - \dfrac{T}{\sigma}\left(\dfrac{1}{T_s} + \dfrac{1-\sigma}{T_r}\right) & -\dfrac{1-\sigma}{\sigma}\omega T & \dfrac{1-\sigma}{\sigma}\dfrac{T}{T_r} \\ \dfrac{T}{T_r} & 0 & 1 - \dfrac{T}{T_r} & \left(\omega_s - \omega\right)T \\ 0 & \dfrac{T}{T_r} & -\left(\omega_s - \omega\right)T & 1 - \dfrac{T}{T_r} \end{vmatrix} = \begin{vmatrix} \boldsymbol{\Phi}_{11}^f & \boldsymbol{\Phi}_{12}^f \\ \boldsymbol{\Phi}_{21}^f & \boldsymbol{\Phi}_{22}^f \end{vmatrix}$$

$$\mathbf{H}^f = \begin{vmatrix} \dfrac{T}{\sigma L_s} & 0 \\ 0 & \dfrac{T}{\sigma L_s} \\ 0 & 0 \\ 0 & 0 \end{vmatrix} = \begin{vmatrix} \mathbf{H}_1^f \\ \mathbf{H}_2^f \end{vmatrix}$$

$$\tag{3.54}$$

After rewriting the discrete state model (3.52) in the form with partial matrices:

$$\begin{cases} \mathbf{i}_s^f\left(k+1\right)=\boldsymbol{\Phi}_{11}^f\,\mathbf{i}_s^f\left(k\right)+\boldsymbol{\Phi}_{12}^f\,\boldsymbol{\psi}_r^{f\prime}\left(k\right)+\mathbf{H}_1^f\,\mathbf{u}_s^f\left(k\right) \\ \boldsymbol{\psi}_r^{f\prime}\left(k+1\right)=\boldsymbol{\Phi}_{21}^f\,\mathbf{i}_s^f\left(k\right)+\boldsymbol{\Phi}_{22}^f\,\boldsymbol{\psi}_r^{f\prime}\left(k\right) \end{cases}$$

(3.55)

and considering, that $\mathbf{H}_2^f$ is a zero matrix, the current process model of the IM and the i-ω flux model for the field synchronous coordinate system are obtained like in equations (3.55) and in figure 3.10. The formal similarity of the two discrete state models of the IM, which is recognizable from the equations and from the pictures, can surely be noticed in the stator-fixed as well as in the field synchronous coordinate system. This formal similarity permits a generalization of both cases and their later summarizing into a common controller design.

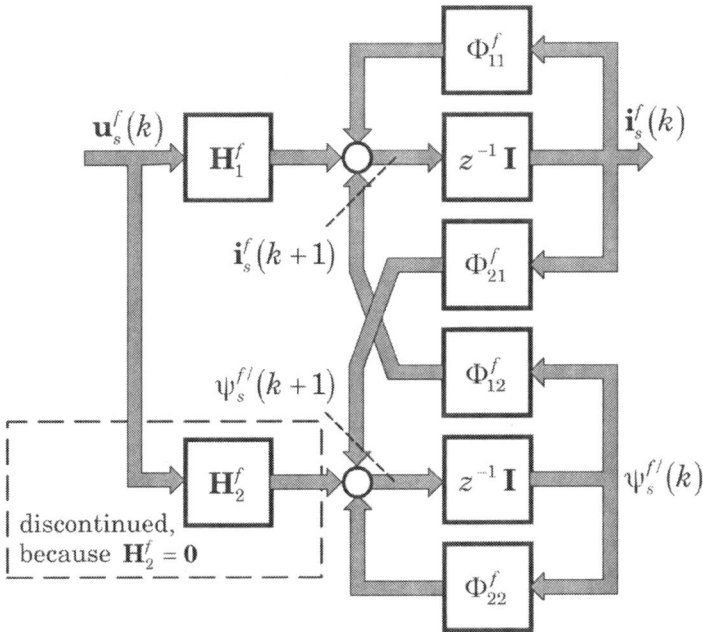

Fig. 3.9   Block structure of the state model of the IM in field synchronous coordinates represented with partial matrices

A decisive difference between the two models can be found in the appearance of ω_s in the transition matrix $\boldsymbol{\Phi}^f$, expressing the coupling between the two current components i_{sd} and i_{sq}. This coupling, as already mentioned in chapter 1, cannot be removed effectively, e.g. by using a decoupling network like indicated in the classical control structure in figure 1.4. This becomes particularly evident if the system is operated constantly with strong field weakening.

The discrete state model (3.52) of the IM in the *dq*-coordinate system was derived by discretization of the continuous model (3.46), which in turn was obtained by transformation from the original $\alpha\beta$-coordinate system into the *dq*-coordinate system, i.e. *the transformation took place before the discretization*.

Another order also may be chosen alternatively: *Discretization before transformation*; i.e. the discrete *dq*-model results from the coordinate transformation of the discrete $\alpha\beta$-model (3.48) (refer to sections 3.1.2 and 12.2). *This way complex eigenvalues of the system matrix or instabilities caused by discretization can be avoided.* In the result a discrete state model is obtained, which provides a larger stable working range for the controller. The parameters of the transition matrix $\mathbf{\Phi}$ will contain sin/cos functions of $\omega_s T$ (e.g. $\omega_s T \rightarrow \sin(\omega_s T)$). Especially for high-speed drives this procedure may yield significant advantages.

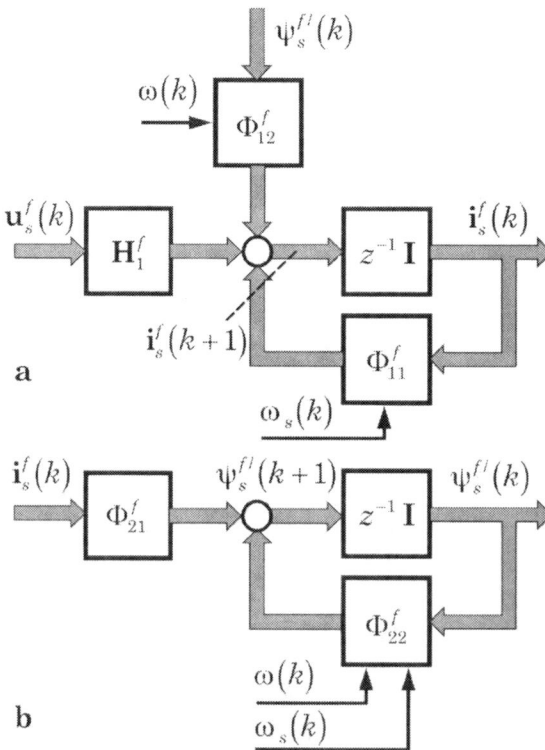

Fig. 3.10 Structure of the current process model (a) and of the i-ω flux model (b) of the IM in field synchronous coordinates

3.3 Permanent magnet excited synchronous machine (PMSM)

Unlike the IM the PMSM has a permanent and constant rotor flux (also pole flux) with a certain preferred axis. With a simultaneous use of a position sensor (resolver, incremental encoder with zero pulse) the pole position can always be clearly identified, and field orientation (also pole flux orientation) is always ensured. For this reason the system design in the stator-fixed coordinate system will be abstained from, and the field synchronous coordinate system will be immediately chosen for the treatment of the machine.

3.3.1 Continuous state space model of the PMSM in the field synchronous coordinate system

The equation (3.25) is the general stator voltage equation of three-phase AC machines, and valid also for the PMSM. A coordinate system rotating with ω or ω_s is conceivable whose axes are the d and q axis. For the PMSM ω and ω_s are identical which means, that the coordinate system rotates not only field synchronously, but is also fixed to the rotor. If the coordinate system is chosen to match the real d-axis with the preferred axis of the pole flux, this coordinate system represents the desired field or pole flux orientation. If the equation (3.25) of the PMSM (in a similar way as in the case of the IM) is transformed from the three winding system of the stator into the field synchronous system, we obtain:

$$\mathbf{u}_s^f = R_s \, \mathbf{i}_s^f + \frac{d\boldsymbol{\psi}_s^f}{dt} + j\omega_s \, \boldsymbol{\psi}_s^f \tag{3.56}$$

For the flux the following equation holds:

$$\boldsymbol{\psi}_s^f = L_s \, \mathbf{i}_s^f + \boldsymbol{\psi}_p^f \tag{3.57}$$

Here $\boldsymbol{\psi}_p^f$ is the vector of the pole flux. Because the real axis of the coordinate system is directly orientated to the preferred axis of the pole flux, the quadrature component of $\boldsymbol{\psi}_p^f$ is zero. Therefore the pole flux vector has only the real direct component $\boldsymbol{\psi}_p$. From that follows:

$$\boldsymbol{\psi}_p^f = \psi_{pd} + j\psi_{pq} = \psi_p \text{ with } \psi_{pq} = 0 \tag{3.58}$$

In addition it has to be taken into account that due to the construction dependent pole gaps on the rotor surface, the stator inductance assumes different values L_{sd}, L_{sq} in the real and quadrature axis, respectively. For

PMSM with cylindrical (non-salient) rotor both inductances are nearly identical and therefore usually equalized in classical control structures. The difference is not pronounced unlike in the case of salient-pole machines and to amounts approx. 3...12%. To obtain an effective decoupling between the current components i_{sd} and i_{sq}, this difference should be and will be taken into account in the following. Application to the stator flux equations thus yields:

$$\begin{cases} \psi_{sd} = L_{sd}\, i_{sd} + \psi_p \\ \psi_{sq} = L_{sq}\, i_{sq} \end{cases} \tag{3.59}$$

Substituting equations (3.57), (3.59) into the equation (3.56) then yields:

$$\begin{cases} u_{sd} = R_s\, i_{sd} + L_{sd}\dfrac{di_{sd}}{dt} - \omega_s L_{sq}\, i_{sq} \\ u_{sq} = R_s\, i_{sq} + L_{sq}\dfrac{di_{sq}}{dt} + \omega_s L_{sd}\, i_{sd} + \omega_s\, \psi_p \end{cases} \tag{3.60}$$

From the general torque equation (3.28) or (3.29) of three-phase AC machines we obtain:

$$m_M = \frac{3}{2} z_p \left(\psi_{sd}\, i_{sq} - \psi_{sq}\, i_{sd} \right) \tag{3.61}$$

After inserting (3.59) into (3.61) the following torque equation results:

$$m_M = \frac{3}{2} z_p \left[\psi_p\, i_{sq} + i_{sd}\, i_{sq} \left(L_{sd} - L_{sq} \right) \right] \tag{3.62}$$

The torque of the PMSM consists of two parts: the main and the reaction torque. With pole flux orientated control of the PMSM the stator current usually will be controlled to obtain a right angle between stator current and pole flux ($i_{sd} = 0$) and therefore not to contribute to magnetization, but only to torque production. Therefore a similar equation as (3.45) for the IM can be obtained:

$$m_M = \frac{3}{2} z_p\, \psi_p\, i_{sq} \tag{3.63}$$

Now equation (3.60) will be rewritten as follows:

$$\begin{cases} \dfrac{di_{sd}}{dt} = -\dfrac{1}{T_{sd}} i_{sd} + \omega_s \dfrac{L_{sq}}{L_{sd}} i_{sq} + \dfrac{1}{L_{sd}} u_{sd} \\ \dfrac{di_{sq}}{dt} = -\omega_s \dfrac{L_{sd}}{L_{sq}} i_{sd} - \dfrac{1}{T_{sq}} i_{sq} + \dfrac{1}{L_{sq}} u_{sq} - \omega_s \dfrac{\psi_p}{L_{sq}} \end{cases} \tag{3.64}$$

The PMSM is completely described by equations (3.62) and (3.64) in field synchronous coordinates (figure 3.11). The equations (3.62), (3.64) are summarized to the following state space model.

$$\frac{d\mathbf{i}_s^f}{dt} = \mathbf{A}_{SM}^f\,\mathbf{i}_s^f + \mathbf{B}_{SM}^f\,\mathbf{u}_s^f + \mathbf{N}_{SM}\,\mathbf{i}_s^f\,\omega_s + \mathbf{S}\,\psi_p\,\omega_s \qquad (3.65)$$

$$\mathbf{A}_{SM}^f = \begin{bmatrix} -\dfrac{1}{T_{sd}} & 0 \\ 0 & -\dfrac{1}{T_{sq}} \end{bmatrix};\ \mathbf{B}_{SM}^f = \begin{bmatrix} \dfrac{1}{L_{sd}} & 0 \\ 0 & \dfrac{1}{L_{sq}} \end{bmatrix};\ \mathbf{N}_{SM} = \begin{bmatrix} 0 & \dfrac{L_{sq}}{L_{sd}} \\ -\dfrac{L_{sd}}{L_{sq}} & 0 \end{bmatrix};\ \mathbf{S} = \begin{bmatrix} 0 \\ -\dfrac{1}{L_{sq}} \end{bmatrix}$$

$$(3.66)$$

$\mathbf{A}_{SM}^f =$	System matrix;	$\mathbf{S} =$	Disturbance vector
$\mathbf{B}_{SM}^f =$	Input matrix;	$T_{sd} = L_{sd}/R_s =$	Time constant of d axis
$\mathbf{N}_{SM} =$	Nonlinear coupling matrix;	$T_{sq} = L_{sq}/R_s =$	Time constant of q axis

The figure 3.12 illustrates the model (3.65) of the PMSM.

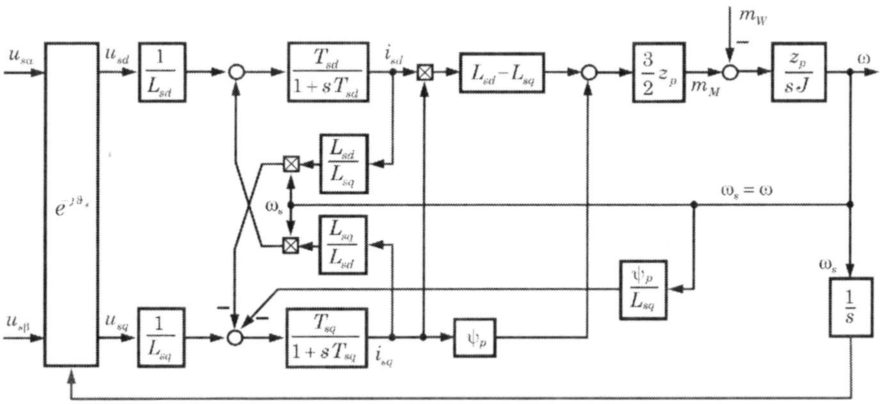

Fig. 3.11 Model of the PMSM in field synchronous or pole flux orientated coordinate system

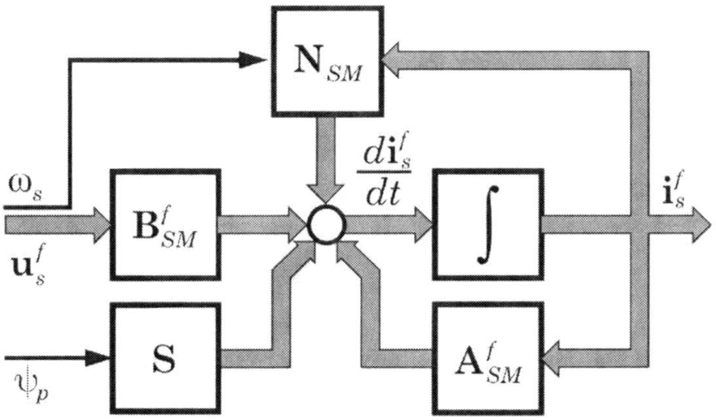

Fig. 3.12 Continuous state model of the PMSM in field synchronous coordinates

The bilinear characteristic of the model is recognizable like in the case of the IM by the matrix $\mathbf{N}_{SM}$. The disturbance, acting on the system through the pole flux ψ_p, does not depend on the stator current but is constant unlike for the IM. The constant excitation shows some advantages for the further treatment:

- The system model is a model of 2nd order (i_{sd}, i_{sq}) – the IM has a model of 4th order (i_{sd}, i_{sq}, ψ_{rd}, ψ_{rq} or $i_{s\alpha}$, $i_{s\beta}$, $\psi_{r\alpha}$, $\psi_{r\beta}$) – and immediately yields the current control process model. For the IM the system of 4th order must be split into partial models to obtain the current process model and the flux model.

- The constant flux ψ_p may be regarded as a system parameter.

- The constant disturbance ψ_p is documented by the machine manufacturer and can, similarly as for the IM, later be separately compensated by a disturbance feed-forward.

3.3.2 Discrete state model of the PMSM

To show clearly that the pole flux ψ_p represents only a constant disturbance variable, ψ_p was introduced into the system by a separate term through the disturbance vector in equation (3.65) and in figure 3.12. However, a discretization of the model is hardly possible in this form. To advance the situation, ψ_p will be viewed as a constant system parameter. The equation (3.65) must be rewritten as follows:

$$\frac{d\mathbf{i}_s^f}{dt} = \mathbf{A}_{SM}^f \mathbf{i}_s^f + \mathbf{B}_{SM}^{f*} \mathbf{v}^f + \mathbf{N}_{SM} \mathbf{i}_s^f \omega_s \tag{3.67}$$

with:

$$\mathbf{v}^{fT} = \left[u_{sd}, u_{sq}, \omega_s\right]; \quad \mathbf{B}_{SM}^{f*} = \left[\mathbf{B}_{SM}^f, \mathbf{S}\psi_p\right] \tag{3.68}$$

With (3.67) the formal, complete identity with (3.46) has been achieved, allowing to treat the PMSM in the same way as the IM to derive its discrete state space description.

Also in this case the state space description (3.67) is characterized by a *bilinear characteristic* because of the multiplicative combination between the state vector $\mathbf{i}_s$ and the element ω_s of the input vector $\mathbf{v}^f$. Under the same assumption regarding the input quantities as in the case of the IM, and after an iterative integration of (3.67) the following equivalent, discrete state model of the permanent magnet excited synchronous machine is obtained.

$$\mathbf{i}_s^f\left(k+1\right)=\mathbf{\Phi}_{SM}^f\,\mathbf{i}_s^f\left(k\right)+\mathbf{H}_{SM}^{f*}\,\mathbf{v}^f\left(k\right) \tag{3.69}$$

There are:

$$\mathbf{\Phi}_{SM}^f=e^{\left[\mathbf{A}_{SM}^f+\mathbf{N}_{SM}\,\omega_s(k)\right]T}=\sum_{\nu=0}^{\infty}\left[\mathbf{A}_{SM}^f+\mathbf{N}_{SM}\,\omega_s\left(k\right)\right]^{\nu}\frac{T^{\nu}}{\nu!}$$

$$\mathbf{H}_{SM}^{f*}=\int_{kT}^{(k+1)T}e^{\left[\mathbf{A}_{SM}^f+\mathbf{N}_{SM}\,\omega_s(k)\right]\tau}\,d\tau\,\mathbf{B}_{SM}^{f*}=\sum_{\nu=1}^{\infty}\left[\mathbf{A}_{SM}^f+\mathbf{N}_{SM}\,\omega_s\left(k\right)\right]^{\nu-1}\frac{T^{\nu}}{\nu!}\mathbf{B}_{SM}^{f*}$$

$$\tag{3.70}$$

The approximation of first order for the transition matrix $\mathbf{\Phi}_{SM}^f$ and the input matrix $\mathbf{H}_{SM}^{f*}$ arise from the series expansion (3.70):

$$\mathbf{\Phi}_{SM}^f=\begin{bmatrix}1-\dfrac{T}{T_{sd}}&\omega_s T\dfrac{L_{sq}}{L_{sd}}\\[2mm]-\omega_s T\dfrac{L_{sd}}{L_{sq}}&1-\dfrac{T}{T_{sq}}\end{bmatrix};\;\mathbf{H}_{SM}^{f*}=\begin{bmatrix}\dfrac{T}{L_{sd}}&0&0\\[2mm]0&\dfrac{T}{L_{sq}}&-\dfrac{\psi_p T}{L_{sq}}\end{bmatrix} \tag{3.71}$$

The discrete state model (3.69) simultaneously represents the expected current control system of the PMSM. The input matrix $\mathbf{H}_{SM}^{f*}$ can be split up as follows:

$$\mathbf{H}_{SM}^{f*}=\left[\mathbf{H}_{SM}^f\,,\mathbf{h}\right]\;\text{with}\;\mathbf{H}_{SM}^f=\begin{bmatrix}\dfrac{T}{L_{sd}}&0\\[2mm]0&\dfrac{T}{L_{sq}}\end{bmatrix};\,\mathbf{h}=\begin{bmatrix}0\\[2mm]-\dfrac{\omega_s T}{L_{sq}}\end{bmatrix} \tag{3.72}$$

The figure 3.13 shows the discrete state model or the current process model of the PMSM arrived at so far. The splitting of $\mathbf{H}_{SM}^{f*}$ in equation (3.72) into two partial matrices is necessary, because:

1. Identical structures of the current process model are obtained in both cases. This commonality later allows the summarized treatment of the current control problem for both machine types and spares a repeated representation of similar designs.

2. It is necessary to later invert the input matrix for the compensation of the disturbance quantity ψ_p. This would not be possible if $\mathbf{H}_{SM}^{f*}$ keeps the form of a 3×2 matrix.

With that the final equation of the discrete state model or the current process model of the PMSM is obtained as:

$$\mathbf{i}_s^f(k+1) = \mathbf{\Phi}_{SM}^f\, \mathbf{i}_s^f(k) + \mathbf{H}_{SM}^f\, \mathbf{u}_s^f(k) + \mathbf{h}\,\psi_p \qquad (3.73)$$

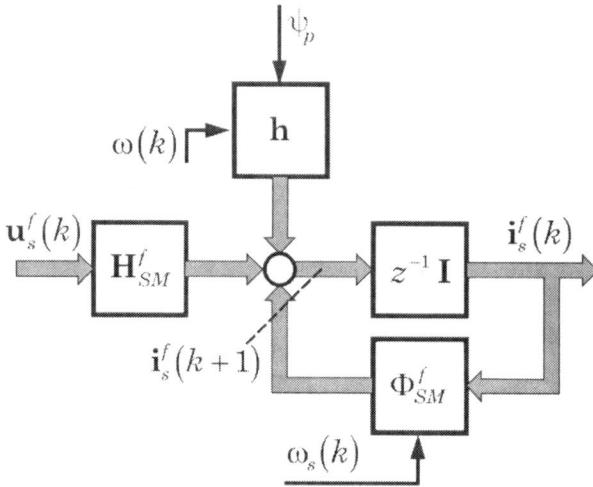

Fig. 3.13 Discrete state model and current process model of the PMSM

3.4 Doubly-fed induction machine (DFIM)

3.4.1 Continuous state space model of the DFIM in the grid synchronous coordinate system

Starting-point for the derivation of the state space model of the DFIM are the voltage equations for stator and rotor winding respectively:

- *Stator voltage in the stator winding system:*

$$\mathbf{u}_s^s = R_s\,\mathbf{i}_s^s + \frac{d\psi_s^s}{dt} \qquad (3.74)$$

- *Rotor voltage in the rotor winding system:*

$$\mathbf{u}_r^r = R_r\,\mathbf{i}_r^r + \frac{d\psi_r^r}{dt} \qquad (3.75)$$

- *Stator and rotor flux:*

$$\begin{cases} \psi_s = L_s\,\mathbf{i}_s + L_m\,\mathbf{i}_r \\ \psi_r = L_m\,\mathbf{i}_s + L_r\,\mathbf{i}_r \end{cases} \text{with} \begin{cases} L_s = L_m + L_{\sigma s} \\ L_r = L_m + L_{\sigma r} \end{cases} \qquad (3.76)$$

All symbols in the formulae (3.74), (3.75) and (3.76) have the same meaning as in the section 3.2.1.

After transforming equations (3.74) and (3.75) to a reference frame rotating with the stator frequency ω_s the following equation is obtained:

$$\begin{cases} \mathbf{u}_s = R_s \mathbf{i}_s + \dfrac{d\psi_s}{dt} + j\omega_s \psi_s \\[2mm] \mathbf{u}_r = R_r \mathbf{i}_r + \dfrac{d\psi_r}{dt} + j\omega_r \psi_r \end{cases} \tag{3.77}$$

Eliminating of stator current $\mathbf{i}_s$ and rotor flux ψ_r from equation (3.77) gives:

$$\begin{cases} \dfrac{d\mathbf{i}_r}{dt} = -\dfrac{1}{\sigma}\left(\dfrac{1}{T_r} + \dfrac{1-\sigma}{T_s}\right)\mathbf{i}_r - j\omega_r \mathbf{i}_r + \dfrac{1-\sigma}{\sigma}\left(\dfrac{1}{T_s} + j\omega\right)\psi_s' \\[3mm] \qquad + \dfrac{1}{\sigma L_r}\mathbf{u}_r - \dfrac{1-\sigma}{\sigma L_m}\mathbf{u}_s \\[3mm] \dfrac{d\psi_s'}{dt} = \dfrac{1}{T_s}\mathbf{i}_r - \left(\dfrac{1}{T_s} + j\omega_s\right)\psi_s' + \dfrac{1}{L_m}\mathbf{u}_s \end{cases} \tag{3.78}$$

with: $\psi_s' = \psi_s / L_m$

After separating both equations into real and imaginary components, we obtain the complete electrical equation system of the DFIM.

$$\begin{cases} \dfrac{di_{rd}}{dt} = -\dfrac{1}{\sigma}\left(\dfrac{1}{T_r} + \dfrac{1-\sigma}{T_s}\right)i_{rd} + \omega_r i_{rq} + \dfrac{1-\sigma}{\sigma}\left(\dfrac{1}{T_s}\psi_{sd}' - \omega\psi_{sq}'\right) \\[3mm] \qquad + \dfrac{1}{\sigma L_r}u_{rd} - \dfrac{1-\sigma}{\sigma L_m}u_{sd} \\[3mm] \dfrac{di_{rq}}{dt} = -\omega_r i_{rd} - \dfrac{1}{\sigma}\left(\dfrac{1}{T_r} + \dfrac{1-\sigma}{T_s}\right)i_{rq} + \dfrac{1-\sigma}{\sigma}\left(\dfrac{1}{T_s}\psi_{sq}' + \omega\psi_{sd}'\right) \\[3mm] \qquad + \dfrac{1}{\sigma L_r}u_{rq} - \dfrac{1-\sigma}{\sigma L_m}u_{sq} \\[3mm] \dfrac{d\psi_{sd}'}{dt} = \dfrac{1}{T_s}i_{rd} - \dfrac{1}{T_s}\psi_{sd}' + \omega_s \psi_{sq}' + \dfrac{1}{L_m}u_{sd} \\[3mm] \dfrac{d\psi_{sq}'}{dt} = \dfrac{1}{T_s}i_{rq} - \omega_s \psi_{sd}' - \dfrac{1}{T_s}\psi_{sq}' + \dfrac{1}{L_m}u_{sq} \end{cases} \tag{3.79}$$

The main control objectives stated above is always the decoupled control of active and reactive current components. This suggests to choose the stator voltage – and respectively grid voltage – orientated reference frame for the further control design.

The realization of the grid voltage orientation requires the accurate and robust acquisition of the phase angle of the grid voltage fundamental wave, considering strong distortions due to converter mains pollution or background grid harmonics. Usually this is accomplished by means of a phase locked loop (PLL).

Summarizing the equation system (3.79) yields the following state space model for the DFIM in the grid voltage orientated reference frame:

$$\frac{dx}{dt} = \mathbf{A}\,\mathbf{x} + \mathbf{B}_s\mathbf{u}_s + \mathbf{B}_r\mathbf{u}_r \qquad (3.80)$$

with:

- State vector $\mathbf{x}^T = \left[i_{rd}, i_{rq}, \psi'_{sd}, \psi'_{sq} \right]$

- Stator voltage vector $\mathbf{u}_s^T = \left[u_{sd}, u_{sq} \right]$ as input vector on stator side

- Rotor voltage vector $\mathbf{u}_r^T = \left[u_{rd}, u_{rq} \right]$ as input vector on rotor side

The system matrix $\mathbf{A}$, the rotor input matrix $\mathbf{B}_r$ and the stator input matrix $\mathbf{B}_s$ may be written as follows:

$$\mathbf{A} = \left[\begin{array}{cc|cc} -\dfrac{1}{\sigma}\left(\dfrac{1}{T_r}+\dfrac{1-\sigma}{T_s}\right) & \omega_r & \dfrac{1-\sigma}{\sigma T_s} & -\dfrac{1-\sigma}{\sigma}\omega \\[2mm] -\omega_r & -\dfrac{1}{\sigma}\left(\dfrac{1}{T_r}+\dfrac{1-\sigma}{T_s}\right) & \dfrac{1-\sigma}{\sigma}\omega & \dfrac{1-\sigma}{\sigma T_s} \\[2mm] \hline \dfrac{1}{T_s} & 0 & -\dfrac{1}{T_s} & \omega_s \\[2mm] 0 & \dfrac{1}{T_s} & -\omega_s & -\dfrac{1}{T_s} \end{array} \right]$$

$$\mathbf{B}_s = \left[\begin{array}{cc} -\dfrac{1-\sigma}{\sigma L_m} & 0 \\[2mm] 0 & -\dfrac{1-\sigma}{\sigma L_m} \\[2mm] \hline \dfrac{1}{L_m} & 0 \\[2mm] 0 & \dfrac{1}{L_m} \end{array} \right] ; \quad \mathbf{B}_r = \left[\begin{array}{cc} \dfrac{1}{\sigma L_r} & 0 \\[2mm] 0 & \dfrac{1}{\sigma L_r} \\[2mm] 0 & 0 \\[2mm] 0 & 0 \end{array} \right]$$

$$(3.81)$$

Doubly-fed induction machine (DFIM) 93

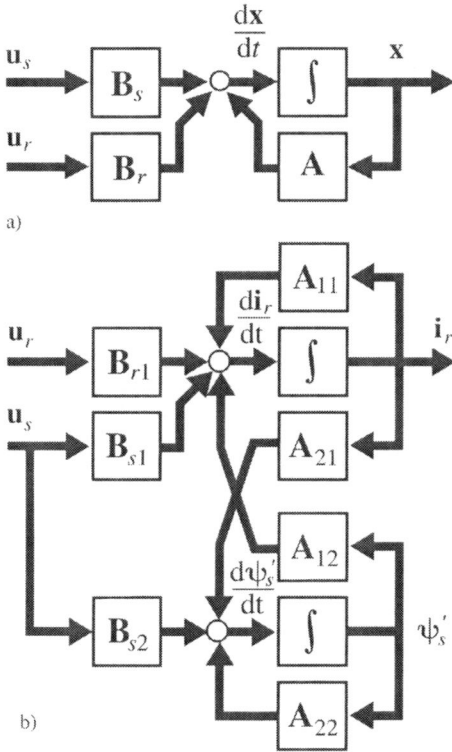

Fig. 3.14 Continuous state space model of the DFIM with stator flux and rotor current as state variables: (**a**) Common representation; (**b**) Split in partial matrices

The state space model of the DFIM is shown in figure 3.14a. The matrices of (3.81) may be split into partial matrices as follows, refer also to figure 3.14b.

$$\mathbf{A} = \begin{bmatrix} \mathbf{A}_{11} & \mathbf{A}_{12} \\ \hline \mathbf{A}_{21} & \mathbf{A}_{22} \end{bmatrix}; \ \mathbf{B}_s = \begin{bmatrix} \mathbf{B}_{s1} \\ \hline \mathbf{B}_{s2} \end{bmatrix}; \ \mathbf{B}_r = \begin{bmatrix} \mathbf{B}_{r1} \\ \hline \mathbf{0} \end{bmatrix} \qquad (3.82)$$

The state space model in partial matrices according to figure 3.14b shows that the rotor voltage $\mathbf{u}_r$ does not influence the stator flux ψ_s directly, but only in an indirect way through the rotor current $\mathbf{i}_r$. The stator flux is determined mainly by the stator voltage. The influence of $\mathbf{u}_s$ to $\mathbf{i}_r$ is like a constant disturbance, and therefore may be compensated by simple feedforward compensation.

3.4.2 Discrete state model of the DFIM

Like in sections 3.2.2 and 3.3.2 the time discrete state model of the DFIM may be obtained by iterative integration of (3.80), yielding the following matrix equation system as base model for the controller design:

$$\mathbf{x}(k+1) = \boldsymbol{\Phi}\,\mathbf{x}(k) + \mathbf{H}_s\mathbf{u}_s(k) + \mathbf{H}_r\mathbf{u}_r(k) \tag{3.83}$$

Transition matrix $\boldsymbol{\Phi}$, stator input matrix $\mathbf{H}_s$ and rotor input matrix $\mathbf{H}_r$ are given by:

$$
\boldsymbol{\Phi} =
\left[
\begin{array}{cc|cc}
1 - \dfrac{T}{\sigma}\left(\dfrac{1}{T_r} + \dfrac{1-\sigma}{T_s}\right) & \omega_r T & \dfrac{1-\sigma}{\sigma}\dfrac{T}{T_s} & -\dfrac{1-\sigma}{\sigma}\omega T \\[3mm]
-\omega_r T & 1 - \dfrac{T}{\sigma}\left(\dfrac{1}{T_r} + \dfrac{1-\sigma}{T_s}\right) & \dfrac{1-\sigma}{\sigma}\omega T & \dfrac{1-\sigma}{\sigma}\dfrac{T}{T_s} \\[3mm]
\dfrac{T}{T_s} & 0 & 1 - \dfrac{T}{T_s} & \omega_s T \\[3mm]
0 & \dfrac{T}{T_s} & -\omega_s T & 1 - \dfrac{T}{T_s}
\end{array}
\right]
= \left[\begin{array}{c|c} \boldsymbol{\Phi}_{11} & \boldsymbol{\Phi}_{12} \\ \hline \boldsymbol{\Phi}_{21} & \boldsymbol{\Phi}_{22} \end{array}\right]
$$

$$
\mathbf{H}_s =
\left[
\begin{array}{cc}
-\dfrac{1-\sigma}{\sigma}\dfrac{T}{L_m} & 0 \\[3mm]
0 & \dfrac{1-\sigma}{\sigma}\dfrac{T}{L_m} \\[3mm]
\dfrac{T}{L_m} & 0 \\[3mm]
0 & \dfrac{T}{L_m}
\end{array}
\right]
= \left[\begin{array}{c} \mathbf{H}_{s1} \\ \mathbf{H}_{s2} \end{array}\right];
\quad
\mathbf{H}_r =
\left[
\begin{array}{cc}
\dfrac{T}{\sigma L_r} & 0 \\[3mm]
0 & \dfrac{T}{\sigma L_r} \\[3mm]
0 & 0 \\[3mm]
0 & 0
\end{array}
\right]
= \left[\begin{array}{c} \mathbf{H}_{r1} \\ \mathbf{0} \end{array}\right]
\tag{3.84}
$$

The discrete state space model is shown in partial matrix form in figure 3.15a. Figure 3.15b shows the rotor current system, being the starting-point for the rotor current controller design. Due to the stiff mains system stator voltage $\mathbf{u}_s$ and stator flux ψ_s can be recognized as almost constant disturbances.

The figure 3.15a was produced by splitting of the equation (3.83) as follows:

$$
\begin{cases}
\mathbf{i}_r(k+1) = \boldsymbol{\Phi}_{11}\mathbf{i}_r(k) + \boldsymbol{\Phi}_{12}\boldsymbol{\psi}_s'(k) + \mathbf{H}_{s1}\mathbf{u}_s(k) + \mathbf{H}_{r1}\mathbf{u}_r(k) \\[2mm]
\boldsymbol{\psi}_s'(k+1) = \boldsymbol{\Phi}_{21}\mathbf{i}_r(k) + \boldsymbol{\Phi}_{22}\boldsymbol{\psi}_s'(k) + \mathbf{H}_{s2}\mathbf{u}_s(k)
\end{cases}
\tag{3.85}
$$

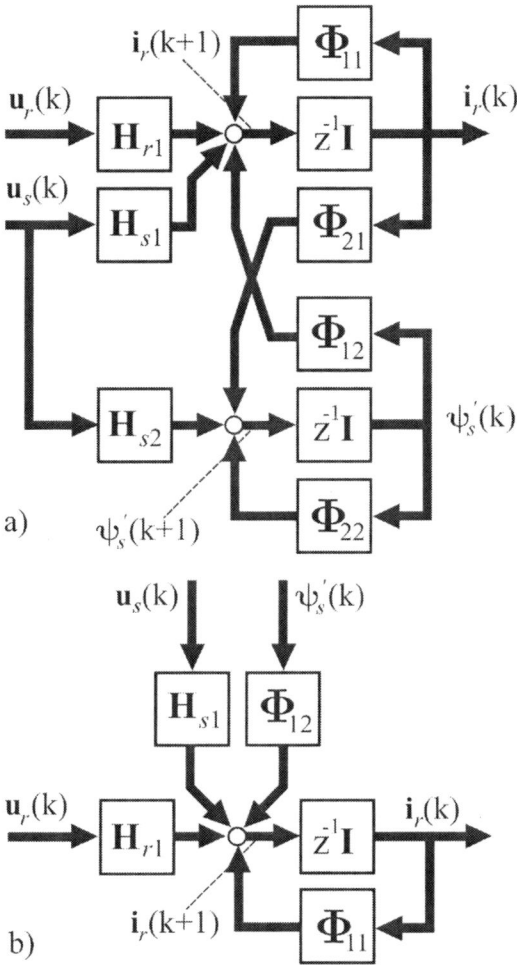

Fig. 3.15 Discrete state model of the DFIM: (a) in grid voltage orientated reference frame; (b) rotor current process model

3.5 Generalized current process model for the two machine types IM and PMSM

In evaluation of the equations (3.51), (3.55) and (3.73) as well as the figures 3.8, 3.10 and 3.13 the formal identity of the two machine types IM and PMSM regarding system structure and system order becomes clearly visible. Therefore it can be regarded theoretically proven, that with respect to hardware and software an identical concept for the stator current impression may be applied in the stator-fixed as well as in the field synchronous coordinate system. In this section a uniform description for all system elements will be derived to support a parallel treatment of both current process models under investigation. The following symbols are defined:

- $\boldsymbol{\Phi}$: Transition matrices $\boldsymbol{\Phi}_{11}^{s}$ or $\boldsymbol{\Phi}_{11}^{f}$ or $\boldsymbol{\Phi}_{SM}^{f}$

- $\mathbf{H}$: Input matrices $\mathbf{H}_{1}^{s}$ or $\mathbf{H}_{1}^{f}$ or $\mathbf{H}_{SM}^{f}$

- $\mathbf{h}$: Disturbance matrices or vector $\boldsymbol{\Phi}_{12}^{s}$ or $\boldsymbol{\Phi}_{12}^{f}$ or $\mathbf{h}$, which represent the intervention of the flux dependent disturbance quantity.

$\mathbf{h}$ is a 2×2 matrix in the case of the IM and only a simple vector in the case of the PMSM. The input vector $\mathbf{u}_{s}$ and the state vector $\mathbf{i}_{s}$ will be written without the subscripts „s" (for stator-fixed) or „f" (for field synchronous coordinate system). This index can be attached later in the concrete choice of the coordinate system to be used. For the rotor and pole flux the symbol ψ is used instead of ψ_{r}^{s} or ψ_{r}^{f} or ψ_{p}. With these arrangements the following common equation results for the current process models:

$$\mathbf{i}_{s}(k+1) = \boldsymbol{\Phi}\,\mathbf{i}_{s}(k) + \mathbf{H}\,\mathbf{u}_{s}(k) + \mathbf{h}\,\psi(k) \tag{3.86}$$

and in the z domain:

$$z\,\mathbf{i}_{s}(z) = \boldsymbol{\Phi}\,\mathbf{i}_{s}(z) + \mathbf{H}\,\mathbf{u}_{s}(z) + \mathbf{h}\,\psi(z) \tag{3.87}$$

with the characteristic equation:

$$\det[z\,\mathbf{I} - \boldsymbol{\Phi}] = 0 \text{ with } \mathbf{I} = \text{ unity matrix} \tag{3.88}$$

The figure 3.16 shows the current process models for the following three cases in the overview:

1. IM in the stator-fixed,
2. IM in the field synchronous and
3. PMSM in the field synchronous coordinate system.

The equation (3.87) as well as the characteristic equation (3.88) are given in the z domain which is advisable for the treatment of discrete systems.

Here the similarity between the model in the figure 3.16 and the current process model of the DFIM in equation (3.85), shown in the figure 3.15b, can also be easily recognized. Because the three models represent linear and time-variant processes, linear current controllers using:

- output feedback or
- state feedback

can be designed. The method to derive these three linear and time-variant process models can be called the linearization within the sampling period. This is possible because the models have been derived under the conditions that:

- the stator-side angular velocity ω_s in the case of the IM or PMSM, and
- the rotor-side angular velocity ω_r in the case of the DFIM are constant within one sampling period.

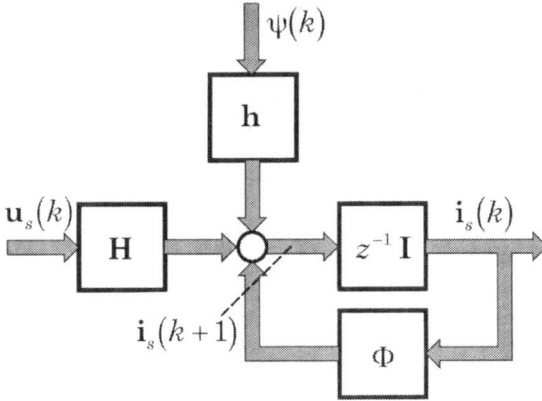

Fig. 3.16 General current process model of IM and PMSM

3.6 Nonlinear properties of the machine models and the way to nonlinear controllers

Electrical 3-phase AC machines exhibit different nonlinearities because of the mechanical construction of their magnetic paths with slots and air-gaps. But only two types of nonlinearities are relevant for the controller design:

- *The nonlinear structure of the process models*: This nonlinearity is caused by products between states variables like current components and input variables ω_s (in cases IM and PMSM), ω_r (in the case DFIM). This structural nonlinearity can only be mastered completely by nonlinear controllers designed – for example – using methods like exact linearization or backstepping based concepts.
- *The nonlinear parameters*: Some parameters like the mutual inductance depends on the rotor flux which is a state variable. The problem with parametric nonlinearities can be solved by identification and adaptation methods.

Because the backstepping based design still is not a mature method for using in the practice, the section 3.6 only deals with the idea of the exact linearization which can be used to master the structural nonlinearities of the process models of the IM, DFIM and PMSM, and to design nonlinear controllers for improving the control performance in difficult operation situations.

3.6.1 Idea of the exact linearization

For the understanding, at first the idea of the relative difference order of a linear system without dead time – the SISO process – shall be explained. If the linear SISO process is represented by the following transfer function:

$$G(s) = \frac{y(s)}{u(s)} = \frac{b_0 + b_1 s + \cdots + b_p s^p}{a_0 + a_1 s + \cdots + a_q s^q}; \quad p < q \tag{3.89}$$

then the pole surplus r with:

$$r = q - p \geq 1 \tag{3.90}$$

can be called the *relative difference order* of the process model described by equation (3.89). If the linear process model is a MISO system with m inputs and only one output, i.e. a system with m transfer functions in a form similar to equation (3.89), then the integer number r:

$$r = \min_i r_i \quad \text{with} \quad 1 \leq i \leq m \tag{3.91}$$

means the relative difference order of the MISO system, in which r_i is the pole surplus of the ith transfer function. If the definition according to the formula (3.91) is applied to a linear process with m inputs and m outputs, the following vector $\mathbf{r}$ of the relative difference orders is obtained:

$$\mathbf{r} = [r_1, r_2, \cdots, r_m] \tag{3.92}$$

with m natural numbers r_j (j = 1, 2, ... , m), and r_j the relative difference order of the jth output. Because the process described by the model (3.89) can be represented in the state space, the relative difference order r and respectively the vector $\mathbf{r}$ of relative difference orders can also be calculated using state space models.

Some classes of nonlinear systems with m inputs and m outputs, the so called nonlinear MIMO systems, can be described by the following equations:

$$\begin{cases} \dfrac{d\mathbf{x}}{dt} = \mathbf{f}(\mathbf{x}) + \mathbf{H}(\mathbf{x})\mathbf{u} \\ \mathbf{y} = \mathbf{g}(\mathbf{x}) \end{cases} \tag{3.93}$$

with:

$$\mathbf{x} = \begin{pmatrix} x_1 \\ \vdots \\ x_n \end{pmatrix}; \mathbf{u} = \begin{pmatrix} u_1 \\ \vdots \\ u_m \end{pmatrix}; \mathbf{g}(\mathbf{x}) = \begin{pmatrix} g_1(\mathbf{x}) \\ \vdots \\ g_m(\mathbf{x}) \end{pmatrix}$$

$$\mathbf{H}(\mathbf{x}) = (\mathbf{h}_1(\mathbf{x}), \quad \mathbf{h}_2(\mathbf{x}), \quad \cdots \quad, \mathbf{h}_m(\mathbf{x}))$$

(3.94)

Similarly to the linear systems, for the system in the equation (3.93) a vector of relative difference orders like (3.92) can also be derived.

The basic idea of the *exact linearization* can be summarized as follows: If the nonlinear MIMO system in the form (3.93) contains a vector of relative difference orders like equation (3.92), which fulfills the following condition:

$$r = r_1 + r_2 + \cdots + r_m = n$$

(3.95)

then the system (3.93) can be transformed using the coordinate transformation:

$$\mathbf{z} = \begin{pmatrix} z_1 \\ \vdots \\ z_n \end{pmatrix} = \mathbf{m}(\mathbf{x}) = \begin{pmatrix} m_1^1(\mathbf{x}) \\ \vdots \\ m_{r_1}^1(\mathbf{x}) \\ \vdots \\ m_1^m(\mathbf{x}) \\ \vdots \\ m_{r_m}^m(\mathbf{x}) \end{pmatrix} = \begin{pmatrix} g_1(\mathbf{x}) \\ \vdots \\ L_f^{r_1-1} g_1(\mathbf{x}) \\ \vdots \\ g_m(\mathbf{x}) \\ \vdots \\ L_f^{r_m-1} g_m(\mathbf{x}) \end{pmatrix}$$

(3.96)

into the following linear MIMO system:

$$\begin{cases} \dfrac{d\mathbf{z}}{dt} = \mathbf{A}\mathbf{z} + \mathbf{B}\mathbf{w} \\ \mathbf{y} = \mathbf{C}\mathbf{z} \end{cases}$$

(3.97)

The original input $\mathbf{u}$ is then controlled by the coordinate transformation law:

$$\mathbf{u} = \mathbf{a}(\mathbf{x}) + \mathbf{L}^{-1}(\mathbf{x})\mathbf{w}$$

(3.98)

The vector $\mathbf{a}(\mathbf{x})$ and the matrix $\mathbf{L}^{-1}(\mathbf{x})$ in (3.98) look as follows:

$$\mathbf{L}(\mathbf{x}) = \begin{pmatrix} L_{h_1} L_f^{r_1-1} g_1(\mathbf{x}) & \cdots & L_{h_m} L_f^{r_1-1} g_1(\mathbf{x}) \\ \vdots & \ddots & \vdots \\ L_{h_1} L_f^{r_m-1} g_m(\mathbf{x}) & \cdots & L_{h_m} L_f^{r_m-1} g_m(\mathbf{x}) \end{pmatrix}; \mathbf{a}(\mathbf{x}) = -\mathbf{L}^{-1}(\mathbf{x}) \begin{pmatrix} L_f^{r_1} g_1(\mathbf{x}) \\ \vdots \\ L_f^{r_m} g_m(\mathbf{x}) \end{pmatrix}$$

(3.99)

Formula (3.99) also requires the ability, with respect to the coordinate transformation or to the exact linearization, to invert the matrix $\mathbf{L(x)}$. In equations (3.96) and (3.99), the term

$$L_f g(\mathbf{x}) = \frac{\partial g(\mathbf{x})}{\partial \mathbf{x}} \mathbf{f}(\mathbf{x}) \qquad (3.100)$$

notifies the Lie derivation of the function $g(\mathbf{x})$ along the trajectory $\mathbf{f}(\mathbf{x})$. The details of the complicated general expressions for the matrices $\mathbf{A}$, $\mathbf{B}$ and $\mathbf{C}$ are abandoned here. The figure 3.17 illustrates the explained facts so far.

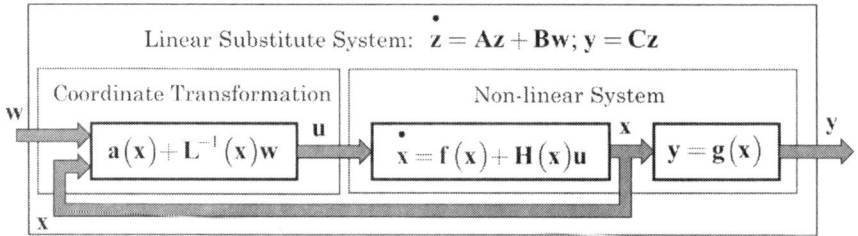

Fig. 3.17 Transformation of a nonlinear system into a linear substitute system

Here it must be highlighted that the coordinate transformation requires exact knowledge of the complete state vector $\mathbf{x}$, which can not always be assumed for 3-phase AC machines.

3.6.2 Nonlinearities of the IM model

The nonlinearity of the IM is clearly represented by the equation (3.46). The two first equations of the system (3.44), which represent the current process model, will be separated and extended by the field angle ϑ_s.

$$\begin{cases} \dfrac{di_{sd}}{dt} = -\left(\dfrac{1}{\sigma T_s} + \dfrac{1-\sigma}{\sigma T_r}\right) i_{sd} + w_s\, i_{sq} + \dfrac{1-\sigma}{\sigma T_r}\psi'_{rd} + \dfrac{1}{\sigma L_s} u_{sd} \\[3mm] \dfrac{di_{sq}}{dt} = -w_s\, i_{sd} - \left(\dfrac{1}{\sigma T_s} + \dfrac{1-\sigma}{\sigma T_r}\right) i_{sq} - \dfrac{1-\sigma}{\sigma}w\psi'_{rd} + \dfrac{1}{\sigma L_s} u_{sq} \quad (3.101) \\[3mm] \dfrac{d\vartheta_s}{dt} = w_s \end{cases}$$

For better understanding temporary parameters and variables are introduced:

- Parameters: $a = 1/\sigma L_s\,; b = 1/\sigma T_s\,; c = (1-\sigma)/\sigma T_r\,; d = b + c$

- State variables: $\quad x_1 = i_{sd};\ x_2 = i_{sq};\ x_3 = \vartheta_s$
- Input variables: $\quad u_1 = u_{sd};\ u_2 = u_{sq};\ u_3 = \omega_s$
- Output variables: $\quad y_1 = i_{sd};\ y_2 = i_{sq};\ y_3 = \vartheta_s$

Now the current process model looks as follows:

$$\begin{cases} \dfrac{dx_1}{dt} = -d\,x_1 + x_2 u_3 + a u_1 + c\psi'_{rd} \\[2mm] \dfrac{dx_2}{dt} = -x_1 u_3 - d\,x_2 + a u_{sq} - c\,T_r\,\omega\,\psi'_{rd} \\[2mm] \dfrac{dx_3}{dt} = u_3 \end{cases} \tag{3.102}$$

or:

$$\begin{cases} \begin{bmatrix} \dot{x}_1 \\ \dot{x}_2 \\ \dot{x}_3 \end{bmatrix} = \begin{bmatrix} -d\,x_1 + c\,\psi'_{rd} \\ -d\,x_2 - c\,T_r\,\omega\,\psi'_{rd} \\ 0 \end{bmatrix} + \begin{bmatrix} a \\ 0 \\ 0 \end{bmatrix} u_1 + \begin{bmatrix} 0 \\ a \\ 0 \end{bmatrix} u_2 + \begin{bmatrix} x_2 \\ -x_1 \\ 1 \end{bmatrix} u_3 \\[6mm] \begin{bmatrix} y_1 \\ y_2 \\ y_3 \end{bmatrix} = \begin{bmatrix} 1 & 0 & 0 \\ 0 & 1 & 0 \\ 0 & 0 & 1 \end{bmatrix} \begin{bmatrix} x_1 \\ x_2 \\ x_3 \end{bmatrix} \end{cases} \tag{3.103}$$

The system (3.103) can now be transferred to the general form with the equation (3.93).

$$\begin{cases} \dot{\mathbf{x}} = \mathbf{f(x)} + \mathbf{h}_1\,u_1 + \mathbf{h}_2\,u_2 + \mathbf{h}_3\,u_3 \\ \mathbf{y} = \mathbf{g(x)} \end{cases} \tag{3.104}$$

with:

$$\mathbf{f(x)} = \begin{bmatrix} -d\,x_1 + c\,\psi'_{rd} \\ -d\,x_2 - c\,T_r\,\omega\,\psi'_{rd} \\ 0 \end{bmatrix};\ \mathbf{h}_1 = \begin{bmatrix} a \\ 0 \\ 0 \end{bmatrix};\ \mathbf{h}_1 = \begin{bmatrix} 0 \\ a \\ 0 \end{bmatrix};\ \mathbf{h}_3 = \begin{bmatrix} x_2 \\ -x_1 \\ 1 \end{bmatrix} \tag{3.105}$$

$$y_1 = g_1(\mathbf{x}) = x_1;\ y_2 = g_2(\mathbf{x}) = x_2;\ y_3 = g_3(\mathbf{x}) = x_3$$

The equation (3.104) represents the new process model and will be used later to design the nonlinear current control loop using exact linearization.

3.6.3 Nonlinearities of the DFIM model

Similar to the case IM, the nonlinearity of the DFIM is represented by the following equation separated from the equation (3.79) and extended by the rotor angle ϑ_r.

$$
\begin{cases}
\dfrac{di_{rd}}{dt} = -\dfrac{1}{\sigma}\left(\dfrac{1}{T_r}+\dfrac{1-\sigma}{T_s}\right)i_{rd}+w_r i_{rq}+\dfrac{1-\sigma}{\sigma}\left(\dfrac{1}{T_s}\psi'_{sd}-w\psi'_{sq}\right) \\
\qquad +\dfrac{1}{\sigma L_r}u_{rd}-\dfrac{1-\sigma}{\sigma L_m}u_{sd} \\
\dfrac{di_{rq}}{dt} = -w_r i_{rd}-\dfrac{1}{\sigma}\left(\dfrac{1}{T_r}+\dfrac{1-\sigma}{T_s}\right)i_{rq}+\dfrac{1-\sigma}{\sigma}\left(\dfrac{1}{T_s}\psi'_{sq}+w\psi'_{sd}\right) \\
\qquad +\dfrac{1}{\sigma L_r}u_{rq}-\dfrac{1-\sigma}{\sigma L_m}u_{sq} \\
\dfrac{d\vartheta_r}{dt}=w_r
\end{cases} \tag{3.106}
$$

After substituting the newly defined temporary parameters:

$$
a=\left(\dfrac{1}{\sigma T_r}+\dfrac{1-\sigma}{\sigma T_s}\right); b=\dfrac{1-\sigma}{\sigma}; c=\dfrac{1}{\sigma L_r}; d=\dfrac{1-\sigma}{\sigma L_m}; e=\dfrac{1-\sigma}{\sigma T_s}
$$

in the partial model of the rotor current in the equation (3.106), the following model is obtained:

$$
\begin{cases}
\dfrac{di_{rd}}{dt} = -ai_{rd}+w_r i_{rq}+e\psi'_{sd}-bw\psi'_{sq}+cu_{rd}-du_{sd} \\
\dfrac{di_{rq}}{dt} = -w_r i_{rd}-ai_{rq}+bw\psi'_{sd}+e\psi'_{sq}+cu_{rq}+du_{sq} \\
\dfrac{d\vartheta_r}{dt}=w_r
\end{cases} \tag{3.107}
$$

New vectors will now be defined as follows:
- Vector of state variables:

$$\mathbf{x}^T=\begin{bmatrix}x_1 & x_2 & x_3\end{bmatrix}; x_1=i_{rd}; x_2=i_{rq}; x_3=\theta_r$$

- Vector of input variables:

$$\mathbf{u}^T=\begin{bmatrix}u_1 & u_2 & u_3\end{bmatrix}; u_1=e\psi'_{sd}-bw\psi'_{sq}+cu_{rd}-du_{sd}$$

$$u_2=bw\psi'_{sd}+e\psi'_{sq}+cu_{rq}+du_{sq}; u_3=w_r$$

- Vector of output variables:

$$\mathbf{y}^T = [y_1 \quad y_2 \quad y_3]; \, y_1 = i_{rd}; \, y_2 = i_{rq}; \, y_3 = \theta_r$$

Finally, the following nonlinear DFIM model in the detailed:

$$
\begin{bmatrix} \dot{x}_1 \\ \dot{x}_2 \\ \dot{x}_3 \end{bmatrix} = \begin{bmatrix} -ax_1 \\ -ax_2 \\ 0 \end{bmatrix} + \begin{bmatrix} 1 \\ 0 \\ 0 \end{bmatrix} u_1 + \begin{bmatrix} 0 \\ 1 \\ 0 \end{bmatrix} u_2 + \begin{bmatrix} x_2 \\ -x_1 \\ 1 \end{bmatrix} u_3
$$

$$
\begin{bmatrix} y_1 \\ y_2 \\ y_3 \end{bmatrix} = \begin{bmatrix} 1 & 0 & 0 \\ 0 & 1 & 0 \\ 0 & 0 & 1 \end{bmatrix} \begin{bmatrix} x_1 \\ x_2 \\ x_3 \end{bmatrix}
$$

(3.108)

or in the generalized form is obtained:

$$
\begin{cases} \dot{\mathbf{x}} = \mathbf{f}(\mathbf{x}) + \mathbf{h}_1(\mathbf{x})u_1 + \mathbf{h}_2(\mathbf{x})u_2 + \mathbf{h}_3(\mathbf{x})u_3 \\ \mathbf{y} = \mathbf{g}(\mathbf{x}) \end{cases}
$$

(3.109)

with:

$$
\mathbf{f}(\mathbf{x}) = \begin{bmatrix} -ax_1 \\ -ax_2 \\ 0 \end{bmatrix}; \mathbf{h}_1(\mathbf{x}) = \begin{bmatrix} 1 \\ 0 \\ 0 \end{bmatrix}; \mathbf{h}_2(\mathbf{x}) = \begin{bmatrix} 0 \\ 1 \\ 0 \end{bmatrix}; \mathbf{h}_3(\mathbf{x}) = \begin{bmatrix} x_2 \\ -x_1 \\ 1 \end{bmatrix}
$$

(3.110)

$$g_1(\mathbf{x}) = x_1; \, g_2(\mathbf{x}) = x_2; \, g_3(\mathbf{x}) = x_3$$

The equations (3.108) and respectively (3.109) are starting points for the later design of the nonlinear controller for DFIM systems.

3.6.4 Nonlinearities of the PMSM model

The model (3.64) of the PMSM will now be extended by the field angle ϑ_s, similarly to the IM in equation (3.101).

$$\begin{cases} \dfrac{di_{sd}}{dt} = -\dfrac{1}{T_{sd}} i_{sd} + \omega_s \dfrac{L_{sq}}{L_{sd}} i_{sq} + \dfrac{1}{L_{sd}} u_{sd} \\[2mm] \dfrac{di_{sq}}{dt} = -\omega_s \dfrac{L_{sd}}{L_{sq}} i_{sd} - \dfrac{1}{T_{sq}} i_{sq} + \dfrac{1}{L_{sq}} u_{sq} - \omega_s \dfrac{\psi_p}{L_{sq}} \\[2mm] \dfrac{d\vartheta_s}{dt} = \omega_s \end{cases} \qquad (3.111)$$

With newly introduced variables and temporary parameters:

- state variables: $\quad x_1 = i_{sd}; x_2 = i_{sq}; x_3 = \vartheta_s$
- input variables: $\quad u_1 = u_{sd}; u_2 = u_{sq}; u_3 = \omega_s$
- output variables: $\quad y_1 = i_{sd}; y_2 = i_{sq}; y_3 = \vartheta_s$
- temporary parameters: $\quad a = 1/L_{sd}; b = 1/L_{sq}; c = 1/T_{sd}; a = 1/T_{sq};$

the equation (3.111) can be transferred to:

$$\begin{bmatrix} \overset{\bullet}{x_1} \\ \overset{\bullet}{x_2} \\ \overset{\bullet}{x_3} \end{bmatrix} = \begin{bmatrix} -cx_1 \\ -dx_2 \\ 0 \end{bmatrix} + \begin{bmatrix} a \\ 0 \\ 0 \end{bmatrix} u_1 + \begin{bmatrix} 0 \\ b \\ 0 \end{bmatrix} u_2 + \begin{bmatrix} \dfrac{a}{b} x_2 \\ -\dfrac{b}{a} x_1 - b\psi_p \\ 1 \end{bmatrix} u_3$$

$$\begin{bmatrix} y_1 \\ y_2 \\ y_3 \end{bmatrix} = \begin{bmatrix} 1 & 0 & 0 \\ 0 & 1 & 0 \\ 0 & 0 & 1 \end{bmatrix} \begin{bmatrix} x_1 \\ x_2 \\ x_3 \end{bmatrix} \qquad (3.112)$$

or to the following generalized form:

$$\begin{cases} \overset{\bullet}{\mathbf{x}} = \mathbf{f}(\mathbf{x}) + \mathbf{h}_1(\mathbf{x})u_1 + \mathbf{h}_2(\mathbf{x})u_2 + \mathbf{h}_3(\mathbf{x})u_3 \\ \mathbf{y} = \mathbf{g}(\mathbf{x}) \end{cases} \qquad (3.113)$$

with:

$$\mathbf{f}(\mathbf{x}) = \begin{bmatrix} -cx_1 \\ -dx_2 \\ 0 \end{bmatrix}; \mathbf{h}_1(\mathbf{x}) = \begin{bmatrix} a \\ 0 \\ 0 \end{bmatrix}; \mathbf{h}_2(\mathbf{x}) = \begin{bmatrix} 0 \\ b \\ 0 \end{bmatrix}; \mathbf{h}_3(\mathbf{x}) = \begin{bmatrix} \dfrac{a}{b} x_2 \\ -\dfrac{b}{a} x_1 - b\psi_p \\ 1 \end{bmatrix} \qquad (3.114)$$

$$y_1 = g_1(\mathbf{x}) = x_1; y_2 = g_2(\mathbf{x}) = x_2; y_3 = g_3(\mathbf{x}) = x_3$$

The equations (3.113) and (3.114) can be used to design nonlinear controllers for systems using 3-phase AC machines of type PMSM.

3.7 References

Ackermann J (1988) Abtastregelung. Springer-Verlag, Berlin Heidelberg New York London Paris Tokyo

Böcker J (1991) Discrete time model of an induction motor. ETEP Journal Vol.1, No.2, March/April

Dittrich JA (1998) Anwendung fortgeschrittener Steuer- und Regelverfahren bei Asynchronantrieben. Habilitationsschrift, TU Dresden

Föllinger O (1982) Lineare Abtastsysteme. R. Oldenbourg Verlag, München Wien

Isermann R (1987) Digitale Regelsysteme. Bd. 2, Springer-Verlag, Berlin Heidelberg New York London Paris Tokyo

Isidori A (1995) Nonlinear Control Systems. 3rd Edition, Springer-Verlag, London Berlin Heidelberg

Jentsch W (1969) Digitale Simulation kontinuierlicher Systeme. R. Oldenbourg Verlag, München

Mayer HR (1988) Entwurf zeitdiskreter Regelverfahren für Asynchronmotoren unter Berücksichtigung der diskreten Arbeitsweise des Umrichters. Dissertation, Uni. Erlangen – Nürnberg

Schönfeld R (1987) Grundlagen der automatischen Steuerung – Leitfaden und Aufgaben aus der Elektrotechnik. Verlag Technik, Berlin

Schwarz H (1979) Zeitdiskrete Regelsysteme. Akademie Verlag, Berlin

Schwarz H (1985) Abtastregelung bilinearer Systeme. 30. Intern. Wiss. Koll. TH Ilmenau , Vortragsreihe „ Techn. Kybernetik / Automatisierungstechnik", H.1

Schwarz H (1991) Nichtlineare Regelungssysteme. R. Oldenbourg Verlag, München Wien

Unbehauen H (1989) Regelungstechnik II: Zustandsregelungen, digitale und nichtlineare Regelsysteme. Vieweg Verlag

Zäglein W (1984) Drehzahlregelung des Asynchronmotors unter Verwendung eines Beobachters mit geringer Parameterempfindlichkeit. Dissertation, Uni. Erlangen – Nürnberg

4 Problems of actual-value measurement and vector orientation

This chapter aims to explain principles of the actual-value measurement, to highlight its problems and to answer some related questions of the field orientation.

The current measurement technique influences decisively the controller design and thus also the dynamic characteristic of the inner current control loop, which in turn is the prerequisite for the superimposed speed control. So the actual-value measurement is an important interface of every drive control system which must be taken into account very carefully for the controller design.

Similarly, for the design of the speed control loop the speed measuring is an important issue to consider. Either an incremental encoder or a resolver can be used to measure the speed. Also the alternative possibility of sensorless capture of the speed will be discussed, and possible ways to solve this problem will be shown.

The second problem of this chapter is the field orientation, which is very closely connected to the speed measurement. Field orientation means namely,

1. that the field angle ϑ_s and respectively the location of the field coordinate system (dq- coordinates) must be calculated, and
2. that the un-measurable rotor flux, which will be used for calculating the rotor frequency or the slip and therefore also the field angle ϑ_s, has to be estimated. The estimated value of the rotor flux can be used as actual value in the flux control loop, which is – for example – of decisive importance for field-weakening operation.

The estimation of the rotor flux, which can be realized either by flux models or by flux observers, and the calculation of the field angle require actual values of current and speed.

4.1 Acquisition of the current

The measurement of the currents can be performed as shown in the figure 1.3. Depending on the coordinate system the inner current control loop is realized in – field synchronous or stator-fixed – actual values $i_{s\alpha}$, $i_{s\beta}$ or i_{sd}, i_{sq} are obtained after the transformation of the measured phase currents i_{su} and i_{sv}. What could not be indicated in this figure are:
1. The technical realization of the measurement and
2. the fact, that *for the current control only the instantaneous value of the fundamental wave is relevant.*
From the technical view, two possibilities to measure the currents exist:
1. The most advanced technique is the measurement of instantaneous values using A/D converters (ADC: Analog to Digital Converter) and
2. The integrating measurement using V/f converters (VFC: Voltage to Frequency Converter).

a) Measurement of instantaneous values using an ADC
This method is frequently applied because of the simplicity of its technical realization and the possibility of a high resolution. The inherent current harmonics have to be suppressed, for example by an additional filter. This however, would result in an additional delay of the measured values. This delay is unwanted, and therefore has to be avoided if possible, to maintain the dynamics of the current control loop, particularly for the new current controller designs in chapter 5.

The *time instant of the current measuring* plays a decisive role for the exact acquisition of the fundamental wave and for the elimination of the pulse frequent harmonics. To achieve this, the measuring instant must be exactly placed in the middle of the zero vector times T_0 or T_7 (using the modulation algorithm in the chapter 2). The figure 4.1 explains the facts.

This measuring strategy has the advantage that the otherwise necessary filter becomes superfluous and the delay connected to it disappears. The obvious disadvantage is, particularly under transient conditions, that the time instant of the measurement sampling will shift (start-up, reversing, field weakening etc.), because the values of the zero vector time T_0 or T_7 are not constant, but depend on the operating state of the motor. The measurement sampling instants, illustrated in the figure 4.1, correspond to the output sequence of $\mathbf{u}_s$ using the time pattern in the figure 2.9a.

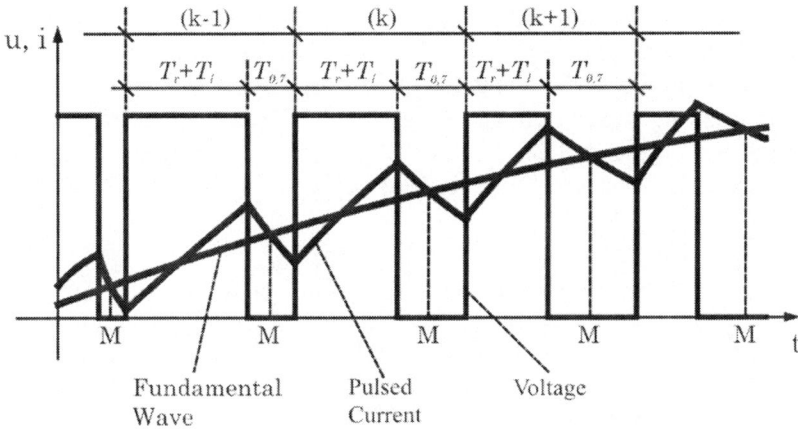

Fig. 4.1 The current measurement sampling instants (*M*) using an A/D converter

The mentioned disadvantage of the shifting sampling instant disappears if the output sequence of the time pattern in figure 2.9b is used. The sampling instant will be exactly in the middle of T_7 and consequently, the pulse and the control sampling frequency will exactly coincide (figure 2.9b).

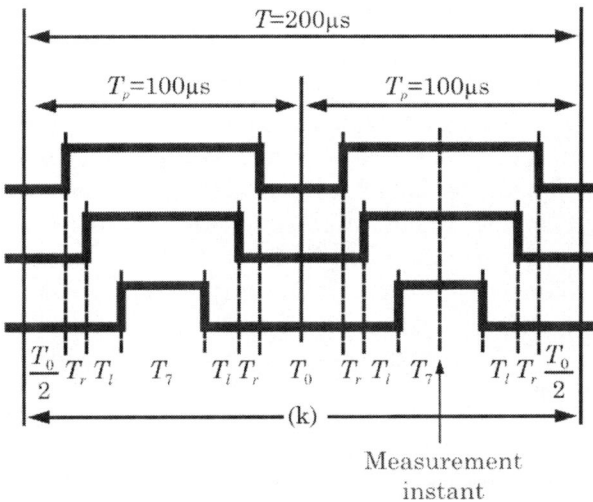

Fig. 4.2 A strict synchronization between pulse period and measurement sampling with sampling in the center of zero vector times

With the tendency toward higher pulse frequencies, the pulse frequency can, however, be a multiple of the sampling frequency. The measurement sampling must always be located either in the center of the last zero vector time T_7 or at the starting points of the sampling periods of the control system. The outlined principle for the realization of the current measurement clarifies the demand for a strict synchronization between pulse periods and measurement sampling which must already be thought through at the hardware design stage. The figure 4.2 presents an example with 10kHz pulse frequency and 5kHz sampling frequency.

b) The integrating measurement using a VFC
This category also includes the method of analog integration with subsequent A/D conversion. The measured signal is converted into a pulse sequence with a frequency which is directly proportional to its amplitude. This pulse sequence is applied to an up/down counter whose counting direction is switched over according to the sign of the measured signal. The impulses are counted over one sampling period. Because of the integrating behavior there is no need for special measures to suppress pulse frequent harmonics. However, the result of the integration does not represent the instantaneous values of the fundamental, which are needed by the control system. They may be back-propagated using an interpolation filter, for example of second order as in equation 4.1.

$$\mathbf{i}_s(k) = 1{,}83\,\bar{\mathbf{i}}_s(k) - 1{,}16\,\bar{\mathbf{i}}_s(k-1) + 0{,}33\,\bar{\mathbf{i}}_s(k-2) \qquad (4.1)$$

$k: 0, 1, 2, ... , \infty; \quad \mathbf{i}_s$: Integrated value

The interpolation filter may be fed with either the phase currents i_{su}, i_{sv} directly or the current components in dq or $\alpha\beta$ coordinates. That means, the back-propagation of the instantaneous values of the fundamental happens before or after processing equations (1.6) and (1.7). The results would show, depending on the sensor resolution largely corresponding feedbacks and actual motor currents, with the restriction that sampling-frequent oscillations cannot be followed. Since only actually measured mean average values of the currents are available, this fact requires special measures for the design of the current controller. Chapter 5 will more deeply deal with these issues.

4.2 Acquisition of the speed

The speed is commonly measured either with a resolver or with an incremental encoder. Because of the pulse counting when using an incremental encoder and the averaging of the speed, over several sampling periods by differentiating the position angle with resolver, the measurement has an integrating characteristic and does not show the instantaneous value of the speed. Similar to the above discussed back-propagation for the current feedback, an interpolation of the measured values might be used to reconstruct the speed instantaneous values (here with filter of first order).

$$w(k) = 1,5\,\overline{w}(k) - 0,5\,\overline{w}(k-1) \tag{4.2}$$

$\overline{w}$ = Integrated actual feedback value

a) Measurement of speed using an incremental encoder (IE)

As is well known, the IE delivers two by 90^0 phase-shifted signals A and B (fig. 4.3) in the form of square pulses, where the impulse number per revolution is given by the construction of the encoder device. The additional channel zero provides once per revolution a zero reference impulse, which is normally congruent with the edges of one of the channels A or B. By measuring of the impulse frequency f_M the speed n can be determined.

$$n = \frac{60\,f_M}{z_{IE}} = [rpm] \tag{4.3}$$

f_M =   Frequency [in Hz] of the impulse sequence

z_{IE} =   Number of impulses per revolution

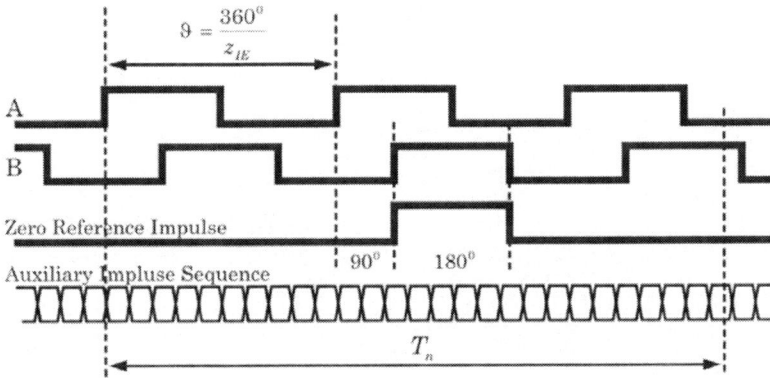

Fig. 4.3 Use of an incremental encoder (IE) to measure the speed

By evaluation of the phase relationship between the signals A and B the direction of rotation can be found. For a given IE and thus defined f_M, the maximal measurable speed n_{max} can be calculated by equation (4.3). Or conversely, with predefined motor and maximal speed n_{max}, the maximal necessary f_M can also be calculated by (4.3).

If the rotor position is described by ϑ, and the sampling period of the speed control by T_n, then the frequency measurement transforms into a counting of the impulses A or B within T_n following the equation:

$$n = \frac{\vartheta(k) - \vartheta(k-1)}{T_n} \tag{4.4}$$

k = 0, 1, 2, ... , ∞ = Sampling time instants

The impulse counting alone does not suffice for a precise measurement at very low-speeds, and here has to be amended by a time measurement where the time for the passing of a certain angular sector is measured with the help of an additional higher-frequent impulse sequence (fig. 4.3). From the equation (4.4) the speed can be obtained as follows:

$$n = \frac{\vartheta}{t(k) - t(k-1)} \quad \text{with } \vartheta = \frac{360^0}{z_{IE}} = \frac{1 \text{ rev}}{z_{IE}} \tag{4.5}$$

The measurement resolution can be found by generalization of the equations (4.4), (4.5). Equation (4.4) can be rewritten as follows:

$$n = \frac{60}{z_{IE} T_n} n_{IE} \quad \text{in } \left[\text{min}^{-1}\right] \tag{4.6}$$

n_{IE} = Number of impulses counted during T_n

The resolution of the speed measurement, using pure impulse counting, can be obtained after derivation of equation (4.6).

$$\frac{dn}{dn_{IE}} = \frac{60}{z_{IE} T_n} \quad \text{or } \Delta n \approx \frac{60}{z_{IE} T_n} \Delta n_{IE} \tag{4.7}$$

With, for example,

$$\Delta n_{IE} = 1 \qquad \text{(Only 1 impulse is counted during } T_n\text{)}$$

$$z_{IE} = 1024 \qquad \text{(IE produces 1024 pulses per revolution)}$$

$$T_n = 1 ms \qquad \text{(Sampling period)}$$

a resolution of $\Delta n \approx 58{,}6$ min^{-1} is obtained. With the help of impulse quadruplication this result can be improved essentially. The equation (4.5) for period measurement can be rewritten as follows:

$$n = \frac{60}{z_{IE}} \frac{1}{\tau} \quad \text{in } \left[\text{min}^{-1}\right] \tag{4.8}$$

τ = in [s] measured time for the passed angle sector $360^0/z_{IE}$

The first derivation of equation (4.8) delivers:

$$\frac{dn}{d\tau} = -\frac{60}{z_{IE} \tau^2} \quad \text{or } \Delta n \approx -\frac{60}{z_{IE} \tau^2} \Delta \tau \tag{4.9}$$

Using a 20 MHz impulse sequence (time resolution $\Delta \tau = 50$ nsec) and a 10-bit counter (maximal measurable time $\tau = 2^{10} \times 50$ ns $= 51{,}2$ μs) a resolution of $\Delta n \approx 1{,}1176$ rpm can be obtained for very low-speeds.

Another possibility to reach a high-resolution result for low-speeds is to use an IE with approximately sinusoidal output signals A and B (fig. 4.4).

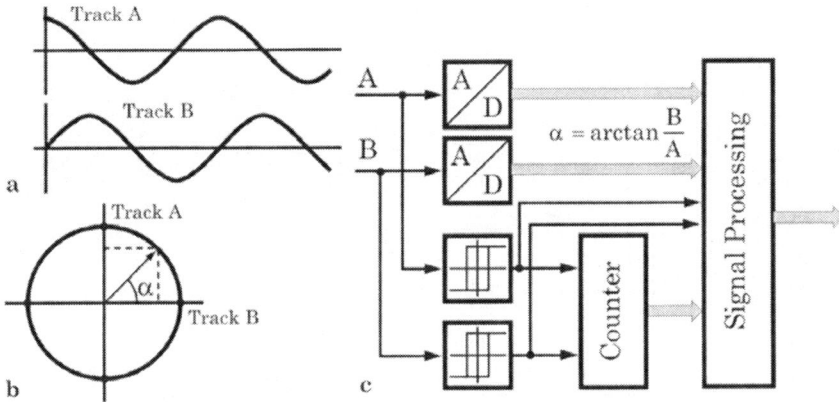

Fig. 4.4 Use of an incremental encoder with sinusoidal output signals for speed measuring

After zero-crossing detection the signals can be processed exactly as in the case of an usual IE. With two additional A/D converters and high-frequency sampling within a signal period, the speed is obtained from:

$$n = \frac{d\alpha}{dt} = \frac{d}{dt}\left(\arctan\frac{B}{A}\right) \tag{4.10}$$

b) Measurement of speed using a resolver

The construction of a resolver is shown in the figure 4.5a. The resolver consists of two parts. The mobile part (the rotor) is fixed to the motor shaft and contains the primary excitation winding, fed by a rotating transformer with an excitation signal of approx. 2...10 kHz. The static part (the stator) contains two secondary (sine, cosine) windings, which are mechanically displaced at 90 deg against each other.

In principle two methods for processing the resolver signals exist. The first one is called angle comparison (fig. 4.6a) and is implemented in integrated circuits like AD2S82 or AD2S90[1]. The angle comparison is carried out with the help of a multiplier (RM: Ratio Multiplier) at the input, which calculates an angle error. After the multiplier (fig. 4.5), the following results are obtained at the outputs of sine/cosine channels:

$$t_c u_0 \sin(\omega t)\sin\vartheta\cos\vartheta_M \quad \text{and} \quad t_c u_0 \sin(\omega t)\cos\vartheta\sin\vartheta_M \tag{4.11}$$

---

[1] from the firm Analog Devices

Substracting the signals of the equation (4.11) from each other yields the error signal:

$$\Delta\vartheta = t_c\, u_0 \sin(\omega t)\left(\sin\vartheta\,\cos\vartheta_M - \cos\vartheta\,\sin\vartheta_M\right)$$
$$= t_c\, u_0 \sin(\omega t)\sin(\vartheta - \vartheta_M) \qquad (4.12)$$

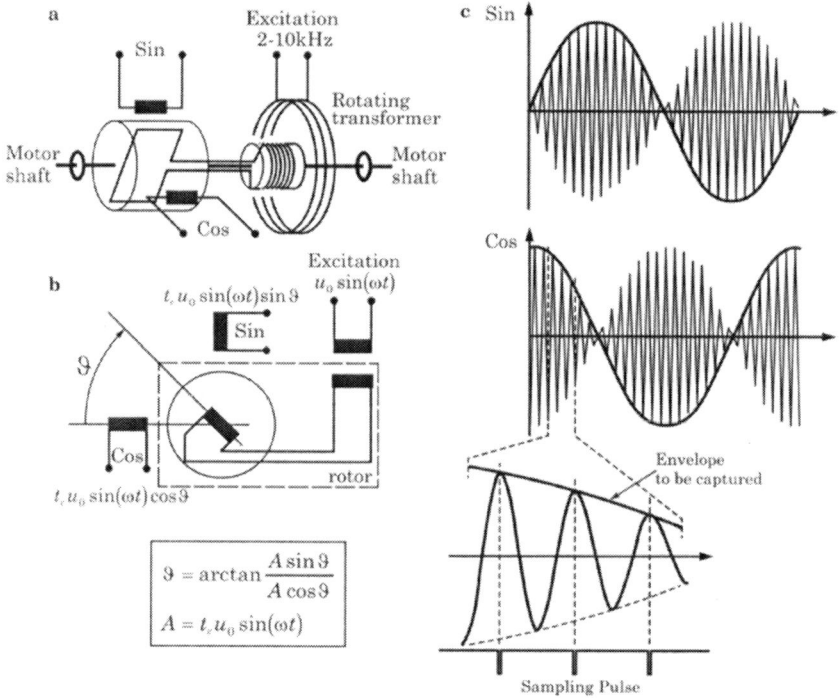

Fig. 4.5 Principle mechanical (**a**) and schematic (**b**) construction of the resolver with its output signals (**c**). t_c: transmission coefficient; u_0: amplitude of excitation signal

According to equation (4.12) the error $\Delta\vartheta$ disappears if the difference $\vartheta - \vartheta_M$ becomes zero. To achieve this, the error signal is fed via a phase sensitive detector (PSD) to an integrator or a PI controller whose output controls a voltage-controlled oscillator (VCO). A up/down counter (UDC) counts the impulses coming from the output of the VCO, in which the counting direction of the UDC depends on the sign of the error signal. This way the phase error will be eliminated with the help of an integrator or a PI controller. The dynamics of the measurement depends on the dynamics of the control loop which poses a considerable disadvantage for the complete system.

The measuring dynamics can be increased significantly if the envelope of the resolver signals (see figure 4.5c) can be captured directly, i.e.

always at the peak value of the curve. With the help of two A/D converters (fig. 4.6b) this can be realized easily provided a *strict synchronization between the measurement sampling, control and modulation,* is observed and taken care of already in the hardware design. Furthermore it has to be considered that the resolver signals are susceptive to noise and distortions through the transmission paths between motor and electronics which may result into loss of the original synchronization and a signal correction (usually by software) becomes necessary.

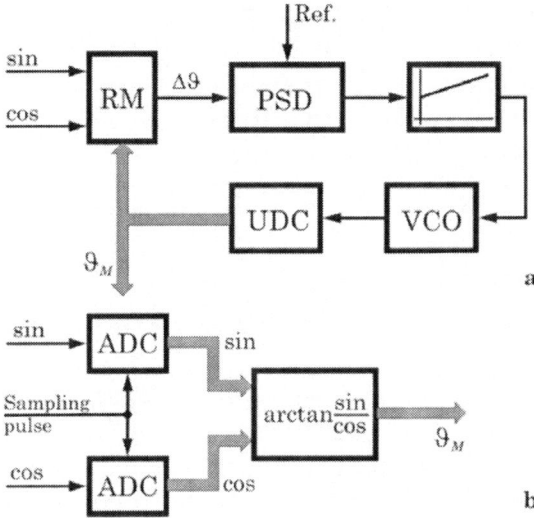

Fig. 4.6 Methods for evaluation of resolver signals: Angle comparison (**a**) and signal sampling with two A/D converters (**b**)

From the relation:

$$\vartheta = \arctan \frac{x}{y} \quad \text{with} \quad x = \sin\vartheta; \, y = \cos\vartheta$$

the total differential of ϑ can be derived easily:

$$d\vartheta = y\,dx - x\,dy \tag{4.13}$$

and:

$$\Delta\vartheta \approx y\,\Delta x - x\,\Delta y \tag{4.14}$$

With $\Delta x = \Delta y = 2^{-z_{ADC}}$, in which z_{ADC} is the resolution of the A/D converter, the corresponding resolution of the measured angle can be assessed.

Considering that the derivatives dx/dt und dy/dt can assume values independent of z_{ADC} at the actual time, and from the relationship for the speed n:

$$n = \frac{60}{2\pi z_p} \omega = \frac{60}{2\pi z_p} \frac{d\vartheta}{dt} \left[\text{min}^{-1}\right] \tag{4.15}$$

the total differential for n can be derived after some conversions:

$$dn = n(x\,dx + y\,dy) \tag{4.16}$$

From (4.16) results:

$$\Delta n \approx n(x\,\Delta x + y\,\Delta y) \tag{4.17}$$

With the equation (4.17) also the speed resolution, depending on the operating point, can be assessed using $\Delta x = \Delta y = 2^{-z_{ADC}}$.

Because the resolver provides the absolute position information, it can be used advantageously in synchronous drives. In addition, the resolver is robust against external influences like high temperatures or magnetic interference fields. With respect to the measuring precision the resolver, however, cannot achieve the high resolution of the IE with analog or sinusoidal output signals.

4.3 Possibilities for sensor-less acquisition of the speed

The idea to save the speed sensor, and to reduce not only the costs but also to increase the reliability, because mechanical parts and the sensitive galvanic connection between sensor and actuator are omitted, was the motivation for numerous research in the last two decades. In principle the developed methods can be divided into three groups:

1. Stator flux orientated methods like Direct Torque Control (DTC), Natural Field Orientation (NFO).

2. Rotor flux orientated methods, following the principle of a Kalman Filter (KF) or a Model Reference Adaptive System (MRAS).

3. Methods which use machine specific effects (unbalance, slots on stator and rotor side etc.).

With respect to the theoretical approaches, the solutions to this problem are very different and partly based on special effects so that not all of them can be discussed in the context of this chapter. Because of the many advantages compared to the stator flux orientation, this chapter exclusively deals with the rotor or pole flux orientated drive systems which are very widespread in practice. Consistently only examples of rotor flux orientated methods will be discussed because the methods to be selected, must be suited for integration into the overall control system. The methods based on the use of machine specific effects can also be used very well in systems controlled with field orientation.

In the area of higher frequencies the speed sensor-less operation works without problems for all methods in the case of an asynchronous drive. The critical area is the area around standstill. The results published in the last decade have led to the conclusion that *zero stator frequency in the case of the IM represents a virtually not observable point and therefore cannot* be controlled correctly *with conventional methods like DTC, NFO, KF, and MRAS.* At set points near zero speed and under influence of a strong load, the rotor can always drift away without the system reacting to it. Thereat the rotor flux vector rotation stops. This is primarily based on the fact, that the magnetization of the slowly rotating rotor of the IM can be easily changed by the (almost) still standing rotor flux vector. Only if the mechanical frequency of motion reaches a certain limit (approximately the slip frequency) and thus the magnetic reversal is no longer possible, the speed can be calculated correctly again. *A clean reversal across the speed zero is always possible, though.* Use of the above mentioned methods, including the already commercialized DTC, always implies theoretically unsolidated detours.

In the case of a PMSM drive, the standstill is less critical because the pole flux is built up permanently. Therefore a magnetic reversal process can not take place, and the moving pole delivers information for the estimation already at low-speeds. Two problems must be solved here:

- The *initial position of the pole flux* must be identified: For the case of an asymmetric rotor build-up (e.g. salient pole machines or full pole machines with only few magnets on the rotor surface), many useable approaches can be found in the literature. The question is still relatively open for machines with exclusive full pole quality (i.e. the magnets are assembled in a larger number, and thus divided up finer) on the rotor surface thanks to the high energy density and the improved construction.
- The development of a *method for the speed sensor-less control* of the drive. The variety of the useable methods is similar as in the case of the IM.

The methods based on the use of the machine specific effects are best suitable both for IM and for PMSM. The unbalances in the mechanical construction and the slots on stator and rotor side are mirrored in the harmonics of the stator currents independent of their fundamental frequency.

In this chapter only two application examples are presented for the speed sensor-less control of the IM and PMSM. For the IM the control system has the principle structure of figure 4.7. It can still be recognized

that the structure of figure 4.7 also applies to the case of the PMSM drive
if the flux controller is dropped and the corresponding algorithm is
implemented in the context of the "speed adaptive observer".

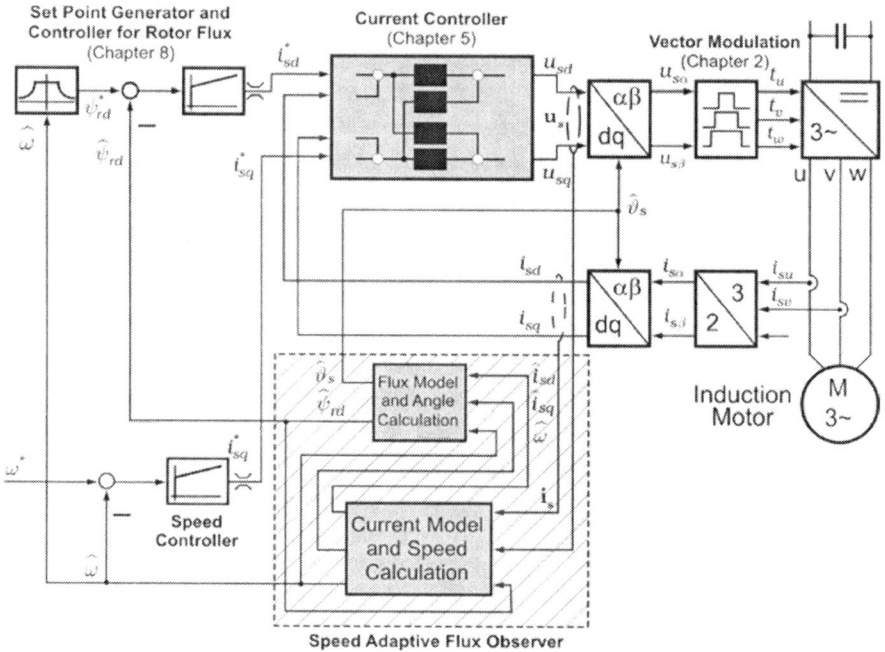

Fig. 4.7 General structure of speed sensor-less and rotor flux orientated control of
an IM drive in *dq* coordinates

4.3.1 Example for the speed sensor-less control of an IM drive

As is well known, the IM can be completely described by the state
model (3.41) (cf. section 3.2) electrically. If we start from the assumption
that the machine parameters are time-invariant, then only **A** in the equation
(3.41) depends on the speed and must be updated on-line with *a measured
or estimated speed*. The estimated quantities are denoted with an index „∧"
in the following.

With the model (3.41) a Luenberger observer can be used to reconstruct
the state vector (fig. 4.8a).

$$\frac{d\hat{\mathbf{x}}}{dt} = \hat{\mathbf{A}}\,\hat{\mathbf{x}} + \mathbf{B}\,\mathbf{u_s} + \mathbf{K}\left(\mathbf{i_s} - \hat{\mathbf{i}_s}\right) \tag{4.18}$$

K = Correction matrix

For the case of measured speed numerous approaches to design $\mathbf{K}$ (e.g. with the help of pole assignment) have been presented. The observer (4.18) then delivers only estimates for the not measurable rotor flux. If the speed, regarded as a system parameter in this model, shall be estimated together with the state quantities, the structure must be extended like shown in figure 4.8b.

Using the definition of the state error $\mathbf{e}$:

$$\mathbf{e} = \mathbf{x} - \hat{\mathbf{x}} \tag{4.19}$$

the following error state equation is obtained after subtracting (3.41) and (4.18):

$$\frac{d\mathbf{e}}{dt} = (\mathbf{A} + \mathbf{KC})\mathbf{e} + \Delta\mathbf{A}\,\hat{\mathbf{x}} \tag{4.20}$$

with:

$$\Delta\mathbf{A} = \mathbf{A} - \hat{\mathbf{A}} = \left[\begin{array}{c|c} \mathbf{0} & \Delta\omega\dfrac{1-\sigma}{\sigma}\mathbf{J} \\ \hline \mathbf{0} & -\Delta\omega\,\mathbf{J} \end{array} \right] \tag{4.21}$$

$$\mathbf{J} = \begin{bmatrix} 0 & 1 \\ -1 & 0 \end{bmatrix}; \; \Delta\omega = \omega - \hat{\omega} = \text{Parameter error}$$

The state estimation techniques using a speed adaptation like in figure 4.8b, are part of the category of methods with model reference adaptive systems (MRAS) in which the motor (the process) plays the role of the reference model.

Because of the nonlinear (process) behavior the stability aspect must be included at the design of such systems from the beginning (cf. [Isermann 1988], chapter 22.3). The stability proof can be carried out either using Popov's method of hyper stability (e.g. [Tajima 1993]) or the direct method of Ljapunov (e.g. [Kubota 1993 and 1994]). The latter will be used in the following.

The Ljapunov function V for the error equation (4.20) is chosen to contain both the state error $\mathbf{e}$ and the parameter error $\Delta\omega$.

$$V = \mathbf{e}^T\mathbf{e} + \frac{[\Delta\omega]^2}{\lambda} \tag{4.22}$$

λ = Positive constant

The first derivation of V yields:

$$\frac{dV}{dt} = \mathbf{e}^T\left[(\mathbf{A} + \mathbf{KC})^T + (\mathbf{A} + \mathbf{KC})\right]\mathbf{e}$$

$$- 2\Delta\omega\left(\tilde{\imath}_{s\alpha}\,\hat{\psi}'_{r\beta} - \tilde{\imath}_{s\beta}\,\hat{\psi}'_{r\alpha}\right)\frac{1-\sigma}{\sigma} + 2\Delta\omega\frac{d\hat{\omega}}{dt}\frac{1}{\lambda} \tag{4.23}$$

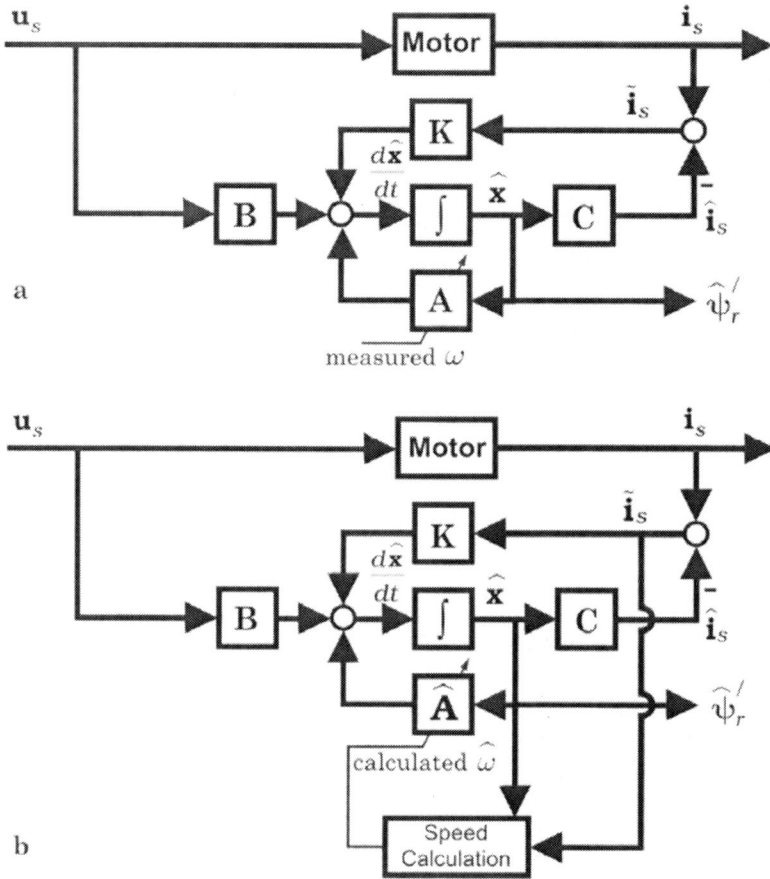

Fig. 4.8 General structure of the rotor flux observer using (a) measured and (b) estimated speed

There $\tilde{i}_{s\alpha} = i_{s\alpha} - \hat{i}_{s\alpha}$; $\tilde{i}_{s\beta} = i_{s\beta} - \hat{i}_{s\beta}$ are components of the state error vector **e**. The right side of the equation (4.23) contains 3 terms. In order for the system to remain stable, the following conditions must be fulfilled.

1.   **K** must be chosen to ensure the negative definiteness of the first term.

2.   The estimation algorithm must be designed so that the second and third terms compensate each other, i.e. the sum of the two terms is zero.

In the references [Kubota 1993 and 1994], a frequency dependent correction matrix **K** in the following form is suggested.

$$\mathbf{K} = \begin{bmatrix} k_1 & k_2 & k_3 & k_4 \\ -k_2 & k_1 & -k_4 & k_3 \end{bmatrix} \tag{4.24}$$

$$k_1 = -\frac{k-1}{\sigma}\left(\frac{1}{T_r}+\frac{1}{T_s}\right); k_2 = (k-1)\widehat{\omega}; k_3 = \frac{k-1}{1-\sigma}\left(\frac{1}{T_r}-\frac{k}{T_s}\right); k_4 = \frac{(k-1)\widehat{\omega}\sigma}{1-\sigma}$$

$$\tag{4.25}$$

The idea behind is to fix the poles of the observer (using the constant $k>0$) proportionally to those of the motor so that the observer (exact like the process or the motor) remains stable. It was found experimentally that k must be chosen in the range of 1 ... 1.5. With this choice the method was tested successfully on different motors, even when only parameterized from name plate data.

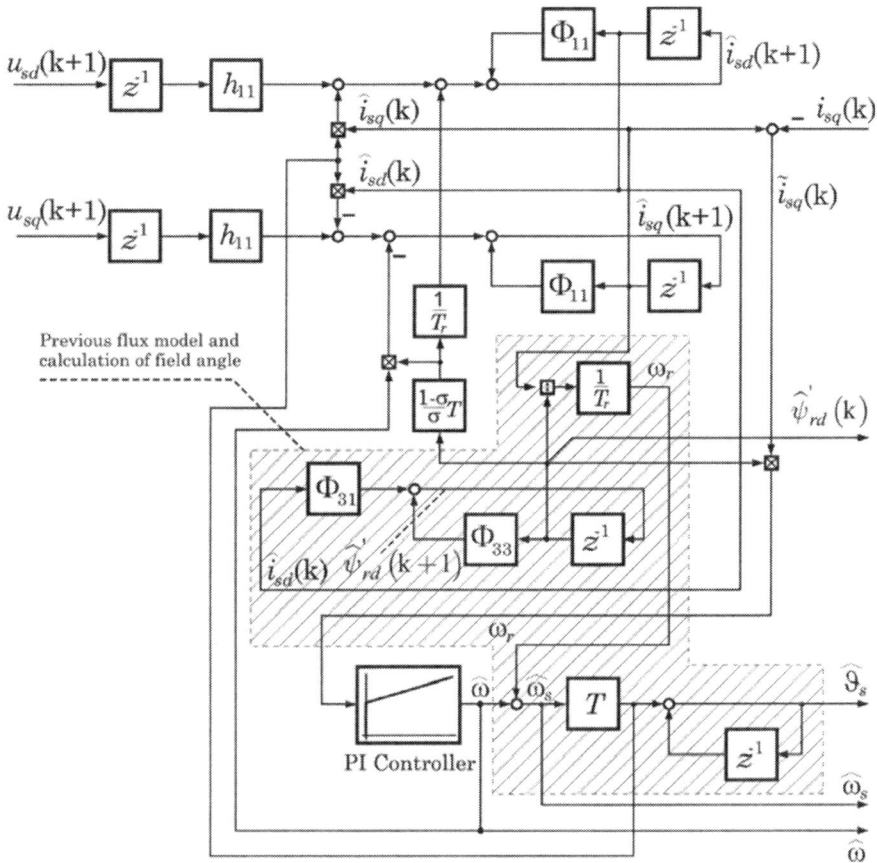

Fig. 4.9 Block circuit diagram of the speed adaptive observer for the calculation of the mechanical angular speed $\widehat{\omega}$ and the rotor flux $\widehat{\psi}'_{rd}$

In order to fulfil the above second condition concerning the stability, the estimated angular speed $\hat{\omega}$ must fulfill the following equation:

$$\frac{d\hat{\omega}}{dt} = \lambda \frac{1-\sigma}{\sigma} \left(\tilde{i}_{s\alpha} \hat{\psi}'_{r\beta} - \tilde{i}_{s\beta} \hat{\psi}'_{r\alpha} \right) \tag{4.26}$$

Considered that the speed can change fast, equation (4.26) can be augmented to the following PI algorithm to calculate $\hat{\omega}$.

$$\hat{\omega} = K_P e_\omega + K_I \int e_\omega dt \quad \text{with} \quad e_\omega = \tilde{i}_{s\alpha} \hat{\psi}'_{r\beta} - \tilde{i}_{s\beta} \hat{\psi}'_{r\alpha} \tag{4.27}$$

$K_P, K_I:$ Gain factors

For the implementation of the described method the speed error must be transformed into dq coordinates first:

$$e_\omega = -\tilde{i}_{sq} \hat{\psi}'_{rd} \quad \text{with} \quad \tilde{i}_{sq} = i_{sq} - \hat{i}_{sq} \tag{4.28}$$

For a step-by-step design we assume first the value $k = 1$ (i.e. flux model without correction). That means, the speed adaptive observer contains the calculation of the current and flux model according to the equation (3.55) (cf. figure 4.7: the hatched area) as well as the PI algorithm (4.27). The angular speed ω_s on the stator side or the stator frequency arises from the following equation:

$$\omega_s = \hat{\omega} + \frac{i_{sq}(k)}{T_r \hat{\psi}'_{rd}(k)} \tag{4.29}$$

The hatched area in figure 4.7 is represented in detail in figure 4.9. The processing of the current control loop is divided into 4 steps (see fig. 4.10, left half), with the estimation algorithm of ω (see fig. 4.9) integrated into the second step, detailed in the right half of the figure 4.10.

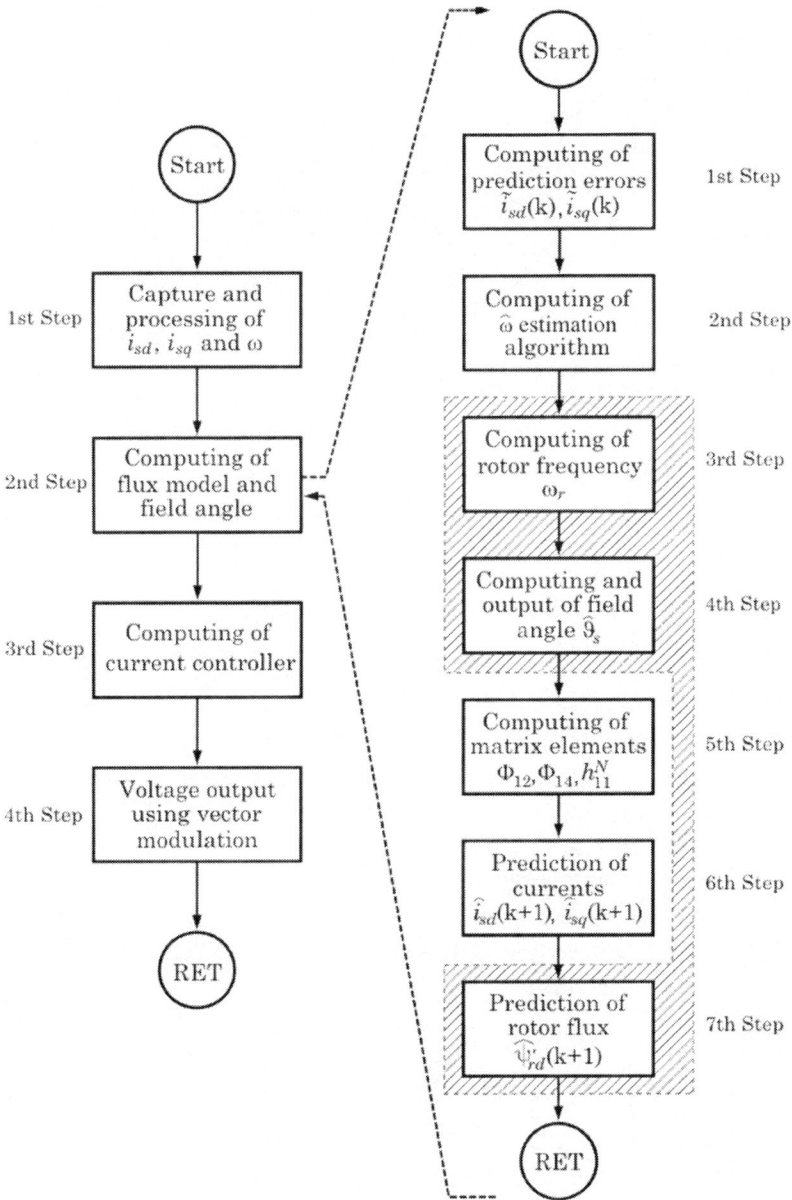

Fig. 4.10 Program flowchart for the implementation of the algorithm from figure 4.9

The speed reversal with field weakening in figure 4.11 illustrates the functionality of the presented method which may be implemented in practical systems easily.

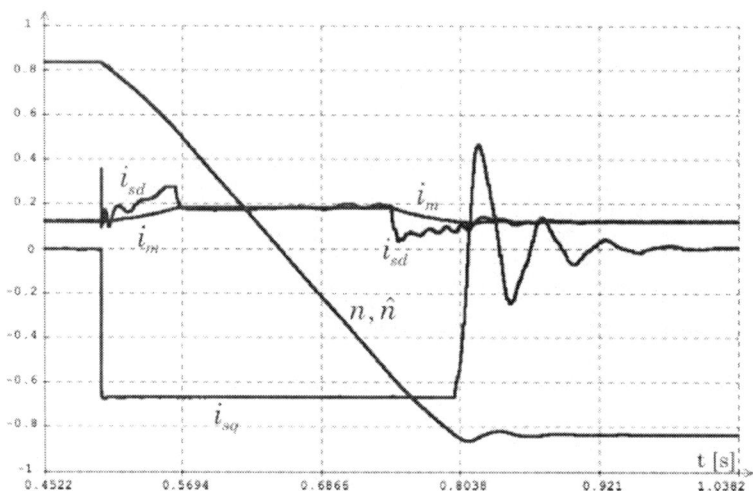

Fig. 4.11 Speed reversal of a sensor-less controlled IM drive: ±3000 min^{-1} with field weakening

The dimensioning of the PI compensation controller (cf. fig. 4.9) is important for the calculation of the speed $\hat{\omega}$. By integrating $\hat{\omega}$ the mechanical angle $\hat{\vartheta}$ is calculated, which forms together with the load angle the transformation angle $\hat{\vartheta}_s$ for the voltage vector output and feedback transformation. I.e. the dynamics of the estimation of ω shows up directly in the innermost loop - the current control loop (cf. figure 4.12). To consider the estimation dynamics in the current controller design, the transfer function $G_e(s)$ is needed.

The transfer function $G_e(s)$ can be derived under the conditions, that:

- the speed in small-signal response only effects the q-axis (cf. 2nd equation from (3.44)), and

- $\psi'_{rd} = \hat{\psi}'_{rd} = i_{sd} = \hat{i}_{sd}$ can be assumed for the d-axis.

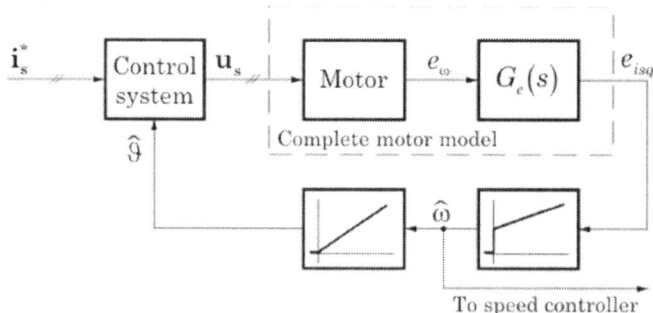

Fig. 4.12 To the necessity of the transfer function $G_e(s)$

Using (3.44) the following equations are obtained in the Laplace domain:

$$\begin{cases} \text{Motor: } s\,i_{sq} = -\omega_s i_{sd} - \frac{1}{\sigma}\left(\frac{1}{T_s}+\frac{1-\sigma}{T_r}\right)i_{sq} - \frac{1-\sigma}{\sigma}\omega\psi'_{rd} + \frac{1}{\sigma L_s}u_{sq} \\[3mm] \text{Model: } s\hat{i}_{sq} = -\hat{\omega}_s \hat{i}_{sd} - \frac{1}{\sigma}\left(\frac{1}{T_s}+\frac{1-\sigma}{T_r}\right)\hat{i}_{sq} - \frac{1-\sigma}{\sigma}\hat{\omega}\hat{\psi}'_{rd} + \frac{1}{\sigma L_s}u_{sq} \end{cases}$$

$$(4.30)$$

With

$$e_{isq} = \tilde{i}_{sq} = i_{sq} - \hat{i}_{sq} \; ; \; e_\omega = \omega - \hat{\omega} \; ; \; \omega_s = \omega + \frac{i_{sq}}{T_r\,\psi'_{rd}} \; ; \; \hat{\omega}_s = \hat{\omega} + \frac{\hat{i}_{sq}}{T_r\,\hat{\psi}'_{rd}}$$

from the subtraction of the two equations in (4.30) and after some remodelling we obtain:

$$\frac{e_{isq}(s)}{e_\omega(s)} = \frac{V_e}{1+sT_e} \quad \text{with } T_e = \frac{\sigma T_s T_r}{T_s+T_r} \; ; \; V_e = -\frac{\psi'_{rd}T_e}{\sigma} \qquad (4.31)$$

K_p, K_I should be chosen to essentially compensate the delay from (4.31) for the closed control loop.

4.3.2 Example for the speed sensor-less control of a PMSM drive

As already mentioned at the beginning of chapter 4.3, two questions must be solved for the speed sensor-less and field orientated operation of a PMSM drive: the *identification of the initial position of the rotor or of the pole flux* and the integration of a *method for the speed sensor-less control*.

The most known publications about the identification of the initial position deal with the case of salient pole machines, where the difference between the direct and the quadrature stator inductance – measured in the d- and q-axis – is relatively large. This makes a relatively simple off-line identification of the rotor position possible either by an indirect measurement of the inductances or by an evaluation of the currents caused by scanning of the total rotor surface with identical voltage transients. Thanks to new magnet materials with a very high energy density and improved construction techniques the magnets of modern machines are finer distributed and fastened on the rotor surface. However, this improvement with respect to the drive quality aggravates the chance to identify the pole flux position. An interesting approach to solve these difficulties with moderate processor power has been presented by [Brunotte 1997].

The PMSM is different from the IM physically only by the way of the magnetization: In the IM $\mathbf{i}_m$ or ψ'_r must be built up, whereas the pole flux is permanently available in the PMSM. Therefore it can be assumed, that the Ljapunov stability approach which led to the error model (4.28) for the IM can be used here as well with the speed error signal:

$$e_\omega = -\tilde{i}_{sq} \frac{\psi_p}{L_{sd}} \quad \text{with} \quad \tilde{i}_{sq} = i_{sq} - \hat{i}_{sq} \tag{4.32}$$

As in the case of the asynchronous drive the current error in the q-axis is used as an input signal for the ω - PI - estimation controller. Because of the preferred axis of the rotor flux, an error signal for the position angle which helps to eliminate the position error from the beginning must be found. The following considerations help to find a solution. From (3.64) the following equations can be obtained:

$$\left| \begin{array}{l} \text{Motor:} \quad s\,i_{sd} = -\dfrac{1}{T_{sd}}i_{sd} + s\vartheta_s \dfrac{L_{sq}}{L_{sd}}i_{sq} + \dfrac{1}{L_{sd}}u_{sd} \\[2em] \text{Model:} \quad s\hat{i}_{sd} = -\dfrac{1}{T_{sd}}\hat{i}_{sd} + s\hat{\vartheta}_s \dfrac{L_{sq}}{L_{sd}}\hat{i}_{sq} + \dfrac{1}{L_{sd}}u_{sd} \end{array} \right. \tag{4.33}$$

After the subtraction of the two above equations and some rewriting the following linear relation arises, under the assumption that the ω - transients have died out and the load is constant:

$$\frac{e_{isd}(s)}{e_\vartheta(s)} = \frac{s\,T_{sq}\,i_{sq}}{1 + s\,T_{sd}} \tag{4.34}$$

The fault model (4.34) means, that the current error in the d- axis can be used as a correction signal for the rotor position. A compensation controller with I behavior, whose output quantity is added to the flux angle, will suffice for this purpose. The figure 4.13 shows the speed reversal of a PMSM drive controlled using this method.

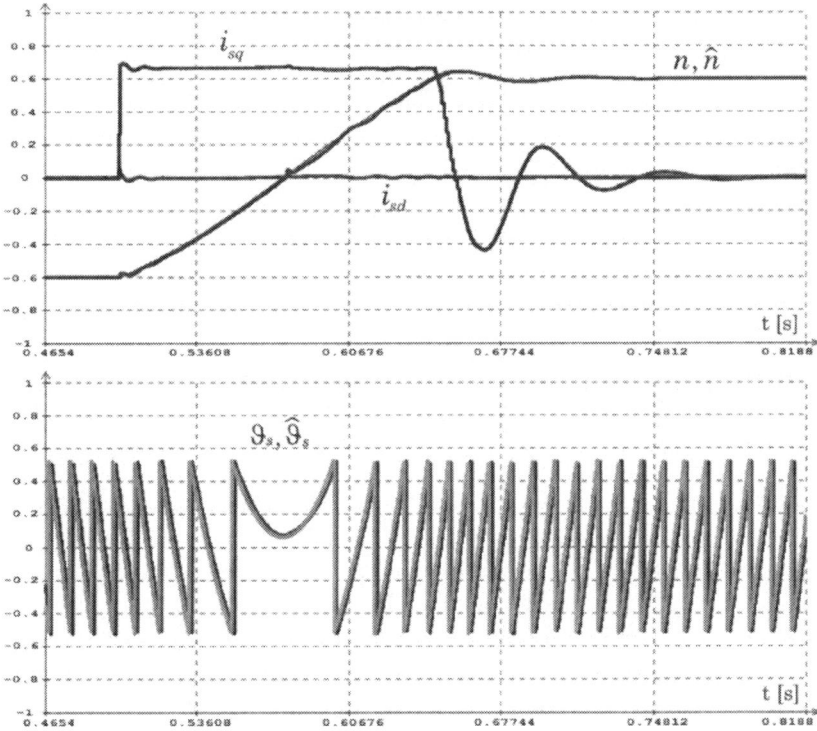

Fig. 4.13 Speed reversal of a speed sensor-less and field-orientated controlled PMSM drive: Currents and speed (**top**), rotor or flux angle (**bottom**)

4.4 Field orientation and its problems

After the essential features of the field-orientated control and important control structures were introduced in chapter 1, some questions of the realization shall be discussed in more detail now. For asynchronous drives the calculation of the rotor flux, which is not measurable without additional costs, is decisive for a successful realization and satisfying control performance. Different models to solve this task will be compared in this chapter. Because control and PWM voltage generation work discontinuously, but the machine, of course, represents a continuous system, a number of issues arise in the context of the interaction of both components. These issues, if disregarded, can have a negative influence on the control accuracy and stability. In the last part of this chapter finally some concrete discretization effects and respective countermeasures will be worked out.

4.4.1 Principle and rotor flux estimation for IM drives

To begin with, some basics of the field orientated control shall be summarized again for the better complete understanding of the matter. As known, the basic idea of the field orientated control is to develop a control structure for the IM similar to that for the DC machine. That means in detail:

1. The process models of torque and flux must be decoupled from each other.
2. At constant flux the torque equation should have a linear characteristic (linear relation between torque and torque-producing quantity).
3. In steady-state, all control variables should be DC quantities.
4. Torque and slip should be proportional. With this proportionality a breakdown-torque caused by the control is avoided, and the maximum torque is expressively determined by the available current or the available voltage.

The requirement 3 is fulfilled by using a reference coordinate system which rotates synchronously with the stator frequency ω_s. The point 2 can be fulfilled, if one axis of the coordinate system is chosen to coincide with the current or the flux vector. It can be shown that under all conceivable variants only the orientation to the vector of the rotor flux (e.g. $\psi_r = \psi_{rd}$, $\psi_{rq} = 0$) fulfils the remaining requirements and at the same time ensures a dynamically exact decoupling between torque and flux. Stator, rotor voltage and torque equations of the IM (cf. chapter 3 and 6) if splitted into their vector components, then may be rewritten as follows:

$$u_{sd} = R_s i_{sd} + \sigma L_s \frac{d i_{sd}}{dt} - \omega_s \sigma L_s i_{sq} + (1-\sigma)L_s \frac{d i_{md}}{dt} \tag{4.35}$$

$$u_{sq} = R_s i_{sq} + \sigma L_s \frac{d i_{sq}}{dt} + \omega_s \sigma L_s i_{sd} + (1-\sigma)L_s \omega_s i_{md} \tag{4.36}$$

$$0 = i_{md} + T_r \frac{d i_{md}}{dt} - i_{sd} \tag{4.37}$$

$$0 = \omega_r T_r i_{md} - i_{sq} \tag{4.38}$$

$$m_M = \frac{3}{2} z_p \frac{L_m}{L_r} \psi_{rd} i_{sq} = \frac{3}{2} z_p \frac{L_m^2}{L_r} i_{md} i_{sq} = \frac{3}{2} z_p (1-\sigma) L_s i_{md} i_{sq} \tag{4.39}$$

with: $i_{md} = \psi_{rd}/L_m = \psi'_{rd}$

In the context of current impression the very simple signal flowchart in figure 4.14 arises. According to (4.37) and (4.39) the current component i_{sd}

works as a control quantity for the rotor flux, and i_{sq} controls the torque at constant rotor flux.

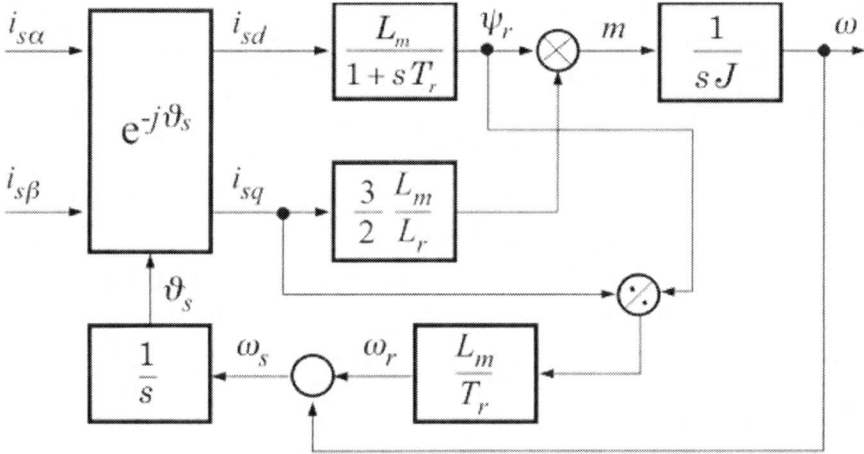

Fig. 4.14 Signal flowchart of the IM in field-orientated coordinates with current impression

Different variants for the implementation of the principle of the rotor flux or field orientation into a control system are conceivable and known. Besides the proposed method using impression of the current vector, all remaining methods differ from each other by their ways to find the modulus and the phase angle of the not directly measurable rotor flux. The exact knowledge of the flux phase angle in stator coordinates

$$\vartheta_s = \arctan\frac{\psi_{r\beta}}{\psi_{r\alpha}} = \vartheta_0 + \int \omega_s \, dt \tag{4.40}$$

is required for the exact transformation of all measured quantities into field coordinates. Using this angle the current vector in field coordinates is obtained to:

$$\mathbf{i}_s^f = \mathbf{i}_s^s \, e^{-j\vartheta_s} \tag{4.41}$$

In the simplest case the needed quantities can be calculated with the help of (4.38) from the reference values (set points) of flux and torque (field orientated feed-forward control [Schönfeld 1987] or indirect field orientation [Vas 1994]). A field orientated (feedback) control or a direct orientation uses a direct measuring of, or an estimation model fed by measured motor quantities, for the flux calculation and therefore has the essential advantage that the current motor state can be captured substantially more exact and independent of the quality of the inner current control loop (e.g. at insufficient voltage reserve). Because a direct flux measuring using Hall sensors or additional coils requires additional incursion in the motor, practically realized systems usually work according to the model method. Typical structures were already introduced and explained in chapter 1.

An analysis of the general voltage equations (ω_k = angular speed of the reference coordinate system)

$$\mathbf{u}_s = R_s \mathbf{i}_s + \sigma L_s \frac{\mathrm{d}\,\mathbf{i}_s}{\mathrm{d}t} + j\omega_k \sigma L_s \mathbf{i}_s + (1-\sigma)L_s\left(\frac{\mathrm{d}\,\mathbf{i}_m}{\mathrm{d}t} + j\omega_k \mathbf{i}_m\right) \quad (4.42)$$

$$0 = \left(1 + j(\omega_k - \omega)T_r\right)\mathbf{i}_m + T_r \frac{\mathrm{d}\,\mathbf{i}_m}{\mathrm{d}t} - \mathbf{i}_s \quad (4.43)$$

shows various possibilities for the calculation or estimation of the rotor flux vector. The different approaches can be distinguished by the used coordinate system and the measured quantities. With regard to the coordinate system characteristically the components of the flux vector are first calculated in stator coordinates, and in the second step modulus and phase angle are derived in field orientated coordinates. Essentially, the following models may be derived (cf. [Verghese 1988], [Zägelein 1984]). Model quantities are indicated with ^.

1. u_s - i_s - *Model in stator coordinates*

$$\frac{\mathrm{d}\,\hat{\mathbf{i}}_m}{\mathrm{d}t} = \frac{\mathbf{u}_s - R_s \mathbf{i}_s}{(1-\sigma)L_s} - \frac{\sigma}{1-\sigma}\frac{\mathrm{d}\,\mathbf{i}_s}{\mathrm{d}t} \quad (4.44)$$

The model immediately results from the suitable rearrangement of the stator voltage equation. From equation (4.44) it can be noticed that because of the open integration any mechanism for the elimination of state errors, caused by wrong initial values or disturbances, is missing. At low rotational speeds considerable precision problems are to expect because of the significant influence of the stator resistance.

From the rotor voltage equation the following

2. i_s - ω - *Model in stator coordinates*

$$\frac{\mathrm{d}\,\hat{\mathbf{i}}_m}{\mathrm{d}t} = (-\frac{1}{T_r} + j\omega)\mathbf{i}_m + \frac{1}{T_r}\mathbf{i}_s \quad (4.45)$$

This model can immediately be derived from the rotor voltage equation. In contrast to (4.44) it can be shown that the state error decays with the rotor time constant [Verghese 1988]. The algorithm contains an integration of sinusoidal input quantities with the corresponding problems at discrete realization, though. This difficulty can be avoided by combination of both models to the following

3. i_s - u_s - ω - *Model in stator coordinates*

The derivation of the flux in (4.42) is eliminated after substituting (4.43):

$$\hat{\mathbf{i}}_m = \frac{-\mathbf{u}_s + (R_s + (1-\sigma)R_r)\mathbf{i}_s + \sigma L_s \dfrac{\mathrm{d}\,\mathbf{i}_s}{\mathrm{d}\,t}}{(1-\sigma)L_s\left(\dfrac{1}{T_r} + j\omega\right)} \tag{4.46}$$

In equation (4.46) no integration is needed anymore, but three measured quantities are to be fed to the model.

The combination of the first two models to *observers* (exactly: observer of reduced order) allows additionally to influence the system dynamics and with dedicated design also the improvement of the parameter sensitivity [Zägelein 1984].

From the rotor voltage equation in field coordinates the following straightforward

4. i_s - ω - *Model in field coordinates*

$$\frac{\mathrm{d}\,\hat{i}_{md}}{\mathrm{d}\,t} = \frac{1}{T_r}\left(-\hat{i}_{md} + \hat{i}_{sd}\right) \tag{4.47}$$

$$\hat{\vartheta}_s = \hat{\omega}_s = \omega + \frac{\hat{i}_{sq}}{T_r\hat{i}_{md}} \tag{4.48}$$

is obtained. Because the currents here are available only after the coordinate transformation, they also were indicated as model quantities. For this model the comparatively low realization effort can particularly be noticed. Special problems in certain speed ranges do not exist. However, all models derived from the rotor equation have the common property that the precision of the phase angle ϑ_s *strongly depends on the temperature-dependent rotor time constant.*

5. *"Natural Field Orientation" [NFO, Jönsson 1991, 1995]*

Stator frequency and flux phase angle may also be calculated directly from the stator equations. In the NFO approach the calculation is divided between stator and field coordinate system. In stator coordinates the EMF voltage is calculated from equation (4.42):

$$\mathbf{e} = \mathbf{u}_s - R_s\mathbf{i}_s - \sigma L_s\frac{\mathrm{d}\,\mathbf{i}_s}{\mathrm{d}\,t} = (1-\sigma)L_s\frac{\mathrm{d}\,\mathbf{i}_m}{\mathrm{d}\,t} \tag{4.49}$$

After transformation into field coordinates the stator frequency can be derived from the quadrature component:

$$\omega_s = \frac{e_q}{(1-\sigma)L_s i_{md}} \tag{4.50}$$

In the original method the flux is controlled in open loop, i.e. the reference value is used for i_{md}. However, it is also possible to include a flux control loop for the flux magnitude (4.47) additionally. Neither rotor quantities nor the speed are needed to calculate the phase angle, at least for feed-forward controlled or constant flux. In this respect a close relationship to the u_s - i_s - model in stator coordinates exists. The voltage integration is transferred to an integration of the stator frequency. Similar precision problems at small rotational speeds may be supposed.

So far, the method does not yet contain any speed model. Such an extension is described in [Jönsson 1995], although within a stator flux related control loop.

Finally the possibility of an

6. u_s - ω - Model

shall yet be mentioned. To its derivation the currents are eliminated in the voltage and flux equations by mutual substitution, and the stator flux is kept as an auxiliary variable.

In all described models, the rotor magnetizing current i_m was used as an equivalent for the rotor flux magnitude in the first place. The actual rotor flux magnitude would have to be calculated from $\psi_{rd} = L_m(|\mathbf{i}_\mu|)i_{md}$ and

$$|\mathbf{i}_\mu| = \sqrt{\left(\frac{L_{r\sigma}}{L_r}i_{sd} + \frac{L_m}{L_r}i_{md}\right)^2 + \left(\frac{L_{r\sigma}}{L_r}i_{sq}\right)^2} \tag{4.51}$$

Because of the proved positive properties over the whole relevant range of stator frequencies, and because of the simple feasibility, the i_s - ω - model (4.47), (4.48) in field coordinates is often preferred in the practice. Additionally to that, the current measurements are anyway available and high-quality speed controlled drives are equipped with speed measuring facilities.

Consideration of the magnetic saturation is required for high dynamics requirements at operation with variable rotor flux. Corresponding control approaches will be discussed in detail in chapter 6.2. A relatively simple and often satisfactorily used model shall be presented here. For the rotor voltage equation in the arbitrary orientated (rotation frequency ω_k) *coordinate system* the following equation can be obtained:

$$0 = \mathbf{i}_m - \mathbf{i}_s + \frac{L'_m(|\mathbf{i}_m|) + L_{r\sigma}}{R_r}\frac{d\,\mathbf{i}_m}{d\,t} + j(\omega_k - \omega)\frac{L_m(|\mathbf{i}_m|) + L_{r\sigma}}{R_r}\mathbf{i}_m \tag{4.52}$$

Split into components, the following equations are arrived at

$$\hat{i}_{md} = \hat{i}_{sd} + \frac{L_m'(\hat{i}_{md}) + L_{r\sigma}}{R_r}\frac{d\,\hat{i}_{md}}{dt} \tag{4.53}$$

$$\overset{\bullet}{\omega}_s = \hat{\omega}_s = \omega + \frac{\hat{i}_{sq}}{\hat{i}_{md}}\frac{R_r}{L_m(\hat{i}_{md}) + L_{r\sigma}} \tag{4.54}$$

with $L_m'(|\mathbf{i}_\mu|) = \dfrac{d|\psi_\mu|}{d|\mathbf{i}_\mu|} = L_m + \dfrac{dL_m}{d|\mathbf{i}_\mu|}|\mathbf{i}_\mu|^{\,1)}$

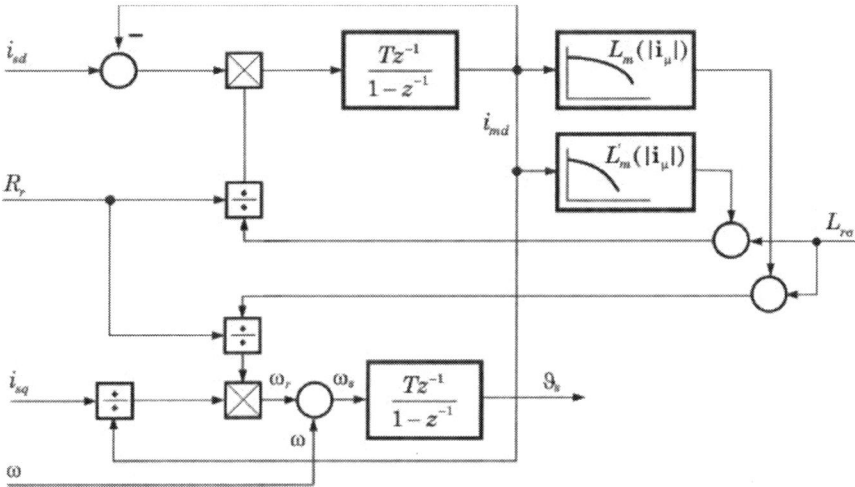

Fig. 4.15 i_s - ω - flux model in field orientated coordinates with saturation of main inductances

Now the i_s - ω - flux model in discrete representation can be rewritten in the following form:

$$\hat{i}_{md}(k+1) = \hat{i}_{md}(k) + \frac{R_r T}{L_m'\left(\hat{i}_{md}(k)\right) + L_{r\sigma}}\left(\hat{i}_{sd}(k) - \hat{i}_{md}(k)\right) \tag{4.55}$$

$$\hat{\vartheta}_s(k+1) = \hat{\vartheta}_s(k) + \omega(k)T + \frac{\hat{i}_{sq}(k)}{\hat{i}_{md}(k)}\frac{R_r T}{L_m\left(\hat{i}_{md}(k)\right) + L_{r\sigma}} \tag{4.56}$$

$$\hat{\vartheta}_s(k+1) = \hat{\vartheta}_s(k) + \hat{\omega}_s(k)T$$

---

1) more about the differential main inductance $L_m'\left(|\mathbf{i}_\mu|\right)$ in chapter 6.2.3

The flux model in this form is represented in figure 4.15. Because of the temperature dependence of the rotor resistance, an on-line tracking of this parameter is additionally required for high-quality drives. The chapter 7 will deal with this problem in greater detail.

4.4.2 Calculation of current set points

Field and torque producing components of the stator current can only be controlled independent of each other as long as the maximum current magnitude is not reached. At this point a suitable strategy for the vectorial current limitation becomes necessary.

The function of the field orientated control relies on the precision or constancy (in the basic speed range) of the impressed rotor flux. Therefore it seems reasonable to give the flux producing component the priority, if the maximum realizable current magnitude is exceeded. A limitation of the maximal reference value i_{sd} to $i_{max}/2$ at the same time allows an adequate control reserve for torque impression. Figure 4.16a illustrates the outlined vector limitation strategy.

If applying a flux control strategy which uses the set point of i_{sq} for calculating the rotor flux set point (cf. chapter 8) an arithmetic loop would arise. This can be avoided by a two-step limitation of i_{sq}^* in the speed controller. Figure 4.16b shows the details.

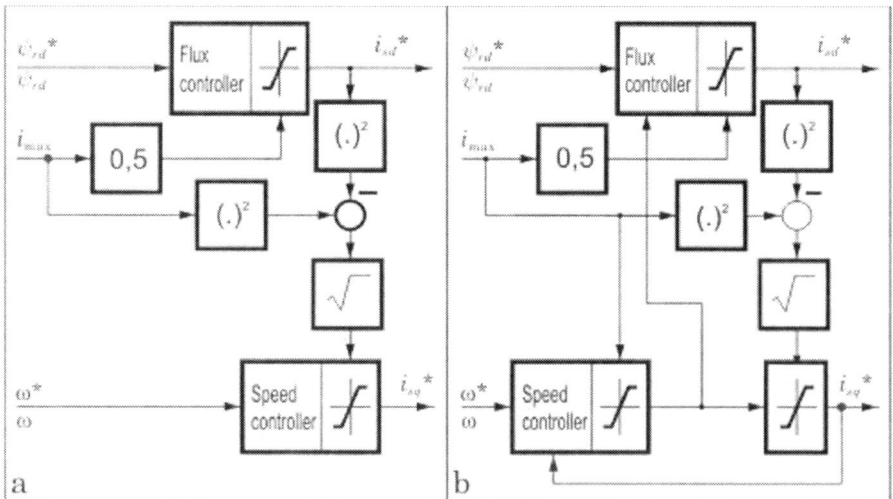

Fig. 4.16 Limitation strategy for calculating the current set points

4.4.3 Problems of the sampling operation of the control system

The problems already discussed in chapter 3.1 for the discretization of the continuous state equations, require a renewed critical assessment at the application within the field orientated control. The system of state equations of the asynchronous machine is the starting point for the design of current controller and flux model, both in field orientated coordinates. In chapter 3.1.2 it was worked out, that for the derivation of an equivalent discrete process model from the continuous state equations the following idealizing prerequisites or approximations must be met:

1. Constancy of the time variant and state-dependent process parameters (frequencies, machine parameters) within a sampling period.
2. Constancy of the input quantities within a sampling period (sampling over zero order hold).

These prerequisites naturally are not fulfilled in the real system, and in the result consequences for precision, control accuracy and stability of a control system, designed on the basis of the mentioned simplifications, have to be expected. Some of these consequences and possible countermeasures shall be examined here. Hereby, we will concentrate on the following main emphases:

1. Validity of the simplifying assumptions for time-variant parameters and input quantities of the continuous system.
2. Choice of the discretization method.
3. Choice of the sampling time.

Thereby the issue (3) must be regarded in closed relation to (1) and (2).

a) Time-variant system parameters

The problem definition can be inverted in a pragmatic way with regard to the time-variant parameters of the system matrix: In order to obtain an equivalent time-discrete system, the boundary conditions of the discretization must be chosen to allow for the system matrix to be considered approximately time-invariant over one sampling period. This primarily has consequences for the selection of the sampling time T which must be chosen adequately small. Time-variant parameters of the system matrix are the speed or mechanical angular velocity ω, the stator angular velocity ω_s and variable (e.g. saturation dependent) machine parameters.

With regard to speed, the prerequisite of approximate time-invariance within a sampling period is fulfilled for usual sampling times of 0.1 ... 1ms. Rotor and stator frequency can change with the dynamics of the impressed torque producing current. The technologically existing limitation of the current allows only a restricted maximum slip and therefore a limited frequency change, though. In the end the assessment of this point will, however, have to be reserved for the detailed investigation of the concrete application, where stability and performance characteristics have to be investigated. The same is valid for time-variant motor parameters to be taken into account. For the incorporation of the main flux saturation it is advantageous that this quantity also can be seen as depending on a slowly varying state variable (rotor flux).

The input variable of the system is the stator voltage vector $\mathbf{u}_s(t)$. The reference value of $\mathbf{u}_s(t)$ is constant over one sampling period, but for the actual motor voltage this is, however, not the case. For a machine model in stator coordinates the stator voltage is piece-wise constant due to the pulse width modulation, and for a machine model in field coordinates it is piece-wise sinusoidal because of the continuously changing phase angle. Thus the used approximation of zero order will cause errors in every case. A workaround could be the consideration of the actual voltage curve or an approximation of higher order for the calculation of the integral in the output equation (3.9):

$$\mathbf{x}((k+1)T) = e^{\mathbf{A}T}\,\mathbf{x}(kT) + \int_0^T e^{\mathbf{A}(T-\tau)}\,\mathbf{B}\mathbf{u}(kT+\tau)\mathrm{d}\tau \qquad (4.57)$$

(cf. chapter 3.1.2). If the zero-order approximation is retained, it has to be made sure that the mean average value of the model input quantity actually matches the effective mean average value at the machine terminals over a sampling period.

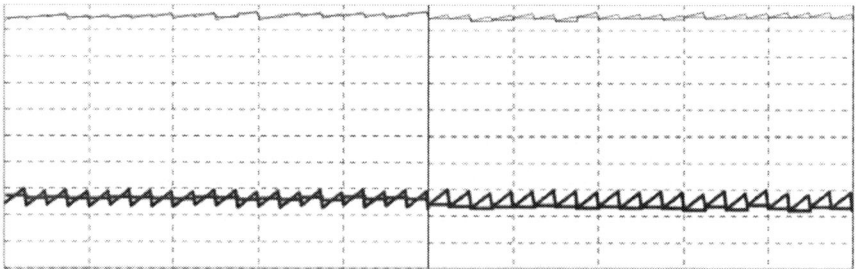

Fig. 4.17 References and actual values of u_{sd} (**bottom**) and u_{sq} (**top**) in field orientated coordinates: with (**left**) and without (**right**) compensation

For a model in field orientated coordinates this can be achieved by a feed-forward compensation of the transformation angle ϑ_s. The equation (4.56) can be amended as follows

$$\hat{\vartheta}_{s1}(k) = \hat{\vartheta}_s(k) + k_c\hat{\omega}_s(k)T \qquad (4.58)$$

with the new transformation angle ϑ_{s1}. The factor $k_c = 1.5$ takes additionally into account the dead time of one sampling period between calculation and output of the control variable. The figure 4.17 shows reference and actual values of the stator voltage in field orientated coordinates with and without the feed-forward compensation of the transformation angle. The pulse width modulation was not simulated in this case.

b) Discretization method
Discretization method and the used approximations decide essentially the stability of the discrete model. This holds particularly when the continuous system matrix contains complex or frequency dependent eigenvalues. To get an opinion about the dimension of the influences and model errors to be expected and also of the differences of the several discretization methods, a series of simulations was carried out whose results are represented in figure 4.18. All parameters of the continuous model are regarded as time-invariant. The state controller with field orientation, described in chapter 5.4, the flux model (4.55), (4.56), (4.58) and a sampling time of $T = 0.5$ms form the basis of the system under investigation. Because an unstable behavior of the system is recognizable by increasing oscillations, the low-pass filtered norm of the current difference vector of two successive sampling periods was chosen as a performance criterion:

$$Q = \frac{\left\| \mathbf{i}_s(k) - \mathbf{i}_s(k-1) \right\|}{1 + sT_F} \tag{4.59}$$

The criterion was recorded during a speed start-up at maximum acceleration. The following methods were simulated (cf. chapter 3.1.2 and the corresponding example in chapter 12.2):

1. Series expansion of the time-discrete system matrix $\mathbf{\Phi}$ with truncation after the linear term (Euler discretization) (field coordinates).
2. Series expansion of $\mathbf{\Phi}$ with truncation after the quadratic term (field coordinates).
3. Euler discretization in stator coordinates and then transformation into field coordinates.
4. Discretization using substitute function in stator coordinates and subsequent transformation into field coordinates.
5. Discretization using substitute function in field coordinates.
6. Current controller in stator coordinates with integration part and voltage limitation in field coordinates, discretization using substitute function.

For methods 1 - 5 the complete current controller was realized in field coordinates. It shall be emphasized that the simulation did not intend to give any statements about control accuracy, current wave form and the like, but exclusively aims at the investigation of the control system stability.

The results show significant differences between the methods. It strikes specifically that *a stable operation up to the maximal theoretically possible frequency following the Shannon theorem* can be achieved, *if the method 5 is used.* The method 1 allows a stable operation to just below a stator frequency of 300Hz. This corresponds to the theoretical stability limit of the Euler method with the condition $\left|\lambda_i + 1/T\right| < 1/T$ for the eigenvalues λ_i of the continuous system matrix (cf. chapter 3.1). Since the chosen sampling time rather lies in the upper range of the usually realized values, it is also confirmed that the bigger part of applications – in terms of the implemented control structures – can already be covered with Euler discretization in field orientated coordinates.

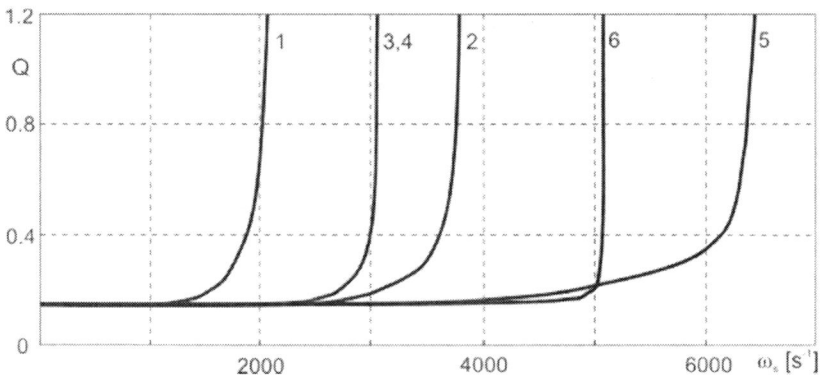

Fig. 4.18 Stability of discretization methods

c) Choice of the sampling time

The choice of the sampling time is one of the most complex problems for the design of a digital control system. To an important part it is a question of the required and available computer power and therefore the hardware costs, in which an optimum is given, because on one hand a minimization of sampling time requires a higher computer power, but on the other hand increasing the sampling time causes the same effect because of the required more sophisticated discretization algorithms.

From the control point of view, stability considerations play a decisive role for the discretization of the continuous model. With regard to the reproducibility of the continuous signals, the absolute lower limit of the sampling frequency is defined by the Shannon theorem. However, for the motor control the opposite case is significant as well: The production of a continuous signal (voltage, current) from a sequence of discrete control signals (signal reconstruction). For the reconstruction of the continuous signal $f(t)$, using a simple D/A converter (zero order hold), the following maximum amplitude error is obtained, provided steady differentiability of $f(t)$ (cf. [Aström 1984]):

$$e_{A,max} = \max_k |f(k+1) - f(k)| \le T \max_t \frac{df(t)}{dt} \qquad (4.60)$$

For a sinusoidal signal $f(t) = a\sin\omega_s t$ the error results to $e_{A,max} \le a\omega_s T$. As is easily to comprehend, the maximum amplitude error is reduced for a continuous signal with assumed linear characteristic between two sampling instants and reconstruction by the mean average value, to half of the value for simple sampling at the sampling instants.

Further approaches for the choice of the sampling time can be obtained from the demanded transient response of the closed control system. For a quasi-continuous design, a value of $T \le (0.25...0.5)t_r$ is recommended in [Aström 1984] from the control response time. The relation between sampling and response time is given for the dead-beat design by the system structure. For a small response time the sampling time has to be chosen as small as possible, which however, on other hand, increases the control gain as well as the amplitude of the control variable, and increases the sensitivity to high-frequency disturbances.

The use of fast pulse-width-modulated inverters with constant switching frequency as control equipment yields another influencing factor: Because of the necessary synchronicity between current control and voltage output, the PWM frequency will be chosen as an integer multiple of the sampling frequency of the current control, which yields values in the range of about 0.1 ... 1ms.

4.5 References

Al-Tayie JK, Acarnley PP (1997) Estimation of speed, stator temperature and rotor temperature in cage induction motor drive using the extended Kalman filter algorithm. IEE Proc.-Electr. Power Appl., Vol. 144, No. 5, September, pp. 301 – 309

Aström K, Wittenmark B (1984) Computer Controlled Systems. Prentice-Hall Englewood Cliffs

Beineke S, Grotstollen H (1997) Praxisgerechter Entwurf von Kalman-Filtern für geregelte Synchronmotoren ohne mechanischen Sensor. SPS/IPC/DRIVES 97, Tagungsband, S. 482 – 493

Boyes GS (1980) Synchro and Resolver Conversion. Memory Devices Ltd

Brunotte C, Schumacher W (1997) Detection of the starting rotor angle of a PMSM at standstill. EPE '97, Trondheim, pp. 1.250 – 1.253

Brunsbach BJ (1991) Sensorloser Betrieb von permanenterregten Synchronmotoren und Asynchronmaschinen mit Kurzschlußläufern durch Zustandsidentifikation. Dissertation, RWTH Aachen

Eulitz Th (1990) Nutzung der Digitalsimulation bei der Entwicklung von Regelstrukturen für Drehstromantriebssysteme. Dissertation, TU Dresden

Isermann R (1988) Identifikation dynamischer Systeme. Bd. II, Springer-Verlag

Jönsson R (1991) Natural Field Orientation for AC Induction Motor Control. PCIM Europe, May/June, pp. 132 – 138

Jönsson R, Leonhard W (1995) Control of an Induction Motor without a Mechanical Sensor, based on the Principle of "Natural Field Orientation" (NFO). International Power Electronics Conference Yokohama

Kiel E (1994) Anwendungsspezifische Schaltkreise in der Drehstrom-Antriebstechnik. Dissertation, TU Carolo - Wilhemina zu Braunschweig

Kubota H, Matsue K, Nakano T (1993) DSP-Based Speed Adaptive Flux Observer of Induction Motor. IEEE Trans. on IA, Vol. 29, No. 2, March/April, pp. 344 – 348

Kubota H, Matsue K (1994) Speed Sensorless Field-Oriented Control of Induction Motor with Rotor Resistance Adaptation. IEEE Trans. on IA, Vol. 30, No. 5, September/October, pp. 1219 – 1224

Matsui N (1996) Sensorless PM Brushless DC Motor Drives. IEEE Trans. on IE, Vol. 43, No. 2, April, pp. 300 – 308

Östlund S, Brokemper M (1996) Sensorless Rotor-Position Detection from Zero to Rated Speed for an Integrated PM Synchronous Motor Drive. IEEE Trans. on IA, Vol. 32, No. 5, September/October

Profumo F, Pastorelli M, Ferraris P, De Doncker RW (1991) Comparision of Universal Field Oriented (UFO) Controllers in Different Reference Frames. Proceedings EPE 1991 Firenze, pp. 689 – 695

Quang NP (1991) Schnelle Drehmomenteinprägung in Drehstromstellantrieben. Dissertation, TU Dresden

Quang NP (1996) Sensorlose feldorientierte Regelung von Asynchronantrieben. Forschungsbericht, TU Dresden

Quang NP, Schönfeld R (1996) Sensorlose und rotorflußorientierte Drehzahlregelung eines Asynchronantriebs in Feldkoordinaten. SPS/IPC/DRIVES 96, Tagungsband, S. 273 – 282

Rajashekara K, Kawamura A, Matsue K (1996) Sensorless Control of AC Motor Drives: Speed and Position Sensorless Operation. IEEE Press, New York

Saito K (1988) A microprocessor-controlled speed regulator with instantanous speed estimation for motor drive. IEEE Trans. on IE, Vol.35, No.1, Feb., pp. 95 – 99

Schönfeld R (1987) Digitale Regelung elektrischer Antriebe. VEB Verlag Technik Berlin

Stärker K (1988) Sensorloser Betrieb einer umrichtergespeisten Synchronmaschine mittels eines Kalman-Filters. Dissertation, RWTH Aachen

Tajima H, Hori Y (1993) Speed Sensorless Field-Orientation Control of the Induction Machine. IEEE Trans. in IA, Vol. 29, No. 1, January/February

Takeshita T, Matsui N (1996) Sensorless Control and Initial Position Estimation of Salient-Pole Brushless DC Motor. Proc. 4th Intern. Workshop on AMC, Mie University Japan

Texas Instruments (1997) Sensorless Control with Kalman Filter on TMS Fixed-Point DSP. Application Report

Vas P, Drury W (1994) Vector Controlled Drives. Proceedings of PCIM '94 Nürnberg, pp. 213 – 228

Verghese GC, Sanders SR (1988) Observers for Flux Estimation in Induction Machines. IEEE Trans. on Industrial Electronics, Vol. 35, No. 1, Febr., pp. 85 – 94

Zägelein W (1984) Drehzahlregelung des Asynchronmotors unter Verwendung eines Beobachters mit geringer Parameterempfindlichkeit. Dissertation, Universität Erlangen-Nürnberg

5 Dynamic current feedback control for fast torque impression in drive systems

The current control loop plays a decisive role in a 3-phase drive system operated with field orientation. The design of the superimposed mechanical systems (speed and position control) wishes for an inner *current control loop with ideal behaviour: With undelayed impression of the stator current.* The assumption that the ideal current control can be modeled by a dead time simplifies fundamentally the control design for often weekly damped oscillating mechanical systems.

Besides the dead time behaviour, which could be achieved by a design aimed at dead-beat response, the current controller also should ensure *an ideal decoupling between the field* and *torque forming components* i_{sd} and i_{sq}, because the two components are strongly coupled with each other in the field synchronous coordinate system. This problem was not solved convincingly with the classic concept (fig. 1.4). From the view of the modern control engineering the current process model of IM or PMSM represents a multivariable process – a MIMO[1] process – which can be mastered only by a multivariable controller. The multivariable controller contains besides controllers in the main (direct) path also cross (decoupling) controllers, so that the difficulties of the decoupling are solved automatically with the controller design.

An important task of the controller design consists in considering a number of implementation dependent issues in controller approach and feedback. With conventional PI controllers such issues are usually neglected.

- The delay of the control variable output of typically one sampling period: The stator voltage calculated by the current controller can only have an effect in the next sampling period.
- The technique of the actual-value measurement: After all, different possibilities like instant value measuring (by ADC) or integrating measuring (by VFC, resolver and incremental encoder) are considered.

---

[1] MIMO: Multi-Input – Multi-Output

Like all control equipment, the inverter can realize only a limited control variable because of the fixed DC-link voltage. To avoid possible oscillations and wind-up effects caused by the implicit integrating part after entering or leaving output limitation (at start-up, speed reversals, magnetization, field weakening), the controller must have the ability to take the limitation of control variables into account effectively.

After discussing the discrete system models in the former chapters, new controllers will now be introduced with uniform and easily comprehensible design and which fulfill all mentioned requirements. But before the controller design is discussed a survey about the existing current control methods shall be given.

5.1 Survey about existing current control methods

The interested reader will find an overview in abbreviated form also in [Quang 1990]. Altogether, the known methods can generally be divided into two groups: nonlinear and linear current controllers.

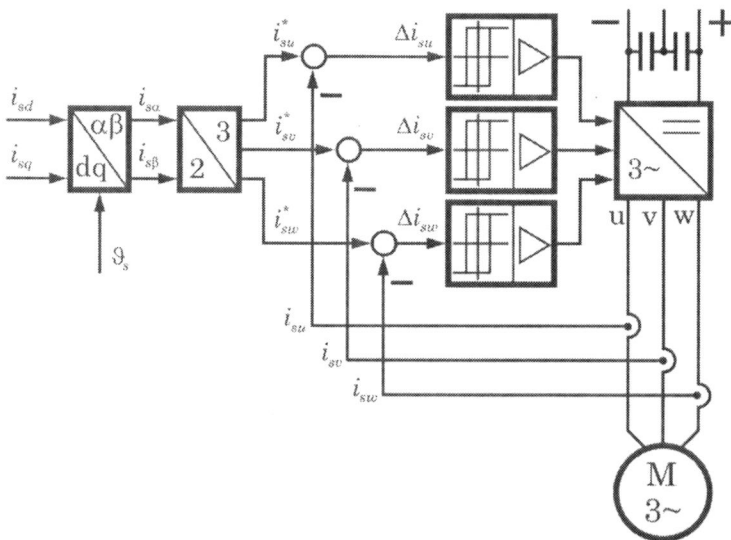

Fig. 5.1 Stator current control with three separate hysteresis (bang-bang) controllers

a) Nonlinear current control
Controllers of this group can show two- or three-point behaviour. A special method is the intelligent predictive control which reacts to the

stator current vector leaving a predefined tolerance circle with a pre-calculated optimal firing pulse and therefore has also two-point behaviour. The most simple version of a current controller with two-point behaviour is to use three separate on-off controllers, refer to [Peak 1982], [Pfaff 1983], [Hofmann 1984], [Brod 1985], [Le-Huy 1986], [Malesani 1987] and [Kazmierkowski 1988]. The principle is shown in the figure 5.1.

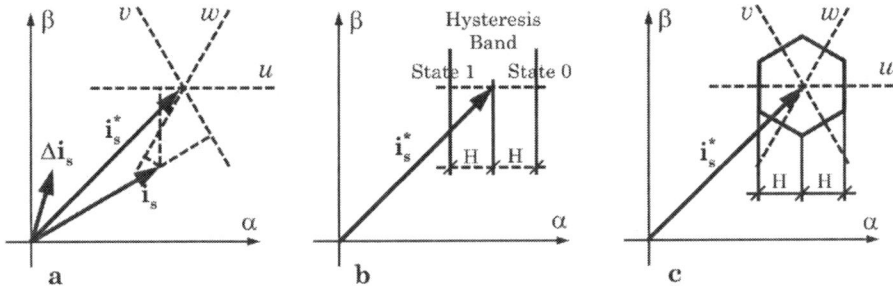

Fig. 5.2 On-off controller for phase currents in vector representation: Components of vector of current error (**a**), tolerance range of one phase (**b**) and tolerance hexagon of all three phases (**c**)

The sinusoidal set points of the phase currents are obtained by coordinate transformation from the field synchronous set points. Depending on the sign of the current errors, the corresponding phase is switched to „+" or „–" potential of the DC-link voltage at exceeding of the permitted error. This control variant stands out by the simplicity of its technical realization and by its convincing dynamic properties, but the following backdrops also have to be mentioned:

- The pulse frequency varies with changing fundamental frequency and load which is particularly unwanted.
- With isolated motor star point the current error can reach the double of the tolerance band.
- The control quality directly depends on analogous comparators which are sensitive to offset and drift and could therefore lead to a slight pre-magnetization of motor or transformer.

The figure 5.2a shows the reference vector of the stator current i_s^*, the actual vector i_s and the error vector Δi_s. The phase current differences are obtained by the projection of the error vector to the axes of the corresponding phase windings. Upon the actual current vector leaving the tolerance hexagon the comparators will become active.

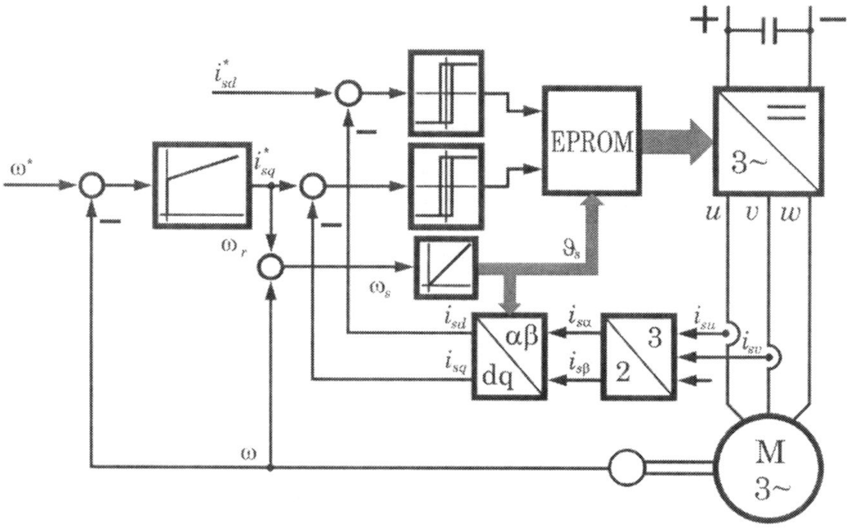

Fig. 5.3 Block structure of the drive system with inner current control loop using two-point controllers in field synchronous coordinates

The figure 5.3 shows the realization of the control with two-point behaviour in field synchronous coordinates (cf. [Pfaff 1983], [Nabae 1985], [Rodriguez 1987] and [Kazmierkowski 1988]). The current error is calculated in field synchronous coordinates. The field angle provides the necessary address to find, depending on the control errors, the fitting pre-defined pulse patterns. The figure 5.4 explains this.

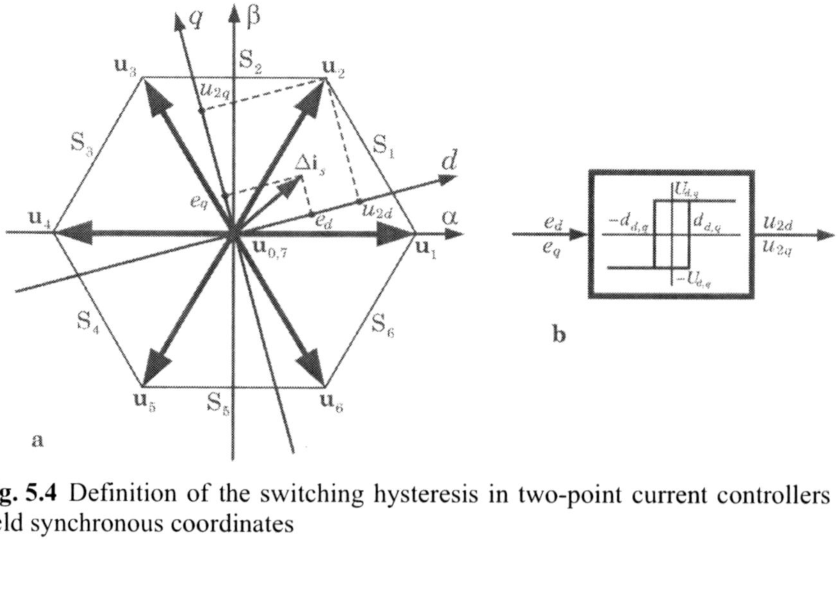

Fig. 5.4 Definition of the switching hysteresis in two-point current controllers in field synchronous coordinates

The actual error vector $\Delta \mathbf{i}_s$ and the position of the coordinate system are shown in figure 5.4a. Following the definition in figure 5.4b the controller behaviour can be summarized as follows:

$$\text{if } \varepsilon_{d,q} > \delta_{d,q}, \text{ then } u_{xd,xq} = U_{d,q} = 1 \text{ and}$$

$$\text{if } \varepsilon_{d,q} \leq \delta_{d,q}, \text{ then } u_{xd,xq} = -U_{d,q} = 0$$

The values 1 and 0 are the logical values which are assigned to the voltages $\pm U_{d,q}$. Index "x" can assume one of the values 0...7 and represents the standard voltage vector to be selected. The projection of $\Delta \mathbf{i}_s = \mathbf{i}_s^* - \mathbf{i}_s$ to the axes dq like in the figure 5.4a yields:

$$\varepsilon_d > \delta_d \text{ , thus } u_{xd} = 1$$

$$\varepsilon_q > \delta_q \text{ , thus } u_{xq} = 1$$

Accordingly, a pulse pattern or a voltage vector has to be chosen whose d and q components minimize these control errors. In the example of figure 5.4 the choice $\mathbf{u}_2$ follows immediately. The assignment

logical values and position of field synchronous coordinate system
→firing pulse

was determined off-line beforehand and then stored in table form in EPROM. [Rodriguez 1987] shows concrete examples.

To control the stator currents, also controllers with three-point behaviour may be used. In [Kazmierkowski 1988] details about this approach can be found which is illustrated in the figure 5.5. In this method the control errors ε_α and ε_β of the stator current are obtained by projection of the error vector to the $\alpha\beta$ axes of the stator-fixed coordinate system. The way to choose the required pulse pattern is similar as in the figure 5.3.

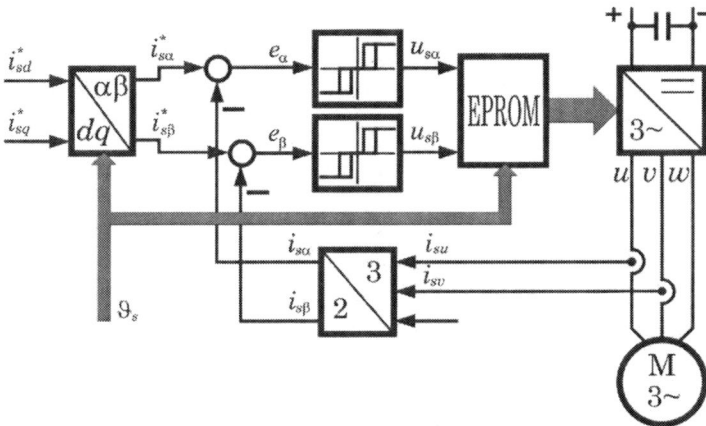

Fig. 5.5 Three-point current controller in stator-fixed coordinate system

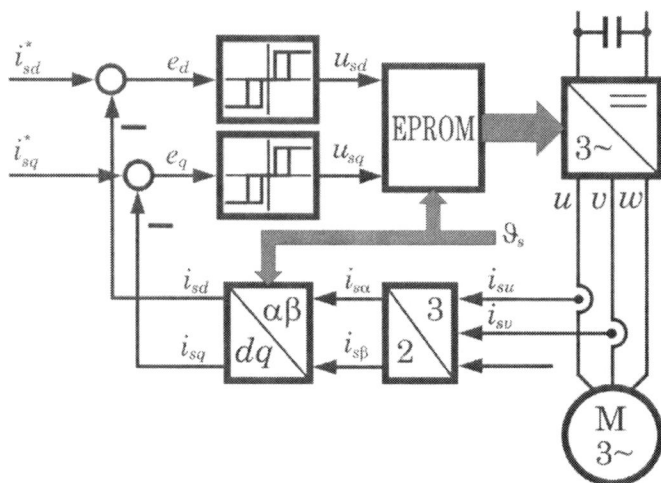

Fig. 5.6 Three-point current controller in field synchronous coordinate system

[Kazmierkowski 1988] further introduced a structure with three-point controllers in the field synchronous coordinate system as shown in the figure 5.6. In principle this variant works exactly like the one in figure 5.3. The only difference between both versions consists in aiming at a higher precision by a finer division of the overall vector space (figure 5.7) into 24 sectors, combined with three-point behaviour. The EPROM table containing the pulse patterns accordingly gets more extensive. In contrast, Rodriguez keeps the six original sectors (figure 5.4).

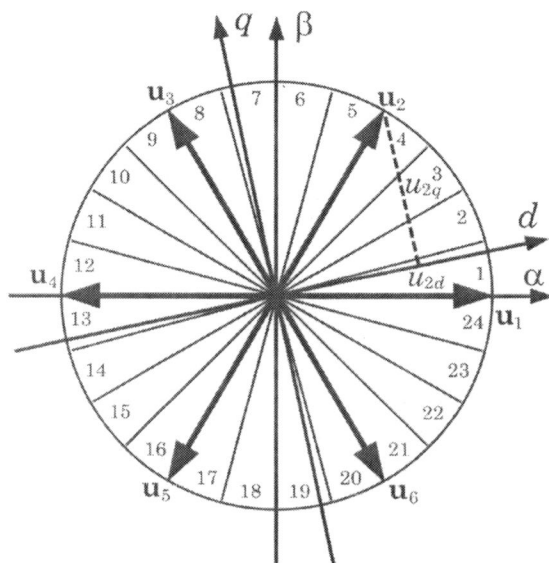

Fig. 5.7 Division of the vector space into 24 sectors

The most intelligent version in the family of the nonlinear current controllers is the predictive control (more in [Holtz 1983, 1985]). This control reacts (figure 5.8) on the actual current vector leaving the tolerance-circle by a predictive calculation of the following, optimized voltage vector. Therefore, it also shows two-point behaviour. The method can be used in field synchronous as well as in stator-fixed coordinates. The principle block structure is shown in the figure 5.9.

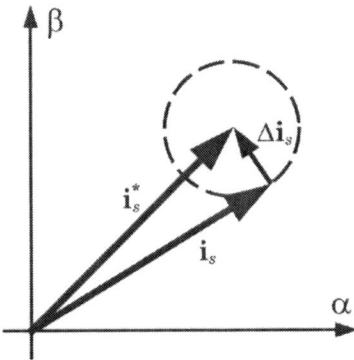

Fig. 5.8 Tolerance-circle of the predictive current controller

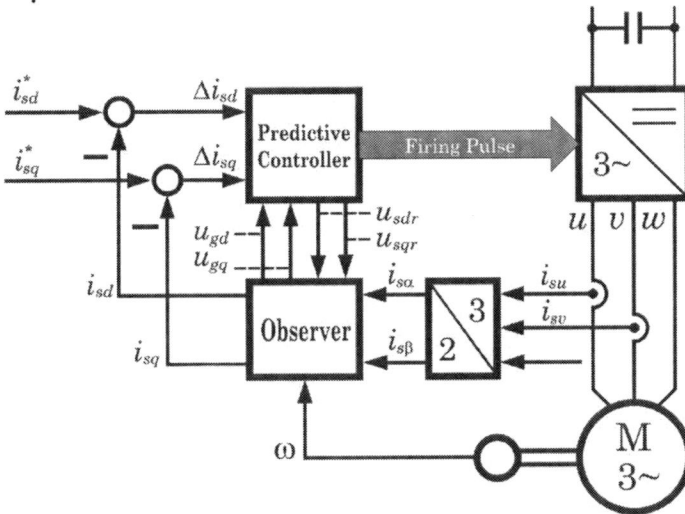

Fig. 5.9 Block structure of the predictive current control

If the actual vector $\mathbf{i}_s$ overlaps the tolerance-circle at the time t_0, the predictive controller must, using the information provided by the observer,

- calculate all possible trajectories of the current vector (figure 5.10a) for each of the seven possible standard voltage vectors, and
- following a certain criterion determine the optimal voltage vector for the chosen current trajectory.

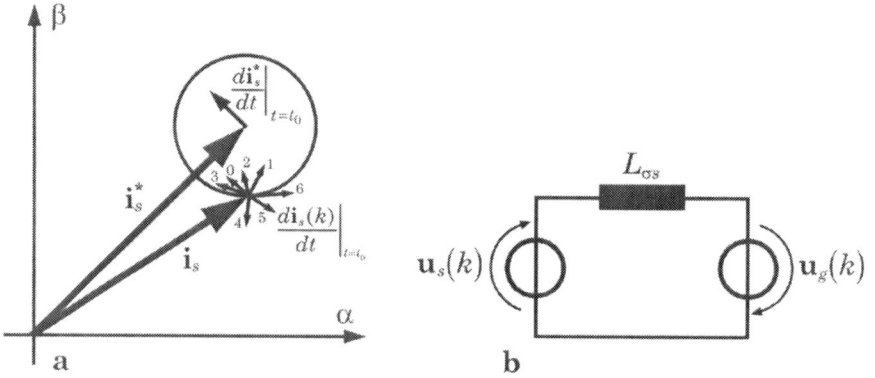

Fig. 5.10 Possible current trajectories on output of all possible standard voltage vectors (**a**) and simplified equivalent circuit of the IM (**b**)

The trajectories can be calculated as follows:

$$\mathbf{i}_s^*(t) = \mathbf{i}_s^*(t_0) + \left.\frac{d\mathbf{i}_s^*}{dt}\right|_{t=t_0}(t-t_0)$$

$$\mathbf{i}_s(t) = \mathbf{i}_s(t_0) + \left.\frac{d\mathbf{i}_s}{dt}\right|_{t=t_0}(t-t_0)$$

(5.1)

In the equation (5.1) the currents $\mathbf{i}_s^*(t_0)$ and $\mathbf{i}_s(t_0)$ are known. The numerical derivation of $\mathbf{i}_s^*$ produces $d\mathbf{i}_s^*/dt$, and for calculation of $d\mathbf{i}_s/dt$ the following equation is used:

$$\frac{d\mathbf{i}_s(k)}{dt} \approx \frac{\mathbf{u}_s(k) - \mathbf{u}_g(t=t_0)}{L_{\sigma s}}$$

(5.2)

with:

k	= 0, 1, ... , 7
$\mathbf{u}_s(k)$	= one of the seven possible standard voltage vectors
$\mathbf{u}_g(t=t_0)$	= the induced e.m.f. at instant t_0
$L_{\sigma s}$	= leakage inductance on the stator side

The formula (5.2) follows from the figure 5.10b in which the stator resistance is neglected. The induced e.m.f. is calculated by a machine model in the observer. Depending on the chosen trajectory ($k = 0, 1, ... , 7$) the following error vector:

$$\Delta\mathbf{i}_s(t,k) = \mathbf{i}_s^*(t) - \mathbf{i}_s(t,k)$$

(5.3)

can be calculated. For a detailed derivation the interested reader is referred to the mentioned literature. Here only the final equation (5.4), which shows the different error trajectories (figure 5.10c) in dependency on the chosen voltage vectors, is given.

$$|\Delta \mathbf{i}_s|^2 (t,k) = |\Delta \mathbf{i}_s|^2 (t=t_0) + a_1 (t-t_0) + a_2 (t-t_0)^2 \qquad (5.4)$$

The error trajectories have the form of a parabola. From the figures 5.10a and 5.10c it can be seen, that the firing pulses corresponding to the voltage vectors $\mathbf{u}_4$ and $\mathbf{u}_5$ would increase the error, while all others would decrease it.

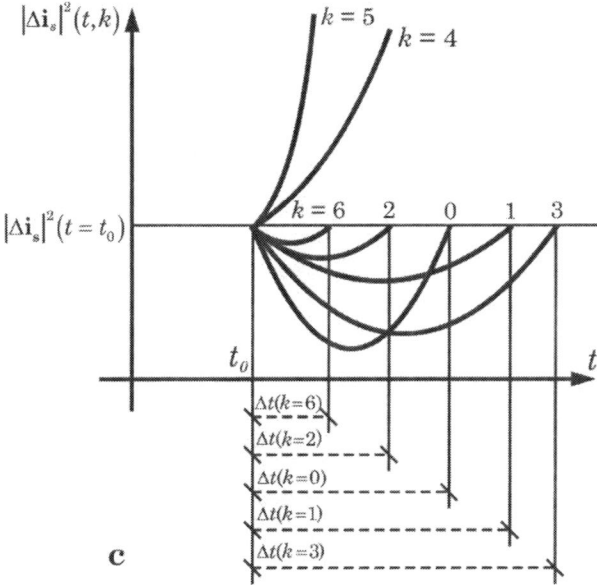

Fig. 5.10c Possible current trajectories at instant t_0

But naturally only one of the five vectors $\mathbf{u}_{0,1,2,3,6}$ can be used. The choice is made according to one of the following criteria:

1. For slow change of current (stationary operation): In this case the actual vector has to be kept within the tolerance circle as long as possible. In addition, the number of necessary switchovers of the semiconductor switches should be as small as possible. Therefore the following criterion is appropriate:

$$\frac{\Delta t(k)}{n(k)} = \text{max} \qquad (5.5)$$

2. For fast change of current (dynamic operation): This case produces very fast changes of the set point vector $\mathbf{i}_s^*$, and it requires that the actual vector $\mathbf{i}_s$ follows the set point vector exactly and as fast as possible. $\mathbf{u}_s(k)$ will then be chosen according to the following criterion:

$$\Delta t(k) = \text{min} \qquad (5.6)$$

For the example in figure 5.10c, using the first criterion would result in choosing vectors $\mathbf{u}_1$ or $\mathbf{u}_3$, whereas the second criterion yields vector $\mathbf{u}_6$.

The predictive control is predominantly used in high power drives, where the assumption of a negligible stator resistance is fulfilled widely and where a very large rotor time constant allows the choice of a relatively large sampling period, what is necessary because of the extensive required calculations.

The disadvantage of all nonlinear current control methods consists in the bad current impression in the area of inverter over-modulation, resulting in a certain orientation error and corresponding torque deviation.

b) Linear current control

Relevant references for this method are [Mayer 1988], [Meshkat 1984], [Rowan 1987] and [Seifert 1986]. The first classical version of linear current controllers was the application of three or two separate PI-controllers to independently control the phase currents (see fig. 5.11). The sinusoidal output signals of the PI controllers would be forwarded to pulse width modulators (PWM) and compared with a sawtooth-shaped pulse sequence. The firing pulses are the immediate result of this comparison.

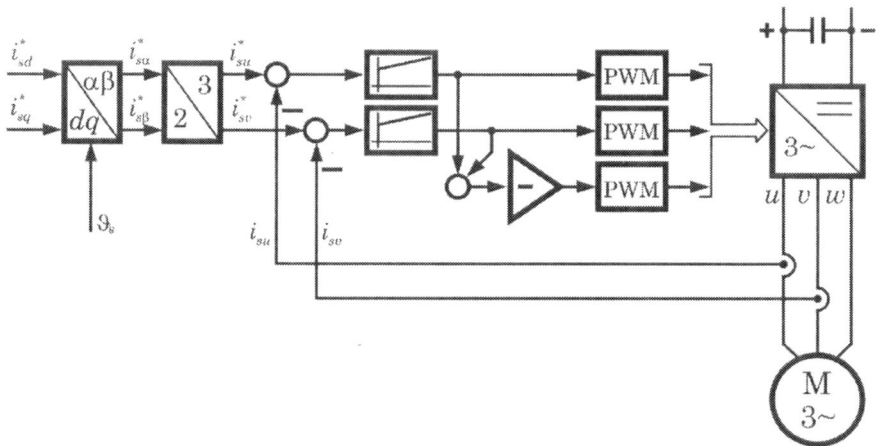

Fig. 5.11 Phase current PI controllers with pulse width modulation

The pulse width modulation was for long time the most widely used control method for inverters. Like all methods in stator-fixed coordinates, the control method shown in the figure 5.11 has the tracking error as a main disadvantage, because the PI controllers permanently have to work in dynamic operation due to the sinusoidal current set points. It was shown by [Rowan 1987] that an *abrupt reduction of the PWM gain* arises if the inverter control goes close to the maximum voltage amplitude (transition mode). This effect of the control variable limitation could not be taken into account effectively with this control method. An essential improvement could be obtained by transforming the control algorithm into the field synchronous coordinate system (figure 1.4) in which the variables to be controlled represent DC quantities in stationary operation.

This control version is very widely applied and possesses the following advantages:

1. The precision is considerably higher because the controller does not have to work in dynamic operation, particularly the current phase error can be controlled to zero.
2. The response near the transition mode is improved.
3. The decoupling of the current components is improved, and therefore a higher accuracy of the field orientation is obtained.

This method however still has a number of disadvantages which shall be mentioned here to motivate the development of improved algorithms in the following chapters.

1. The response time (or the dynamics) of the control strongly depends on the stator leakage time constant. Therefore, a nearly undelayed current or torque impression as ideally required by the speed control loop is hardly achievable.
2. The current components i_{sd} and i_{sq} are strongly coupled to each other in field synchronous coordinates. Can an adequate decoupling be ensured?
3. Can the transfer characteristic of the current measuring technique actually used (measurement of instantaneous values, integrating measurement) be taken into account with this control concept effectively to guarantee a wide application range?
4. Can the one-step delay of the control variable $\mathbf{u}_s$, calculated by PI controllers, effectively be integrated into the control equations?
5. How does the controller react to the control variable limitation, and can switching-off of the integral part (anti-reset wind up) be regarded as a sufficient method in the PI controllers?

These questions will be answered in context with new designs of the current controller in this chapter. However, it has already to be highlighted that this variant represented a considerable progress to formerly applied methods.

A last method shall be mentioned yet, being a mixture between a linear and nonlinear regulation. This is the method introduced in [Enjeti 1988] and [Zhang 1988] with current modulus and current phase control (fig. 5.12a). The current modulus and current phase control loops are designed separately and have in principle linear characteristics. The decoupling of the two quantities, however, is of nonlinear nature. The references and the actual values are rectified and then compared with each other. The current deviation is supplied to a PI controller. The phase angles are determined and compared with each other by phase detectors. The phase deviation has the form of a time interval during which a counter counts. The output of the A/D converter, following the PI modulus controller, and the output of the phase counter then build the address word for the corresponding switching pattern stored in a 64 Kbytes EPROM table.

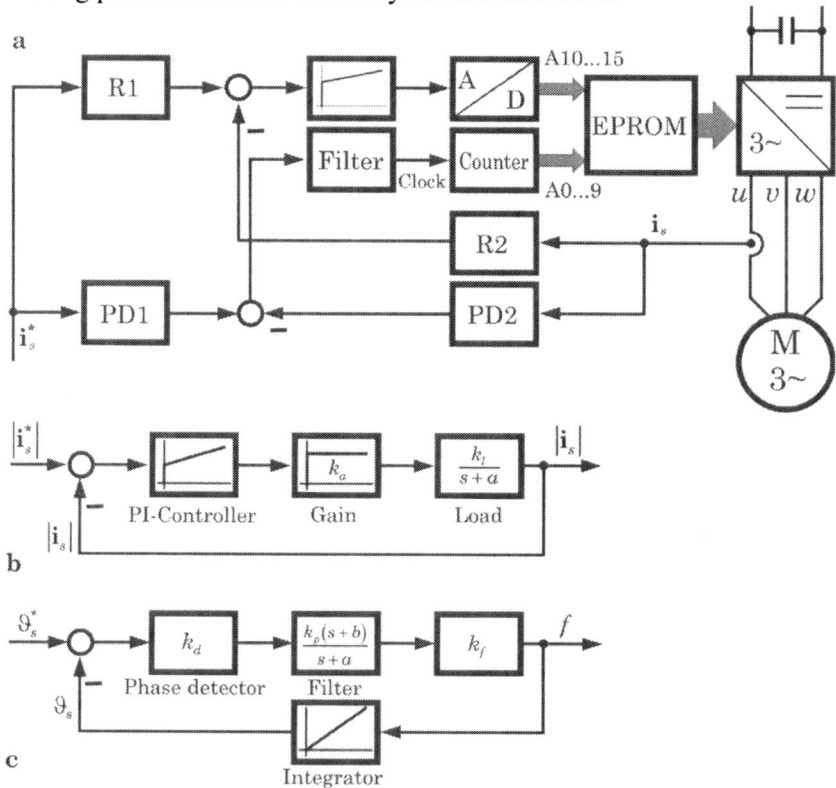

Fig. 5.12 Structure of the concept using current modulus and current phase control (a): the modulus control loop (b) and the phase or frequency control loop (c)

This control concept is mainly used in current source inverters with the control system designed according to the signal diagrams shown in figure 5.12b,c (see [Enjeti 1988]).

c) Closing remark to the overview
This chapter tried to give a summary of the known current control methods. Where possible, the functional principle was outlined. In connection with this, reference sources are included so that the possibility for background investigation is always ensured.

Because of the wide variety of the known methods deviations to the originals are conceivable. The aim of this summary was not to deliver a complete analysis about all methods, but rather to give a stimulus for own study.

5.2 Environmental conditions, closed loop transfer function and control approach

The consideration of all environmental conditions is one of the most important tasks of the controller design. Before the controller approach itself is developed these conditions shall be discussed here. In addition, the final closed loop behaviour to be achieved shall also be outlined.

a) Environmental conditions
The first condition to be considered is the *applied technique for capturing the actual-values of current and speed*. Basically, two main techniques exist: The measurement of instantaneous values using A/D converters, and the integrating measurement using V/f converters for the current and incremental encoder or resolver for the speed. The difficulties connected to this were discussed extensively in the chapter 4, but how they influence the controller design will be subject of this chapter 5.

The second environmental condition is the *one-step delayed output of the control variable* $\mathbf{u}_s$ of the current controller. This delay must be taken into account in the controller approach.

The rotor flux of the IM is, in comparison to other electrical quantities, a slowly changing variable. The pole flux of the PMSM is constant. Therefore the fluxes can be looked at as disturbance variables and shall be accounted for in the controller approach separately.

c) Closed loop response
The closed loop response is the intended transfer behaviour of the controlled system. In the case of the stator current controller, it is characterized by the following properties:
1. The step rise time, characterizing the control dynamics, and
2. the decoupling between the components in steady-state and dynamic operation.

The ideal dynamic behaviour can be achieved by the so-called *dead beat response* which means that the actual value will match the reference value after one sampling period, or, if the one-step delay of the control output is taken into account, after two sampling periods. Considering that for some systems working with very short sampling times (e.g. T = 100μs) this rise time of 2×100μs would be too small from the viewpoint of the required energy to drive the current, a rise time of 3×100μs or 4×100μs (meaning after three or four periods) could be more useful in these cases. The dynamics does not become worse because a rise time of 300μs or 400μs (still much smaller then 1ms) can only be wished for with conventional PI controllers. To be able to express the demanded behaviour in general terms we start from a closed loop response with n sampling periods (figure 5.13) for the SISO process.

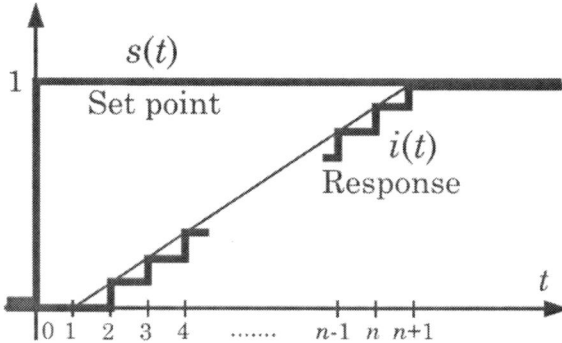

Fig. 5.13 Set point signal and its response of a SISO process controlled with deadbeat behaviour

The discrete control by means of micro computer allows for an exact tracking of the actual value so that it can reach the set point after n sampling periods exactly and without overshots. Such a controller is conceived, as well known, for *finite adjustment time* (FAT response). Considering the one-step delay of the control variable the FAT will then be exact $(n+1)$ periods. Therefore, the approach for the output signal can be written in the z domain as follows (figure 5.13):

$$i(z)=\sum_{\nu=0}^{\infty}c_{\nu}\,z^{-\nu} \quad \text{with} \quad \begin{cases} c_0 = 0 \\ c_1...c_n = c_k = \dfrac{k-1}{n} \quad (k=1...n) \\ c_{n+1}=c_{n+2}=...=c_{\infty}=1 \end{cases} \quad (5.7)$$

With $z<1$ it can be obtained:

$$i(z)=\frac{z^{-(n+1)}}{1-z^{-1}}+\sum_{\nu=1}^{n}\frac{\nu-1}{n}z^{-\nu} \tag{5.8}$$

If a step change of the set point

$$s(z)=\frac{1}{1-z^{-1}} \tag{5.9}$$

is considered as characteristical excitation signal the general transfer function is obtained as:

$$i(z) = \left[z^{-(n+1)} + \left(1 - z^{-1}\right) \sum_{\nu=1}^{n} \frac{\nu-1}{n} z^{-\nu} \right] s(z) \qquad (5.10)$$

for the controlled SISO process with FAT response (figure 5.13). The closed loop response of the vectorial stator current control is obtained from equation (5.10) to:

$$\mathbf{i}_s(z) = \left[z^{-(n+1)} + \left(1 - z^{-1}\right) \sum_{\nu=1}^{n} \frac{\nu-1}{n} z^{-\nu} \right] \mathbf{i}_s^*(z) \qquad (5.11)$$

The closed loop response (5.11) means,

1. that the dynamic as well as the static decoupling between the current components i_{sd} and i_{sq} will be guaranteed, because the transfer matrix is the unity matrix or a diagonal matrix respectively, and
2. that the FAT response with FAT = $(n+1)$ sampling periods will result for the decoupled current components.

It will be shown later that a FAT response with a higher step number is always connected with a complete change of the controller structure or with an increased computing time. It is therefore impractical to increase the number of steps exaggeratedly. The investigation has shown that a FAT response with FAT = 2, 3 or 4 periods, referring to the computation effort which must be handled during a very short sampling period (e.g. 100...200μs), would be realistic and practicable. Therefore, only controller designs for these three cases are offered later on. The reference transfer functions or the closed loop response are obtained as follows for:

1. n=1: FAT = n+1 = 2 (dead beat behaviour)

$$\mathbf{i}_s(z) = z^{-2}\, \mathbf{i}_s^*(z) \qquad (5.12)$$

2. n=2: FAT = n+1 = 3

$$\mathbf{i}_s(z) = \frac{1}{2}\left(z^{-2} + z^{-3}\right) \mathbf{i}_s^*(z) \qquad (5.13)$$

3. n=3: FAT = n+1 = 4

$$\mathbf{i}_s(z) = \frac{1}{3}\left(z^{-2} + z^{-3} + z^{-4}\right) \mathbf{i}_s^*(z) \qquad (5.14)$$

c) Controller approach

It was tried in the subchapter 3.5 to agree on a common representation for the current control processes for IM and PMSM, resulting in the general process models (3.86) or (3.87) and the block structure in the figure 3.16. The equations represent the control process both in the field synchronous and in the stator fixed coordinate system. They are repeated here in favor of a better overview.

$$\mathbf{i}_s(k+1) = \mathbf{\Phi}\mathbf{i}_s(k) + \mathbf{H}\mathbf{u}_s(k) + \mathbf{h}\psi(k) \tag{5.15}$$

In z domain:

$$z\mathbf{i}_s(z) = \mathbf{\Phi}\mathbf{i}_s(z) + \mathbf{H}\mathbf{u}_s(z) + \mathbf{h}\psi(z) \tag{5.16}$$

Using these equations the controller design shall be carried out first in general and then applied for concrete cases. Under the assumption that $\mathbf{y}$ is the actual controller output quantity the following general controller approach arises.

$$\mathbf{u}_s(k) = \mathbf{H}^{-1}\left[\mathbf{y}(k-1) - \mathbf{h}\psi(k)\right] \text{ or}$$
$$\mathbf{u}_s(k+1) = \mathbf{H}^{-1}\left[\mathbf{y}(k) - \mathbf{h}\psi(k+1)\right] \tag{5.17}$$

The term $\mathbf{y}(k-1)$ takes into account by the time shift (k-1), that in the current calculation the value of the controller output quantity $\mathbf{y}$ calculated in the period before is used. With that the one-step delayed output of the control variable is included in the approach. The 2nd term with $-\mathbf{h}\psi(k)$ compensates the flux dependent part. After inserting equation (5.17) into the equation (5.15) this immediately becomes recognizable, and the compensated general current process model (5.18) arises for the IM as well as the PMSM:

$$\mathbf{i}_s(k+1) = \mathbf{\Phi}\mathbf{i}_s(k) + \mathbf{y}(k-1) \tag{5.18}$$

In the z domain the following equation holds:

$$\left[z\mathbf{I} - \mathbf{\Phi}\right]\mathbf{i}_s(z) = z^{-1}\mathbf{y}(z) \tag{5.19}$$

The figure 5.14 illustrates the compensated current process model which serves as a starting point subsequently for all controller designs. In the following the methodical procedure will always be to address the general design first. After that the design will be specified to the concrete case: IM or PMSM, in field synchronous or in stator fixed coordinates. For this purpose the designs are always represented both in the form of equations and by circuit diagrams so that programming will be made easier.

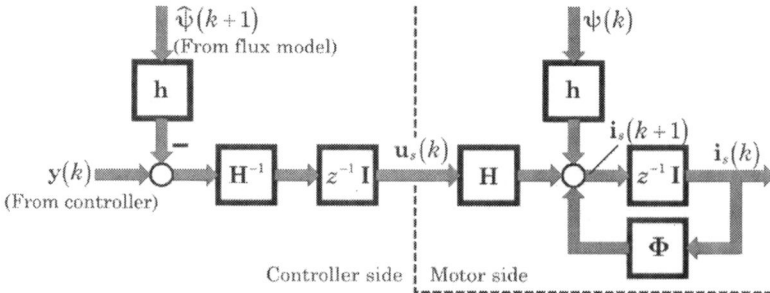

Fig. 5.14 General compensated current process model of the IM and PMSM

5.3 Design of a current vector controller with dead-beat behaviour

The designs in this chapter were introduced repeatedly in different papers [Quang 1991, 1993 and 1996].

5.3.1 Design of a current vector controller with dead-beat behaviour with instantaneous value measurement of the current actual-values

The figure 5.15 shows the principle block structure of the current vector controller with instantaneous value measurement for the example of the measuring strategy in figure 4.1. The controller equation is for this case:

$$\mathbf{y}(z) = \mathbf{R_1}\left[\mathbf{i}_s^*(z) - \mathbf{i}_s(z)\right] \tag{5.20}$$

$\mathbf{i}_s^*(z) =$ Reference or set point vector of the current

After substituting equation (5.20) into the equation (5.19) the following transfer function of the current controlled IM or PMSM can be obtained:

$$\mathbf{i}_s(z) = z^{-1}\left[z\mathbf{I} - \mathbf{\Phi} + z^{-1}\mathbf{R_1}\right]^{-1}\mathbf{R_1}\,\mathbf{i}_s^*(z) \tag{5.21}$$

The approach (5.12) is valid for the closed loop response and respectively for the reference transfer function. The equation (5.12) will be identical with (5.21), if the following equation holds for $\mathbf{R_1}$:

$$\mathbf{R_1} = \frac{\mathbf{I} - z^{-1}\mathbf{\Phi}}{1 - z^{-2}} \tag{5.22}$$

The transfer function (5.12) illustrates by the diagonal matrix whose elements are z^{-2} a both statically and dynamically good decoupling between the current components. The controller $\mathbf{R_1}$ (5.22) in figure 5.15 shows that a decoupling network in the classical presentation (figure 1.4) can be abandoned.

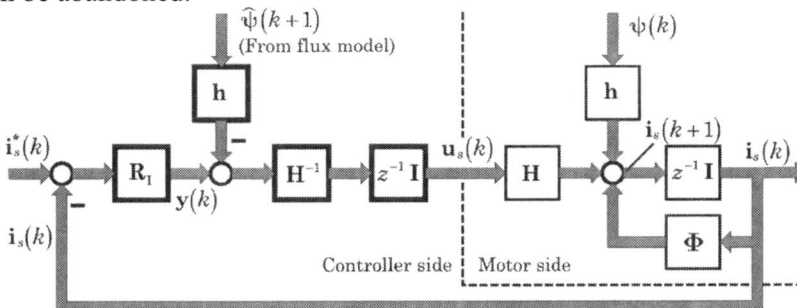

Fig. 5.15 Block structure of the current vector controller for IM or PMSM

With the current control error:

$$\mathbf{x}_w(z) = \mathbf{i}_s^*(z) - \mathbf{i}_s(z) \tag{5.23}$$

it will be obtained:

$$\mathbf{y}(z) = \mathbf{R}_1\,\mathbf{x}_w(z) \tag{5.24}$$

In the time domain the following controller equation results from equation (5.22):

$$\mathbf{y}(k) = \mathbf{x}_w(k) - \mathbf{\Phi}\mathbf{x}_w(k-1) + \mathbf{y}(k-2) \tag{5.25}$$

After inserting equation (5.25) into the equation (5.17) the control variable and respectively the stator voltage, which must be applied on the motor by the vector modulation, is obtained.

$$\mathbf{u}_s(k+1) = \mathbf{H}^{-1}\left[\mathbf{x}_w(k) - \mathbf{\Phi}\mathbf{x}_w(k-1) + \mathbf{y}(k-2) - \mathbf{h}\widehat{\psi}(k+1)\right] \tag{5.26}$$

With the equation (5.26) the design is complete. Two notes, however, are still necessary here.

1. The estimated rotor flux $\widehat{\psi}(k+1)$ (by equations 3.51, 3.55; in detail cf. subchapter 4.4) is used to compensate its disturbance effect. It is constant in the constant flux area and perhaps can be neglected in the practical implementation. The implicit I part in the controller is able to compensate for the missing flux compensation. However, the slowly variable flux in the field weakening area is exposed to permanent changes. It is therefore more advantageous to include the compensation into the equation (5.26).
2. The voltage or the control variable $\mathbf{u}_s$ will be calculated by processing the equation (5.26) always one sampling period ahead. With that the delay of the control variable $\mathbf{u}_s$ by one sampling period is taken into account.

a) Use of the controller for the IM in field synchronous coordinates

To be able to use the design (5.22) and respectively the equation (5.26), the following matrix elements must be replaced corresponding to the models derived in chapter 3:

$$\mathbf{\Phi} \text{ by } \mathbf{\Phi}_{11}^f,\ \mathbf{H} \text{ by } \mathbf{H}_1^f \text{ and } \mathbf{h} \text{ by } \mathbf{\Phi}_{12}^f$$

From equation (3.54) it will be obtained:

$$\mathbf{\Phi}_{11}^f = \begin{bmatrix} \Phi_{11} & \Phi_{12} \\ -\Phi_{12} & \Phi_{11} \end{bmatrix} = \begin{bmatrix} 1 - \dfrac{T}{\sigma}\left(\dfrac{1}{T_s} + \dfrac{1-\sigma}{T_r}\right) & \omega_s T \\[3mm] -\omega_s T & 1 - \dfrac{T}{\sigma}\left(\dfrac{1}{T_s} + \dfrac{1-\sigma}{T_r}\right) \end{bmatrix} \tag{5.27}$$

$$\Phi_{12}^{f} = \begin{bmatrix} \Phi_{13} & \Phi_{14} \\ -\Phi_{14} & \Phi_{13} \end{bmatrix} = \begin{bmatrix} \dfrac{1-\sigma}{\sigma}\dfrac{T}{T_r} & \dfrac{1-\sigma}{\sigma}\omega T \\ -\dfrac{1-\sigma}{\sigma}\omega T & \dfrac{1-\sigma}{\sigma}\dfrac{T}{T_r} \end{bmatrix} \qquad (5.28)$$

$$\mathbf{H}_{1}^{f} = \begin{bmatrix} h_{11} & 0 \\ 0 & h_{11} \end{bmatrix} = \begin{bmatrix} \dfrac{T}{\sigma L_s} & 0 \\ 0 & \dfrac{T}{\sigma L_s} \end{bmatrix} \qquad (5.29)$$

If the matrix elements from (5.27), (5.28) and (5.29) are used in the equation (5.26) now, the following controller equations will be obtained considering that the cross component ψ_{rq} of the rotor flux is zero because of an exact field orientation:

$$\begin{cases} u_{sd}(k+1) = h_{11}^{-1}\Big[x_{wd}(k) - \Phi_{11} x_{wd}(k-1) - \Phi_{12} x_{wq}(k-1) \\ \qquad\qquad + y_d(k-2) - \Phi_{13}\,\psi_{rd}'(k+1) \Big] \\ u_{sq}(k+1) = h_{11}^{-1}\Big[x_{wq}(k) + \Phi_{12} x_{wd}(k-1) - \Phi_{11} x_{wq}(k-1) \\ \qquad\qquad + y_q(k-2) + \Phi_{14}\,\psi_{rd}'(k+1) \Big] \end{cases} \qquad (5.30)$$

Because of the necessary storage of the temporary variable **y** through several sampling periods a direct programming of the equation (5.30) is impractical. The following sequence is more advantageous:

1. Calculation of the vector y(k) using (5.25):

$$\begin{cases} y_d(k) = x_{wd}(k) - \Phi_{11} x_{wd}(k-1) - \Phi_{12} x_{wq}(k-1) + y_d(k-2) \\ y_q(k) = x_{wq}(k) + \Phi_{12} x_{wd}(k-1) - \Phi_{11} x_{wq}(k-1) + y_q(k-2) \end{cases}$$

$$(5.31)$$

2. Then calculation of the stator voltage using (5.17):

$$\begin{cases} u_{sd}(k+1) = h_{11}^{-1}\Big[y_d(k) - \Phi_{13}\,\psi_{rd}'(k+1) \Big] \\ u_{sq}(k+1) = h_{11}^{-1}\Big[y_q(k) + \Phi_{14}\,\psi_{rd}'(k+1) \Big] \end{cases} \qquad (5.32)$$

Now the equations (5.31) and (5.32) can be used for programming provided that the axis-related deviations x_{wd}, x_{wq}, and the accumulated quantity **y** still must be corrected at stator voltage limitation to avoid instabilities. The subchapter 5.5 will deal with the problem of the control variable limitation later in detail.

b) Use of the controller for the IM in stator-fixed coordinates

Φ, $\mathbf{H}$ and $\mathbf{h}$ are replaced by Φ_{11}^s, $\mathbf{H}_1^s$ and Φ_{12}^s from the equation (3.50):

$$\Phi_{11}^s = \begin{bmatrix} \Phi_{11} & 0 \\ 0 & \Phi_{11} \end{bmatrix} = \begin{bmatrix} 1 - \dfrac{T}{\sigma}\left(\dfrac{1}{T_s} + \dfrac{1-\sigma}{T_r}\right) & 0 \\ 0 & 1 - \dfrac{T}{\sigma}\left(\dfrac{1}{T_s} + \dfrac{1-\sigma}{T_r}\right) \end{bmatrix} \tag{5.33}$$

$$\Phi_{12}^s = \begin{bmatrix} \Phi_{13} & \Phi_{14} \\ -\Phi_{14} & \Phi_{13} \end{bmatrix} = \begin{bmatrix} \dfrac{1-\sigma}{\sigma}\dfrac{T}{T_r} & \dfrac{1-\sigma}{\sigma}\omega T \\ -\dfrac{1-\sigma}{\sigma}\omega T & \dfrac{1-\sigma}{\sigma}\dfrac{T}{T_r} \end{bmatrix} \tag{5.34}$$

$$\mathbf{H}_1^s = \begin{bmatrix} h_{11} & 0 \\ 0 & h_{11} \end{bmatrix} = \begin{bmatrix} \dfrac{T}{\sigma L_s} & 0 \\ 0 & \dfrac{T}{\sigma L_s} \end{bmatrix} \tag{5.35}$$

If the matrix elements of (5.33), (5.34) and (5.35) are inserted into the equation (5.26), then the following voltage components in $\alpha\beta$ coordinates will be obtained.

$$\begin{cases} u_{s\alpha}(k+1) = h_{11}^{-1}\big[x_{w\alpha}(k) - \Phi_{11}x_{w\alpha}(k-1) + y_\alpha(k-2) \\ \qquad\qquad - \Phi_{13}\psi_{r\alpha}'(k+1) - \Phi_{14}\psi_{r\beta}'(k+1)\big] \\ u_{s\beta}(k+1) = h_{11}^{-1}\big[x_{w\beta}(k) - \Phi_{11}x_{w\beta}(k-1) + y_\beta(k-2) \\ \qquad\qquad + \Phi_{14}\psi_{r\alpha}'(k+1) - \Phi_{13}\psi_{r\beta}'(k+1)\big] \end{cases} \tag{5.36}$$

The next steps are again useful to support programming:

1. Calculation of the vector $\mathbf{y}(k)$ according to (5.25):

$$\begin{cases} y_\alpha(k) = x_{w\alpha}(k) - \Phi_{11}x_{w\alpha}(k-1) + y_\alpha(k-2) \\ y_\beta(k) = x_{w\beta}(k) - \Phi_{11}x_{w\beta}(k-1) + y_\beta(k-2) \end{cases} \tag{5.37}$$

2. Then the calculation of the voltage using (5.17):

$$\begin{cases} u_{s\alpha}(k+1) = h_{11}^{-1}\big[y_\alpha(k) - \Phi_{13}\psi_{r\alpha}'(k+1) - \Phi_{14}\psi_{r\beta}'(k+1)\big] \\ u_{s\beta}(k+1) = h_{11}^{-1}\big[y_\beta(k) + \Phi_{14}\psi_{r\alpha}'(k+1) - \Phi_{13}\psi_{r\beta}'(k+1)\big] \end{cases} \tag{5.38}$$

c) Use of the controller for the PMSM in field synchronous coordinates

Instead of $\boldsymbol{\Phi}$, $\boldsymbol{H}$ and $\boldsymbol{h}$, the matrices $\boldsymbol{\Phi}_{SM}^{f}$, $\boldsymbol{H}_{SM}^{f}$ and $\boldsymbol{h}$ from equations (3.71) and (3.72) are used for the PMSM:

$$\boldsymbol{\Phi}_{SM}^{f} = \begin{bmatrix} \Phi_{11} & \Phi_{12} \\ \Phi_{21} & \Phi_{22} \end{bmatrix} = \begin{bmatrix} 1 - \dfrac{T}{T_{sd}} & \omega_s T \dfrac{L_{sq}}{L_{sd}} \\ -\omega_s T \dfrac{L_{sd}}{L_{sq}} & 1 - \dfrac{T}{T_{sq}} \end{bmatrix} \tag{5.39}$$

$$\boldsymbol{H}_{SM}^{f} = \begin{bmatrix} h_{11} & 0 \\ 0 & h_{22} \end{bmatrix} = \begin{bmatrix} \dfrac{T}{L_{sd}} & 0 \\ 0 & \dfrac{T}{L_{sq}} \end{bmatrix} ; \quad \boldsymbol{h} = \begin{bmatrix} 0 \\ -\dfrac{\omega_s T}{L_{sq}} \end{bmatrix} = \begin{bmatrix} 0 \\ h_2 \end{bmatrix} \tag{5.40}$$

After replacing the matrix elements of (5.39), (5.40), the *dq* components of the stator voltage result to:

$$\begin{cases} u_{sd}(k+1) = h_{11}^{-1}\left[x_{wd}(k) - \Phi_{11}x_{wd}(k-1) - \Phi_{12}x_{wq}(k-1) + y_d(k-2)\right] \\ u_{sq}(k+1) = h_{22}^{-1}\left[x_{wq}(k) - \Phi_{21}x_{wd}(k-1) - \Phi_{22}x_{wq}(k-1) + y_q(k-2) - h_2\psi_p\right] \end{cases}$$
$$\tag{5.41}$$

and the following programming equations will be obtained:

1. **y**(k) is calculated by using (5.25):

$$\begin{cases} y_d(k) = x_{wd}(k) - \Phi_{11}x_{wd}(k-1) - \Phi_{12}x_{wq}(k-1) + y_d(k-2) \\ y_q(k) = x_{wq}(k) - \Phi_{21}x_{wd}(k-1) - \Phi_{22}x_{wq}(k-1) + y_q(k-2) \end{cases} \tag{5.42}$$

2. then the voltage calculation using (5.17) follows:

$$\begin{cases} u_{sd}(k+1) = h_{11}^{-1}y_d(k) \\ u_{sq}(k+1) = h_{22}^{-1}\left[y_q(k) - h_2\psi_p\right] \end{cases} \tag{5.43}$$

5.3.2 Design of a current vector controller with dead-beat behaviour for integrating measurement of the current actual-values

In principle the process equations (5.18) and (5.19) are only valid for processes with instantaneous value measurement of the current values. In case of an integrating measurement (cf. subchapter 4.1) the measuring equipment is modeled by using the averaging function:

$$\mathbf{i}_s^M(k) = \frac{1}{2}\left[\mathbf{i}_s(k) + \mathbf{i}_s(k-1)\right] \tag{5.44}$$

raised index M: average value

and the result $\mathbf{i}_s^M(k)$ is available for the control as actual value of the stator current. The final process equation in case of integrating measurement results from (5.44) by using (5.18):

$$\mathbf{i}_s^M(k+1) = \mathbf{\Phi}\mathbf{i}_s^M(k) + \frac{1}{2}\left[\mathbf{y}(k-1) + \mathbf{y}(k-2)\right] \tag{5.45}$$

and in the z domain:

$$\left[z\mathbf{I} - \mathbf{\Phi}\right]\mathbf{i}_s^M(z) = \frac{1}{2}\left[z^{-1} + z^{-2}\right]\mathbf{y}(z) \tag{5.46}$$

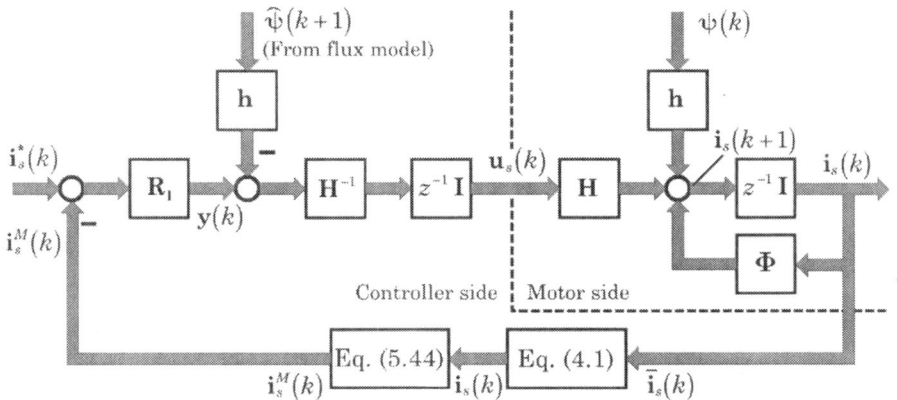

Fig. 5.16 Block structure of the current vector controller for IM or PMSM with integrating measurement

The controller equation starts from:

$$\mathbf{y}(z) = \mathbf{R_1}\left[\mathbf{i}_s^*(z) - \mathbf{i}_s^M(z)\right] \tag{5.47}$$

After eliminating $\mathbf{y}(z)$ in (5.46) and (5.47) we obtain the transfer function:

$$\mathbf{i}_s^M(z) = \frac{1}{2}\left(z^{-1} + z^{-2}\right)\left[z\mathbf{I} - \mathbf{\Phi} + \frac{1}{2}\left(z^{-1} + z^{-2}\right)\mathbf{R_1}\right]^{-1}\mathbf{R_1}\mathbf{i}_s^*(z) \tag{5.48}$$

The approach for the closed loop response and respectively the reference transfer function is:

$$\mathbf{i}_s^M(z) = \frac{1}{2}\left(z^{-2} + z^{-3}\right)\mathbf{i}_s^*(z) \tag{5.49}$$

This is equivalent to a dead beat response. The equations (5.48) and (5.49) are identical, if the following controller is chosen:

$$\mathbf{R_1} = \frac{\mathbf{I} - z^{-1}\mathbf{\Phi}}{1 - \frac{1}{2}\left(z^{-2} + z^{-3}\right)} \tag{5.50}$$

The equation (5.50) looks as follows in the time domain:

$$\mathbf{y}(k) = \mathbf{x}_w(k) - \mathbf{\Phi}\mathbf{x}_w(k-1) + \frac{1}{2}\left[\mathbf{y}(k-2) + \mathbf{y}(k-3)\right] \tag{5.51}$$

$$\mathbf{x}_w(k) = \mathbf{i}_s^*(k) - \mathbf{i}_s^M(k)$$

The derivation of the equations for the controller application in figure 5.16 can similarly be carried out – for the cases IM or PMSM, in field synchronous or stator fixed coordinates – like in the subchapter 5.3.1. For the problem of the control variable limitation it is again referred to the subchapter 5.5.

5.3.3 Design of a current vector controller with finite adjustment time

The controllers introduced in this chapter are derived like in chapter 5.3.1 from the common theoretical approach (5.11) for the closed loop response.

It was shown repeatedly in the literature references mentioned at the beginning of the subchapter 5.3 that the fastest dynamics can be achieved by a dead beat design. This approach provides a virtually undelayed torque impression which is particularly advantageous for the conception of superimposed control loops for mechanical systems (speed, position). Step response times of under 1 ms were reached. The application of fast microprocessors (digital signal processors, high performance microcontrollers) and the tendency toward higher pulse frequencies (10-kHz and more) however result in yet faster sampling of the current control ($T = 100...200$ μs). If the current control was prepared for dead beat behaviour, the inverter could not produce the voltage over time areas necessary to drive the required current step amplitudes (at dynamic processes like magnetization, start up or speed reversal) within the very short demanded rise times of 2×100μs ... 2×200μs = 200...400μs. This is extremely critical for inverters with small control reserve (low DC link voltage). It becomes critical as well if the drive is operated at the voltage limit and dynamic processes (e.g. speed reversal out of the field weakening range) take place simultaneously. Preferably, at these small sampling times and with fast processors like DSP's the current control is not adjusted to dead beat response any more, but to FAT behaviour with more than 2 steps response time. As indicated in the section 5.2 it would be realistic to realize the FAT behaviour with 3 or 4 sampling steps.

For instantaneous value measuring of the stator currents the transfer function (5.21) results for the current-controlled IM in *dq* coordinates. The reference transfer functions for the recommended step number are given in (5.13) and (5.14). The equation (5.21) is identical with either (5.13) or (5.14), if for:

1. **n=2** (FAT = *n*+1 = 3) the following controller:

$$\mathbf{R_1} = \frac{1}{2} \frac{\left(1 + z^{-1}\right)\left[\mathbf{I} - z^{-1}\mathbf{\Phi}\right]}{1 - \frac{1}{2}\left(z^{-2} + z^{-3}\right)}, \text{ and} \tag{5.52}$$

2. **n=3** (FAT = *n*+1 = 4) the following controller:

$$\mathbf{R_1} = \frac{1}{3} \frac{\left(1 + z^{-1} + z^{-2}\right)\left[\mathbf{I} - z^{-1}\mathbf{\Phi}\right]}{1 - \frac{1}{3}\left(z^{-2} + z^{-3} + z^{-4}\right)} \tag{5.53}$$

is valid. The controller designs (5.52) and (5.53) can be used – regarding the available processing capacity – almost without problems by application of digital signal processors with a sampling period of 100 µs, including necessary functions like the vector modulation, the coordinate transformation, the flux model or flux observer and the feedback value processing. The outlined design was carried out assuming an instantaneous value measurement of the stator current.

The current driving voltage over time area is – in comparison with the dead beat design – the same, but distributed over several steps. With that the control voltage $\mathbf{u_s}$ rarely goes into the limitation. This property is seen as an important advantage, especially for inverters with small control reserve (low DC voltage). This also takes effect particularly if the inverter is operated at the limits of the control reserve (e.g. in the field weakening area or at full load). The system stability is fundamentally improved while entering into and recovering from limitation.

5.4 Design of a current state space controller with dead beat behaviour

The main advantage of the *current vector regulators* introduced in the chapters 5.3.1 and 5.3.3 is primarily the practically proven ruggedness when applied to machines whose data are known only inaccurately or calculated only from the name plate. This chapter on the other hand introduces a design in the state space which can produce superior qualities with respect to smooth running and dynamical or decoupling behaviour at higher stator frequencies if exact machine data are available. This allows the particularly advantageous use of the new controller, called *the current state controller* from now on (figure 5.17), in precision drives.

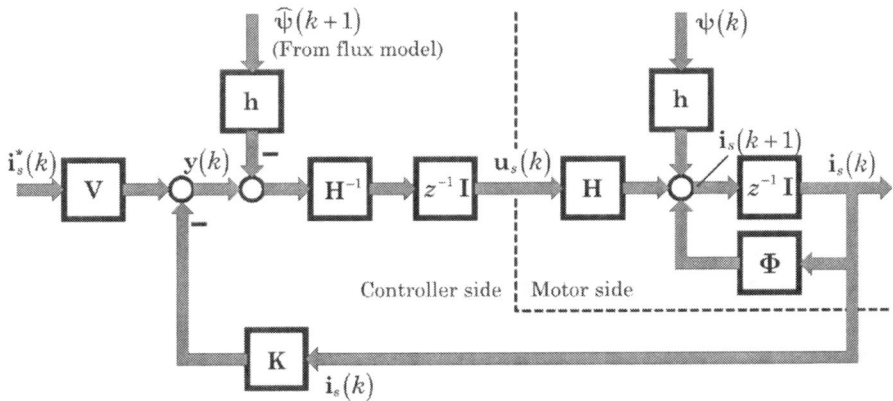

Fig. 5.17 Block structure of the current state controller with pre-filter matrix **V** and feedback matrix **K**

The design starts out as usual from the general approach (5.17) and from the compensated process model (5.18) or (5.19). The controller equation can be written in the z domain as:

$$\mathbf{y}(z) = \mathbf{V}\,\mathbf{i}_s^*(z) - \mathbf{K}\,\mathbf{i}_s(z) \tag{5.54}$$

The equation of the closed loop system is obtained after inserting the equation (5.54) into (5.19):

$$\left[z\mathbf{I} - \left(\mathbf{\Phi} - z^{-1}\mathbf{K} \right) \right] \mathbf{i}_s(z) = z^{-1}\mathbf{V}\,\mathbf{i}_s^*(z) \tag{5.55}$$

Using equation (5.55), the state controller can be designed now, and it has to be noticed that

1. the *feedback matrix* **K** changes the pole positions of the closed loop system, and is therefore decisive for *dynamics* and *stability*. With that, different design strategies, such as the design

- on dead beat behaviour or
- on well damped characteristic, can be derived, and
2. the *pre-filter matrix* **V** serves the adjustment of the demanded working point, and therefore is responsible for the *stationary transfer characteristic.*

This means with respect to the decoupling between the torque and flux forming current components that **K** determines *the dynamic* and **V** *the static decoupling properties.*

5.4.1 Feedback matrix K

By using (5.55) the characteristic equation of the closed loop system is:

$$\det\left[z\,\mathbf{I}-\left(\mathbf{\Phi}-z^{-1}\mathbf{K}\right)\right]=0 \qquad (5.56)$$

The polynomial on the left side of (5.56) has the following general form:

$$\det\left[z\,\mathbf{I}-\left(\mathbf{\Phi}-z^{-1}\mathbf{K}\right)\right]=\sum_{i=0}^{\infty}a_i\,z^i \qquad (5.57)$$

The system has two poles. To achieve dead beat behaviour, both poles must be located (see e.g. Föllinger [1982]) in the coordinate origin. This means that

$$a_i = 0 \ \text{ for } \ i \neq 2\,; \ a_2 = 1$$

Following the Cayley-Hamilton theorem (cf. Föllinger [1982], Isermann [1987]) the matrix $(\mathbf{\Phi}-z^{-1}\mathbf{K})$ fulfils its own characteristic equation. From that, we obtain:

$$\left[\mathbf{\Phi}-z^{-1}\mathbf{K}\right]^2 = \mathbf{0} \ \text{ or } \ \left[\mathbf{\Phi}-z^{-1}\mathbf{K}\right]=\mathbf{0} \qquad (5.58)$$

and then:

$$\mathbf{K}=z\,\mathbf{\Phi} \qquad (5.59)$$

Two remarks shall follow to interpret this result:
1. The equation (5.59) contains a z operator, which means that a *prediction* (one sampling period in advance) of the actual-value of the stator current is necessary.
2. The dead beat behaviour would cause large control amplitudes (as explained in the subchapter 5.3.3) at set point steps, and from this, strong control movements for stochastically disturbed control variables. Therefore the design (5.59) could have an unfavourable effect for inverters with small control reserve (low DC link voltage). A FAT behaviour according to the subchapter 5.3.3 would be sensible

and useful. The application of this reference transfer function in the state space, however, is not possible. A behaviour prepared for a good damping is, on the other hand, practicable. The poles then should not be assigned directly in the coordinate origin but in its near vicinity.

$$\det\left[z\mathbf{I}-\left(\boldsymbol{\Phi}-z^{-1}\mathbf{K}\right)\right]=\left(z-z_1\right)^2 \quad \text{with} \quad z_1 \neq 0 \tag{5.60}$$

From (5.60) it will be obtained:

$$\mathbf{K}=z\left[\boldsymbol{\Phi}-z_1\mathbf{I}\right] \tag{5.61}$$

For practical realization it suffices to determine the satisfactory behaviour by varying z_1 experimentally without having to exaggerate the theory here further.

5.4.2 Pre-filter matrix V

A stationary exact transfer characteristic and good decoupling between the two current components can be expected if the following is valid:

$$\mathbf{i}_s\left(k+1\right)=\mathbf{i}_s\left(k\right)=\mathbf{i}_s^*\left(k\right) \quad \text{für} \quad k\rightarrow\infty$$

or:

$$\mathbf{i}_s\left(z\right)=\mathbf{i}_s^*\left(z\right) \quad \text{für} \quad z\rightarrow 1$$

It follows from (5.55):

$$\mathbf{V}=\mathbf{I}-\left[\boldsymbol{\Phi}-\mathbf{K}\right] \tag{5.62}$$

After using the matrix $\mathbf{K}$ like (5.59) or (5.61) it will be obtained:
1. For the dead beat behaviour:

$$\mathbf{V}=\mathbf{I} \tag{5.63}$$

2. For the design with good damping:

$$\mathbf{V}=\left(1-z_1\right)\mathbf{I} \tag{5.64}$$

With $\mathbf{K}$ and $\mathbf{V}$ calculated by equations (5.59) and (5.63) or (5.61) and (5.64) we obtain from (5.55) the following transfer function of the controlled process:
1. For the dead beat behaviour:

$$\mathbf{i}_s\left(z\right)=z^{-2}\,\mathbf{i}_s^*\left(z\right) \tag{5.65}$$

2. For the design with good damping:

$$\mathbf{i}_s\left(z\right)=\frac{1-z_1}{1-z_1\,z}\mathbf{i}_s^*\left(z\right) \tag{5.66}$$

The two state space designs point to a good dynamic decoupling judging from their diagonal transfer matrices. In contrast to the current vector controller (cf. subchapter 5.3) however, a stationary error has always to be expected because of *the missing integral term*. This stationary error, partly caused by the first order approximation of the discrete state models and partly caused by parameter deviations, shall be eliminated by introducing an additional integral term. Because the current components are dynamically and statically decoupled by the controller design with **K** and **V**, the elimination of the stationary error or deviation can be realized separately for every single current component. Therefore the control structure is extended by two additional integral controllers (figure 5.18).

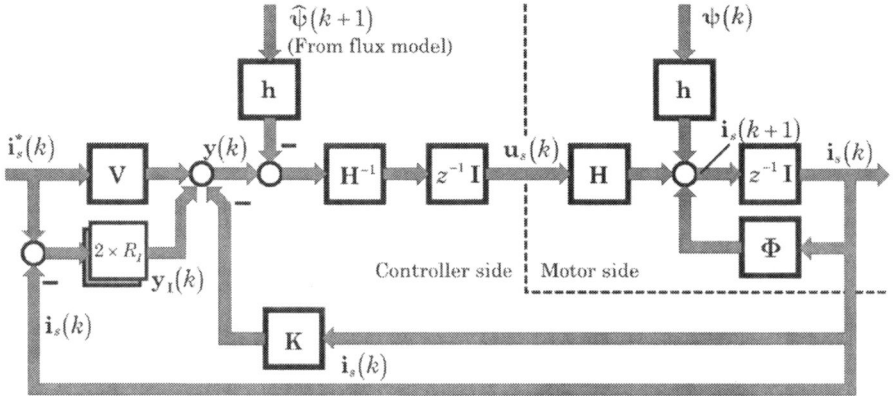

Fig. 5.18 Current state space control with two additional integral controllers

The equation of the additional integral controller R_I will be:

$$\left(1-z^{-1}\right)\mathbf{y_I}\left(z\right)=V_I\,\Delta\mathbf{i}_s\left(z\right)\ \text{ or}$$
$$\mathbf{y_I}\left(k\right)=V_I\,\Delta\mathbf{i}_s\left(k\right)+\mathbf{y_I}\left(k-1\right) \tag{5.67}$$

V_I Controller gain
$\Delta\mathbf{i}_s$ Stationary current error
$\mathbf{y_I}$ Output variable of the integral controller

The controller output variables $\mathbf{y_I}$ have the task to eliminate the stationary errors $\Delta\mathbf{i}_s$ of the stator current. $\mathbf{y_I}$ and $\Delta\mathbf{i}_s$ also fulfill the process equations (5.18) and (5.19):

$$\left[z\mathbf{I}-\mathbf{\Phi}\right]\Delta\mathbf{i}_s\left(z\right)=z^{-1}\mathbf{y_I}\left(z\right)\ \text{ or}$$
$$\Delta\mathbf{i}_s\left(k+1\right)=\mathbf{\Phi}\Delta\mathbf{i}_s\left(k\right)+\mathbf{y_I}\left(k-1\right) \tag{5.68}$$

Since an effective decoupling between the current components is already ensured by the basic structure of the current state space control, the equations (5.67) and (5.68) can be re-written in component notation as follows:

Controller : $y_{Id,q}(z) = \dfrac{V_{Id,q}}{1-z^{-1}} \Delta i_{sd,q}(z)$

(5.69)

Process : $\Delta i_{sd,q}(z) = \dfrac{z^{-2}}{1-\Phi_{11} z^{-1}} y_{Id,q}(z)$

The equation (5.69) is substituted into the closed loop transfer function to calculate the gain factors $V_{Id,q}$ which are usually chosen identical. Because these factors correspond to the ratio T/T_I, (T_I is in comparison with the sampling time T a very big integration time), it suffices in the practice to choose for these factors after the normalizing (about normalizing: subchapter 12.1) a value of approx. 0.05 ... 0.25.

An even better choice would be to feed the integral controller with the current feedback not directly but through a model of the closed loop control system. This would prevent the controller from being invoked at every set point change.

The reader's attention was already drawn to the z operator in equations (5.59), (5.61). The z operator requires a prediction of the stator current. With the actually realized stator voltage, the estimated rotor flux and the measured stator current this prediction can be simply carried out according to the equation (3.74).

$$\hat{\mathbf{i}}_s(k+1) = \boldsymbol{\Phi}\mathbf{i}_s(k) + \mathbf{H}\mathbf{u}_s(k) + \mathbf{h}\hat{\psi}(k) \qquad (5.70)$$

The equation (5.70) has to be adapted to the usage of IM or PMSM and in which coordinate system the motor will be controlled. The complete structure of the current state space control is represented in the figure 5.19.

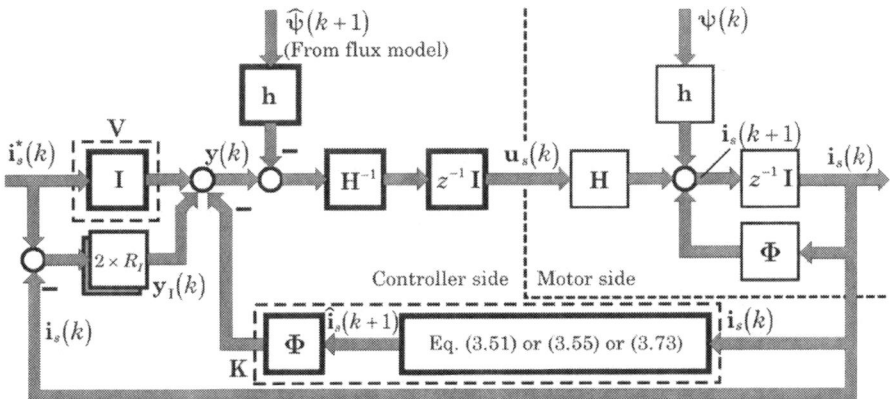

Fig. 5.19 Detailed block structure of the current state space controller for the IM and PMSM

---

[1] Caution: Instead of $\boldsymbol{\Phi}_{11}$, $\boldsymbol{\Phi}_{22}$ is used for q axis in the case PMSM

5.5 Treatment of the limitation of control variables

Generally, the control variable or the stator voltage is limited by the DC link voltage. At uncertain time, e.g. because of a dynamic transient, the current controller requires excessive amplitudes of the control variable which, however, cannot be provided by the inverter. So the control variable hits its maximum consuming all available control reserve. After the current has reached its reference the control variable still stays on its maximum until the integral part has decayed. In this process, oscillations or vibrations of the controlled variable around limitation may develop.

The described process is known and understandable. It is also known that these difficulties can be normally solved by *switching off the integral part* (*anti-reset windup*) once the control variable goes into the limitation. Regarding the new current controllers this strategy could be applied for the additional integral parts of the current state space controller because these parts do obviously exist separately. What would, however, happen with the current vector controllers? The integral part is here not recognizable as part of the design in its own right. Furthermore, being a rather empirical method, turning-off the integrating part does not fit into a fully consistent design and leaves a number of open questions as to the optimal instants to disable and re-enable integration. A better and consistent solution can be provided by reverse-correction of the control deviation (cf. Schönfeld [1985]) which is elaborated on further in this chapter.

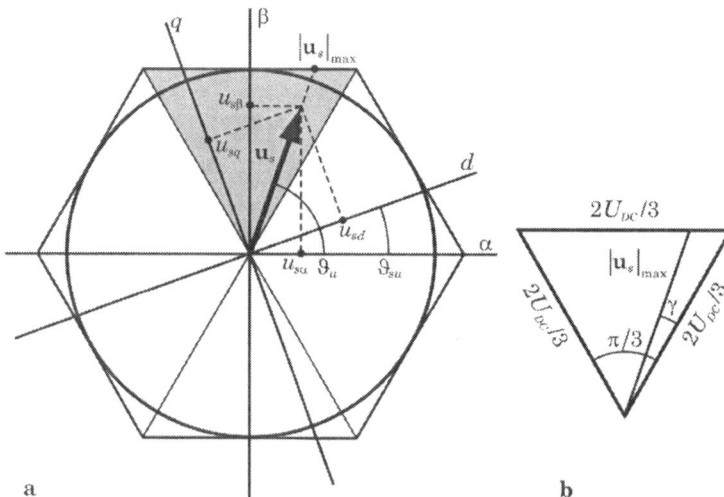

Fig. 5.20 Limitation of control variable: (**a**) Voltage vector $\mathbf{u}_s$ with arbitrary phase angle ϑ_u and (**b**) the maximum modulation ratio $|\mathbf{u}_s|_{max}$ of inverter

Instead of measuring the stator voltage to detect when entering limitation, the stator voltage can be limited intentionally to the maximum modulation ratio.

From the chapter 2 is known, that the maximum usable stator voltage lies within a hexagon (fig. 5.20) and furthermore, that only the limitation of the amplitude of the voltage vector is of importance. However, the stator voltage actually exists in components, u_{sd} and u_{sq} or $u_{s\alpha}$ und $u_{s\beta}$. That means:

> *The voltage limitation must be split into components as well. Suitable methods for this have to be worked out for the chosen coordinate system.*

The voltage limitation itself is completed with its splitting into components. But as mentioned above:

> *A reverse correction strategy, which prevents the vibrations or oscillations caused by the implicitly existing integral part, must be worked out.*

The figure 5.20a has pointed to the possibilities of setting the limitation boundaries on the inner circle touched by the hexagon or on the outer hexagon. The limitation most simply works with the circle, but a loss of control reserve (the area between hexagon and circle) would be the result. The phase angle ϑ_u of the stator voltage then is:

$$\vartheta_u = \vartheta_{su} + \arctan\left(\frac{u_{sq}}{u_{sd}}\right) \tag{5.71}$$

With the help of (5.71) and fig. 5.20b the maximum amplitude of the voltage vector or the maximum modulation ratio (at normalization with $2U_{DC}/3$) depending on the phase angle can be found as:

$$\left|\mathbf{u_s}\right|_{max} = \frac{\sqrt{3}}{2}\frac{1}{\sin\left(\gamma + \dfrac{\pi}{3}\right)} \,^{1} \tag{5.72}$$

The limitation on the outer hexagon according to (5.72) yields the best actuator utilization with respect to deliverable control voltage, however, causes an additional third harmonic in the current. This is unwanted in the stationary operation where the field and torque forming components represent DC quantities. It is therefore recommended for high-quality servo drives to limit on the inner circle. The maximum modulation ratio then is:

---

[1] After normalizing with $2U_{DC}/3$ the voltage is formulated as modulation ratio here; the angle γ is defined in fig. 5.20b.

$$|\mathbf{u}_s|_{max} = \frac{\sqrt{3}}{2} \qquad (5.73)$$

In principle the limitation can be implemented on one of the three following levels (fig. 5.21).

1. *Level of dq components:* This is the mostly applied, most effective variant for the limitation. The decoupling between the *dq* axes or between torque formation and magnetization can be ensured largely with a correct splitting strategy (cf. chapter 5.5.1).

2. *Level of αβ components:* The application of this variant is only possible if the torque impression is implemented using a current controller in the stator-fixed coordinate system. Unfortunately, the decoupling between torque formation and magnetization cannot be ensured any more.

3. *Level of switching times:* This variant is rarely used. The decoupling is not ensured any more. For low-quality drives, where microprocessor power (for splitting and reverse correction) is missing and/or slow semiconductor components are used, the use of this variant could be interesting.

The following chapters only deal with the limitation at the level of the *dq* coordinate system.

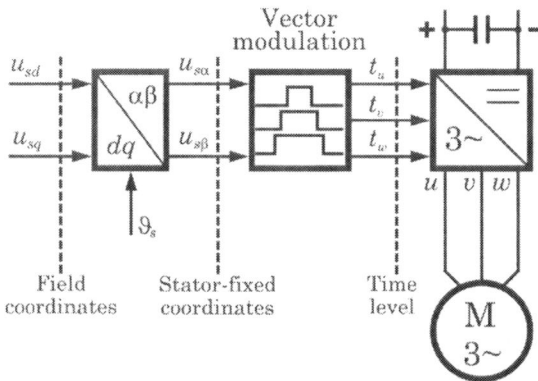

Fig. 5.21 Possible levels for the realization of the control variable limitation

5.5.1 Splitting strategy at voltage limitation

Geometrically the voltage limitation is equivalent to shortening the voltage amplitude. For non-reactive loads, i.e. current and voltage have the same sign, the current gets smaller at reduced voltage. For reactive,

inductive/capacitive or mixed (ohmic-inductive, ohmic-capacitive) loads –
i.e. current and voltage can have different signs – a voltage shortening
would be able to cause the current to increase and duly cause the system to
become unstable. It is known that u_{sd} and i_{sd} as well as u_{sq} and i_{sq} very often
have different signs which indicate the operating state (motor, generator)
of the system.

*These introductory words make already clear that a splitting
strategy, which ensures the system stability, must be able to
recognize priorities for the coordinate axes depending on the
operating state and then perform the limit splitting according to
the geometric possibilities.*

a) Geometric possibilities for limitation

From geometrical point of view and depending on whether the outer
hexagon or the inner circle is chosen as the limitation curve, one of the
three following possibilities (cf. figure 5.22) can be used for the splitting:

1. u_{sd} is cut down, u_{sq} will be kept or has priority:

$$u_{sdr} = sign(u_{sd})\sqrt{|\mathbf{u_s}|^2_{max} - u^2_{sq}} \ ; \ u_{sqr} = u_{sq} \tag{5.74}$$

2. u_{sd} and u_{sq} are truncated in the same proportion (called: the phase correct limitation):

$$u_{sdr} = u_{sd}\sqrt{\frac{|\mathbf{u_s}|^2_{max}}{u^2_{sd}+u^2_{sq}}} \ ; \ u_{sqr} = u_{sq}\sqrt{\frac{|\mathbf{u_s}|^2_{max}}{u^2_{sd}+u^2_{sq}}} \tag{5.75}$$

3. u_{sq} is cut down, u_{sd} will be kept or has priority:

$$u_{sdr} = u_{sd} \ ; \ u_{sqr} = sign(u_{sq})\sqrt{|\mathbf{u_s}|^2_{max} - u^2_{sd}} \tag{5.76}$$

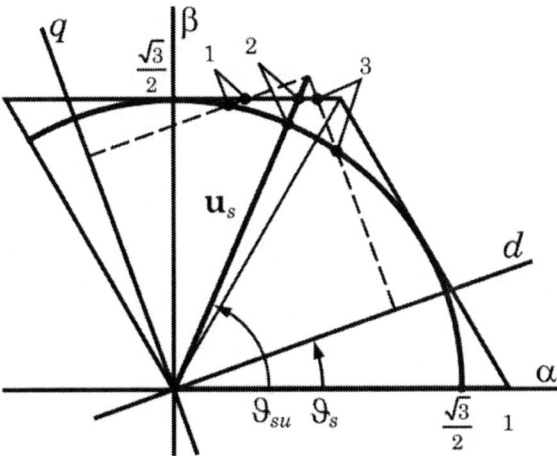

Fig. 5.22 Geometric possibilities for limitation splitting: **(1)** only d component, or **(2)** d and q component in the same proportion, or **(3)** only q component will be truncated

The figure 5.22 clarifies that with the inner circle as limitation curve the value of the maximum modulation ratio $|\mathbf{u}_s|_{max}$ is always given according to (5.73). Unlike this, $|\mathbf{u}_s|_{max}$ using the outer hexagon can adopt different values (cf. fig. 5.22) for the same reference voltage vector, whose complex calculation cannot be handled by every microprocessor and therefore is rarely used. For this reason the limitation on the hexagon will not be further followed here.

The equations (5.74) and (5.76) point to the possible case in which even the component with priority can exceed the value of $|\mathbf{u}_s|_{max}$. In this case the component with priority must also be shortened.

b) Splitting strategy by [Quang 1994]
The strategy starts out from an analysis of the possible operating states of the electrical machine.

Asynchronous drive using IM
In stationary operation the following system of equations is valid for the stator voltage:

$$\begin{cases} u_{sd} = R_s\, i_{sd} - \omega_s\, \sigma\, L_s\, i_{sq} \\ u_{sq} = R_s\, i_{sq} + \omega_s\, L_s\, i_{sd} \end{cases} \tag{5.77}$$

The operating states which lead to the voltage limitation are always connected to higher frequencies so that resistive voltage drops are negligible in the equation (5.77). Therefore they can be reduced to:

$$\begin{cases} u_{sd} \approx -\omega_s\, \sigma\, L_s\, i_{sq} \\ u_{sq} \approx \omega_s\, L_s\, i_{sd} \end{cases} \tag{5.78}$$

The equation (5.78) obviously points to a static coupling between d and q axes and implies that in the area of higher frequencies (where limitations often take place) the components u_{sd} and u_{sq} usually have to provide the greater part for overcoming this coupling than for keeping its own current component. The following facts can be stated in evaluation of equation (5.78):

• The field forming current i_{sd} *always has positive sign in stationary operation.* (Remark: The field forming current i_{sd} could accept negative sign only for feedback-controlled flux for a short time).
• *The product* $m_M \times \omega_s$ *or* $i_{sq} \times \omega_s$ *is always positive in motor operation.* I.e. i_{sq} and ω_s always have the same sign. This means, that:
  $\Rightarrow u_{sd} < 0$, or u_{sd} and i_{sd} have different signs, and
  $\Rightarrow u_{sq}$ and i_{sq} have the same sign.

- *The product $m_M \times \omega_s$ or $i_{sq} \times \omega_s$ is always negative in generator operation.* I.e. i_{sq} and ω_s always have different signs. This means, that:
  $\Rightarrow u_{sd} > 0$, or u_{sd} and i_{sd} have *the same sign*, and
  $\Rightarrow u_{sq}$ and i_{sq} have *different signs*.
  The above short analysis says, that:
- In motor operation the component u_{sd} and
- in generator operation the component u_{sq}
  will get priority. If the priority component already exceeds the value $|\mathbf{u}_s|_{max}$, so approx. 95% of $|\mathbf{u}_s|_{max}$ shall be assigned to this component.

Synchronous drive using PMSM

The following relation will be arrived at for the synchronous drive in stationary operation similar to the case of the asynchronous drive:

$$\begin{cases} u_{sd} = R_s\, i_{sd} - \omega_s\, L_{sq}\, i_{sq} \\ u_{sq} = R_s\, i_{sq} + \omega_s\, L_{sd}\left(i_{sd} + \dfrac{\psi_p}{L_{sd}} \right) \end{cases} \tag{5.79}$$

or

$$\begin{cases} u_{sd} \approx -\omega_s\, L_{sq}\, i_{sq} \\ u_{sq} \approx \omega_s\, L_{sd}\left(i_{sd} + \dfrac{\psi_p}{L_{sd}} \right) \end{cases} \tag{5.80}$$

In the two above equations, the term $\left(i_{sd} + \psi_p / L_{sd} \right)$, in which the current i_{sd} assumes the value zero in the basic operation range and negative values only in the field weakening range, represents the substitute magnetization current with always positive values. The equation (5.80) can be interpreted now similarly to (5.78) of the IM so that the following conclusions can be drawn:

- *The product $m_M \times \omega_s$ or $i_{sq} \times \omega_s$ is always positive in motor operation.* i_{sd} is either zero or negative. This means, that:
  $\Rightarrow u_{sd} < 0$, or u_{sd} and i_{sd} have *the same sign*, and
  $\Rightarrow u_{sq}$ and i_{sq} have *the same sign*.
- *The product $m_M \times \omega_s$ or $i_{sq} \times \omega_s$ is always negative in generator operation.* i_{sd} is either zero or negative. This means, that:
  $\Rightarrow u_{sd} > 0$, or u_{sd} and i_{sd} have *different signs*, and
  $\Rightarrow u_{sq}$ and i_{sq} have *different signs*.
  The analysis has shown the clear difference between the IM and the PMSM: While in the motor operation with the IM the component u_{sd} shall

get the priority obviously, the priority must be assigned to none of the axis voltages in the case PMSM.

The generator operation with PMSM seems to be more problematic than with IM because both couples u_{sd}, i_{sd} and u_{sq}, i_{sq} have different signs. Also this case can be realized exactly as for the IM: I.e. priority for u_{sq}. Amplification of $|i_{sd}|$ for a short time after shortening $|u_{sd}|$ only weakens the permanent magnetization which in turn would increase the control reserve, and the limitation would disappear. With these considerations a simple algorithm outlined in figure 5.23 can be worked out for both types of machines.

$\left|\mathbf{u}_s\right| > \left|\mathbf{u}_s\right|_{max}$			
$sign(\omega_s) \neq sign(i_{sq})$?			
No (motor operation)		Yes (generator operation)	
$\left|u_{sd}\right| > 95\%\left|\mathbf{u}_s\right|_{max}$?		$\left|u_{sq}\right| > 95\%\left|\mathbf{u}_s\right|_{max}$?	
No	Yes	No	Yes
$u_{sdr} := u_{sd}$ $u_{sqr} := sign(u_{sq}) \cdot$ $\sqrt{\left|\mathbf{u}_s\right|^2_{max} - u^2_{sdr}}$	$u_{sdr} := sign(u_{sd}) \cdot$ $\left(95\%\left|\mathbf{u}_s\right|_{max}\right)$ $u_{sqr} := sign(u_{sq}) \cdot$ $\sqrt{\left|\mathbf{u}_s\right|^2_{max} - u^2_{sdr}}$	$u_{sqr} := u_{sq}$ $u_{sdr} := sign(u_{sd}) \cdot$ $\sqrt{\left|\mathbf{u}_s\right|^2_{max} - u^2_{sqr}}$	$u_{sqr} := sign(u_{sq}) \cdot$ $\left(95\%\left|\mathbf{u}_s\right|_{max}\right)$ $u_{sdr} := sign(u_{sd}) \cdot$ $\sqrt{\left|\mathbf{u}_s\right|^2_{max} - u^2_{sqr}}$

Fig. 5.23 Algorithm for voltage limitation by [Quang 1994] (index r: actually realized)

c) Splitting strategy by [Dittrich 1998]

The basic idea of this strategy is ensure decoupling between rotor flux and torque in large-signal behaviour. To achieve this, an intervention in form of a limitation should as much as possible only effect the voltage component, which has caused the maximum voltage vector to exceed its limit, and leave the other component uninfluenced. This concept presumes that such a separation of causes is actually possible and that the voltage vector can be reduced to its maximum value by reduction of one component only. The context is generally more complex and requires a detailed analysis, in particular, if the controlled system must be operated for longer time at the limit of the control variable.

For splitting the voltage limitation after [Dittrich 1998] two questions must be answered:

1. Which component obtains the priority, i.e. which component must remain as unchanged as possible?
2. Which value does the other component get?

The algorithm which is found and realized eventually answers these questions as follows:

Priority decision

Stability considerations are decisive. If current and voltage have different signs in one axis, a limitation of the voltage leads to a temporarily unstable and uncontrollable behavior. If current and voltage signs are different in one axis, this axis must get the priority. If the signs are different in both axes, the axis with the larger current amplitude gets the priority, or the phase correct limitation (using equation (5.75)) is applied. Equal or different signs in the q axis are equivalent to motor or generator operation.

Voltage in the non-priority axis

Two cases have to be distinguished. If the priority component is smaller than the maximum voltage, i.e. the limitation was caused by the non-priority component essentially, the non-priority component results simply from the geometric difference between the maximum voltage and the priority component. In the other case, the non-priority component is assigned the share from the cross-coupling of the current components to support the stationary decoupling of the current components also at control variable limitation.

$\left\|\mathbf{u}_s\right\| > \left\|\mathbf{u}_s\right\|_{max}$			
$\left[sign\left(u_{sd}\right) \neq sign\left(i_{sd}\right)\right] \vee \left\{\left[sign\left(\omega_s\right) = sign\left(i_{sq}^*\right)\right] \wedge \left(i_{sd}^* < 1,5\,i_m\right)\right\}$			
Yes		No	
$\left\|u_{sd}\right\| > \left\|\mathbf{u}_s\right\|_{max}$?		$\left\|u_{sq}\right\| > \left\|\mathbf{u}_s\right\|_{max}$?	
Yes	No	Yes	No
$u_{sqr} := \sigma L_s \omega_s i_{sd}$ $u_{sdr} := sign\left(u_{sd}\right)\cdot$ $\sqrt{\left\|\mathbf{u}_s\right\|_{max}^2 - u_{sqr}^2}$	$u_{sqr} := sign\left(u_{sq}\right)\cdot$ $\sqrt{\left\|\mathbf{u}_s\right\|_{max}^2 - u_{sd}^2}$	$u_{sdr} := -\sigma L_s \omega_s i_{sq}$ $u_{sqr} := sign\left(u_{sq}\right)\cdot$ $\sqrt{\left\|\mathbf{u}_s\right\|_{max}^2 - u_{sdr}^2}$	$u_{sdr} := sign\left(u_{sd}\right)\cdot$ $\sqrt{\left\|\mathbf{u}_s\right\|_{max}^2 - u_{sq}^2}$

Fig. 5.24 Algorithm for voltage limitation by [Dittrich 1998]

The figure 5.24 shows the described algorithm in the overview. A similar approach was attended in [Wiesing 1994].

5.5.2 Correction strategy at voltage limitation

The basic idea of the reverse correction is a correction of the control error $\mathbf{x}_w$ to prevent the windup-integration of the integral part which implicitly exists in the control algorithm.

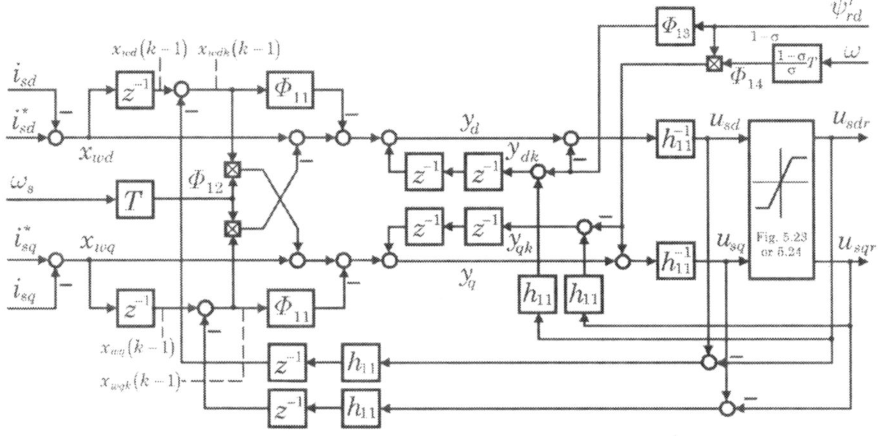

Fig. 5.25 Complete structure of the current vector controller with dead-beat-behaviour

To derive – the design in the chapter 5.3.1 serves as an example – the formula for the reverse correction, the equation (5.17) is re-written as follows:

$$\mathbf{y}(k) = \mathbf{H}\,\mathbf{u}_s(k+1) + \mathbf{h}\,\psi(k+1) \tag{5.81}$$

Assuming a largely constant rotor flux the following result will be obtained after substituting the equation (5.81) into (5.25):

$$\mathbf{H}\,\mathbf{u}_s(k) = \mathbf{x}_w(k-1) - \mathbf{\Phi}\,\mathbf{x}_w(k-2) + \mathbf{H}\,\mathbf{u}_s(k-2) \tag{5.82}$$

Assumed that the voltage goes into the limitation in time instant (k), i.e. instead of the voltage $\mathbf{u}_s(k)$ to be realized only $\mathbf{u}_{sr}(k)$ was realized, (5.82) turns into the equation (5.83).

$$\mathbf{H}\,\mathbf{u}_{sr}(k) = \mathbf{x}_{wc}(k-1) - \mathbf{\Phi}\,\mathbf{x}_w(k-2) + \mathbf{H}\,\mathbf{u}_s(k-2) \tag{5.83}$$

$\mathbf{x}_{wc}$ = Control errors corrected

$\mathbf{u}_{sr}$ = Voltage actually realized after limitation

The subtraction of the equations (5.82) and (5.83) produces for the corrected deviation:

$$\mathbf{x}_{wc}(k-1) = \mathbf{x}_w(k-1) - \mathbf{H}\big[\mathbf{u}_s(k) - \mathbf{u}_{sr}(k)\big] \tag{5.84}$$

Also the accumulated values $\mathbf{y}$ according to the equation (5.25) have to be corrected according to the equation (5.17) with the correct voltage values:

$$\mathbf{y}_k\left(k\right) = \mathbf{H}\mathbf{u}_{sr}\left(k+1\right) + \mathbf{h}\,\psi\left(k+1\right) \tag{5.85}$$

The formulae for the reverse correction for the designs with FAT behavior or for the additional integral controllers of the state space design can also be derived similarly. The figure 5.25 exemplarily illustrates the design with dead beat behaviour.

The implementation of the complete control algorithm in figure 5.25 is outlined by the program flowchart in figure 5.26.

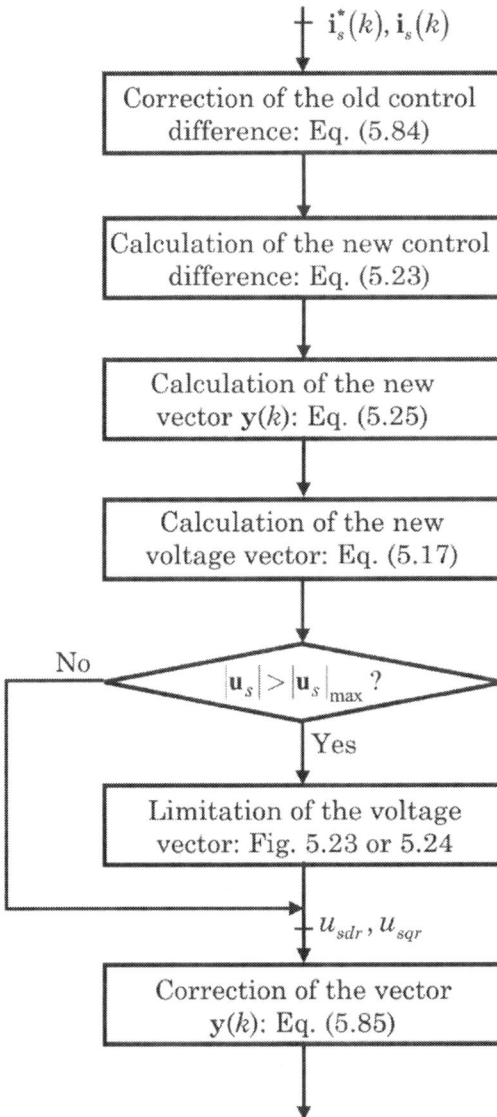

$$\mathbf{i}_s^*(k), \mathbf{i}_s(k)$$

Correction of the old control
difference: Eq. (5.84)

Calculation of the new control
difference: Eq. (5.23)

Calculation of the new
vector $\mathbf{y}(k)$: Eq. (5.25)

Calculation of the new
voltage vector: Eq. (5.17)

No

$\left|\mathbf{u}_s\right| > \left|\mathbf{u}_s\right|_{max}$?

Yes

Limitation of the voltage
vector: Fig. 5.23 or 5.24

u_{sdr}, u_{sqr}

Correction of the vector
$\mathbf{y}(k)$: Eq. (5.85)

Fig. 5.26 Program flowchart of the current vector controller with dead beat behaviour

5.6 References

Brod DM, Novotny DN (1985) Current control of VSI-PWM inverters. IEEE Trans. on IA 21

Dittrich JA (1998) Anwendung fortgeschrittener Steuer- und Regelverfahren bei Asynchronantrieben. Habilitationsschrift, TU Dresden

Enjeti P, Lindsay JF, Ziogas PD, Rashid MH (1988) New current control scheme for PWM inverters. IEE Proceedings, Vol. 135, Pt.B, No. 4, July, pp. 172 – 179

Föllinger O (1982) Lineare Abtastsysteme. R. Oldenbourg Verlag: München Wien

Hofmann W (1984) Entwurf und Eigenschaften einer digitalen Vektorregelung von Asynchronmotoren mit gesteuertem Rotorfluß. Dissertation, TU Dresden

Holtz J, Stadtfeld S (1983) Fieldoriented control by forced motor currents in a voltage fed inverter drive. IFAC Symposium Control in Power Electronics an Electrical Drives, Lausanne, Switzerland, pp. 103 – 110

Holtz J, Stadtfeld S (1983) A predictive controller for the stator current vector of AC machines fed from a switched voltage source. IPEC Tokyo, Conf. Rec., pp. 1665 – 1675

Holtz J, Stadtfeld S (1985) A PWM inverter drive system with on-line optimized pulse patterns. EPE Brussel Conf. Rec., pp. 3.21 – 3.25

Isermann R (1987) Digitale Regelsysteme. Bd. 2, Springer Verlag: Berlin Heidelberg New York London Paris Tokyo

Kazmierkowski MP, Wojciak A (1988) Current control of VSI-PWM inverter-fed induction motor. Warsav Uni. of Technology, Institute of Control and Industrial Electronics, PE 7945

Kazmierkowski MP, Dzieniakowski MA, Sulkowski W (1988) Novel space vector based current controller for PWM-inverters. Summary, Warsav Uni. of Technology, Institute of Control and Industrial Electronics

Le-Huy H, Dessaint LA (1986) An adaptive current controller for PWM-inverters. IEEE PESC Conf., pp. 610 – 616

Malesani L, Tenti P (1987) A novel hysteresis control method for current controlled VSI-PWM inverters with constant modulation frequency. Conf. Rec. of IEEE-IAS Ann. Meet., pp. 851 – 855

Mayer HR (1988) Entwurf zeitdiskreter Regelverfahren für Asynchronmotoren unter Berücksichtigung der diskreten Arbeitsweise des Umrichters. Dissertation, Uni. Erlangen – Nürnberg

Meshkat S, Persson EK (1984) Optimum current vector control of brushless servo amplifier using microprocessors. IEEE IAS Ann. Meet. Conf. Rec., pp. 451 – 457

Nabae A, Oyasawara S, Akagi H (1985) A novel control scheme of current-controlled PWM inverters. IEEE – IAS Ann. Meet. Conf. Rec., pp. 473 – 478

Peak SC, Plunkett AB (1982) Transistorized PWM inverter-induction motor drive system. IEEE – IAS Ann. Meet. Conf. Rec., pp. 892 – 893

Pfaff G, Wick A (1983) Direkte Stromregelung bei Drehstromantrieben mit Pulswechselrichter. Regelungstechnische Praxis (rtp) 24, H. 11, S. 472 – 477

Quang NP (1990) Verfahren zur Stromregelung in Drehstromstellantrieben: Lösungsprinzipien und deren Grenzen. Fachtagung „Steuerung mechanischer Systeme", TU Chemnitz, Februar, S. 102 – 105

Quang NP (1991) Schnelle Drehmomenteinprägung in Drehstromstellantrieben. Dissertation, TU Dresden

Quang NP (1994) Dokumentation zur Regelungssoftware mit TMS 320C25 von REFU 402Vectovar. REFU Elektronik GmbH, Abt. E1, interner Bericht

Quang NP (1996) Mehrgrößenregler löst PI-Regler ab: Von den Parametern zu programmierbaren Reglergleichungen. Elektronik, H.8, S. 112 – 120

Quang NP (1996) Digital Controlled Three-Phase Drives. Education Publishing House: Hanoi (book in vietnamese: Điều khiển tự động truyền động điện xoay chiều ba pha. Nhà xuất bản Giáo dục Hà Nội)

Quang NP, Schönfeld R (1991) Stromvektorregelung für Drehstromstellantriebe mit Pulswechselrichter. atp, Nov., S. 401 – 405 (msr)

Quang NP, Schönfeld R (1991) Stromzustandsregelung: Neues Konzept zur Ständerstromeinprägung für Drehstromstellantriebe mit Pulswechselrichter. atp, Dez., S. 432 – 436 (msr)

Quang NP, Schönfeld R (1993) Dynamische Stromregelung zur Drehmomenteinprägung in Drehstromantrieben mit Pulswechselrichter. Archiv für Elektrotechnik / Archiv of Electrical Engineering 76, S. 317 – 323

Quang NP, Schönfeld R (1993) Eine Stromvektorregelung mit endlicher Einstellzeit für dynamische Drehstromantriebe. Archiv für Elektrotechnik / Archiv of Electrical Engineering 76, S. 377 – 385

Rodriguez J, Kastner G (1987) Nonlinear current control of an inverter-fed induction machine. etz Archiv, Bd. 9, H. 8, S. 245 – 250

Rowan TM, Kerkman RJ, Lipo TA (1987) Operation of naturally sampled current regulators in the transition mode. IEEE Trans. on IA, IA- 23, No. 4, July/August, pp. 586 – 596

Schönfeld R, Krug H, Geitner GH, Stoev A (1985) Regelalgorithmen für digitale Regler von elektrischen Antrieben. msr, Berlin 28 H. 9, S. 390 – 394

Seifert D (1986) Stromregelung der Asynchronmaschine. etz Archiv, Bd. 6, H. 5, S. 151 – 156

Wiesing J (1994) Betrieb der feldorientiert geregelten Asynchronmaschine im Bereich oberhalb der Nenndrehzahl. Dissertation, Uni. Paderborn

Zhang J, Thiagarajan V, Grant T, Barton TH (1988) New approach to field orientation control of a CSI induction motor drive. IEE Proceedings, Vol. 135, Pt.B, No. 1, January, pp. 1 – 7

6 Equivalent circuits and methods to determine the system parameters

For the clear specification of the electromagnetic processes in 3-phase AC machines and as a starting point for control design, equivalent circuits which are based on the representation of the physical quantities as complex space vectors *in a stator-fixed coordinate system* will be a very useful tool. The underlying mathematics is strongly related to the complex calculations known from the AC technology. To abstract the physical operation of the machines, inductances and resistances are represented as concentrated components, and symmetrical conditions are assumed with regard to the 3-phase windings.

For the satisfactory function of a controller designed using equivalent circuits the parameters of the equivalent circuits must be known with sufficient accuracy. From modern drives it will be expected that they fulfill the projected quality parameters without special tuning to be carried out by the customer, and keep the parameters durably. Because frequency converters, particularly in small and medium power ranges, are offered in principle as separate units without motors, parameter pre-setting or measuring the used motor by means of classical methods (no-load or short circuit test) are not practicable. Therefore the second part of this chapter deals with possibilities of the automated computation of the electrical motor parameters.

A first starting-point and also a base for start values of a more exact estimation will be provided by the name plate or by the rated data of the motor. For a more exact parameter setting off-line identification methods which provide estimated values of motor parameters during a test run in standstill are discussed.

6.1 Equivalent circuits with constant parameters

6.1.1 Equivalent circuits of the IM

6.1.1.1 T equivalent circuit

The general voltage and flux-linkage equations in the stator-fixed coordinate system (cf. chapter 3)

$$\mathbf{u}_s = R_s \mathbf{i}_s + \dot{\boldsymbol{\psi}}_s \tag{6.1}$$

$$\mathbf{u}_r^r = R_r^r \mathbf{i}_r^r + \dot{\boldsymbol{\psi}}_r^r - j\omega\,\boldsymbol{\psi}_r^r \tag{6.2}$$

$$\boldsymbol{\psi}_s = L_s \mathbf{i}_s + L_m \mathbf{i}_r \tag{6.3}$$

$$\boldsymbol{\psi}_r^r = L_r^r \mathbf{i}_r^r + L_m^r \mathbf{i}_s^r \tag{6.4}$$

describe a transformer with an additional secondary (rotor-sided) voltage source as represented in figure 6.1. In this case the superscript r means that the so labelled parameters and quantities are related to the rotor side, and therefore correspond to the values measured at rotor terminals physically. Quantities without such index are related to the stator side.

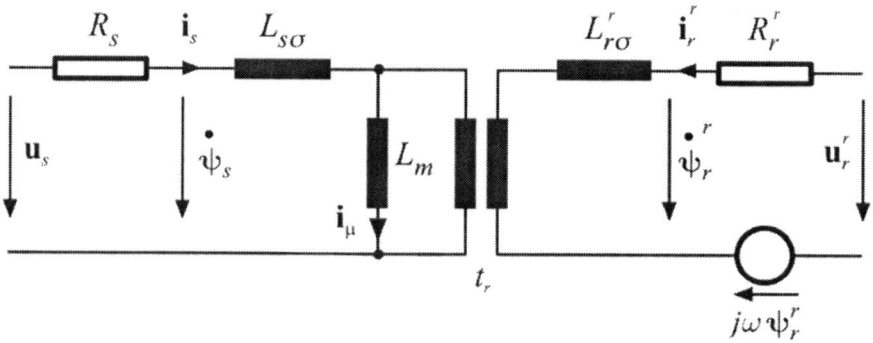

Fig. 6.1   Transformer-equivalent circuit of the induction machine

The actual transformer symbol in the equivalent circuit marks an ideal transformer with the transfer ratio t_r. This contains the turn ratio and winding factors, and can be expressed by the relationship between the no-load nominal voltages.

$$t_r = \frac{U_{rN}}{U_{sN}} \tag{6.5}$$

Because the induction machine is fed normally from either the stator side or the rotor side, it is usual and useful to relate all electrical quantities to either the stator side or the rotor side. Subsequently, on principle, the stator side reference shall be used. For the transformation of the rotor quantities to the stator side the following relations are obtained by using the transfer ratio t_r defined above:

$$\mathbf{u}_r = \frac{\mathbf{u}_r^r}{t_r}$$

$$\mathbf{i}_r = t_r \mathbf{i}_r^r \tag{6.6}$$

$$\psi_r = \frac{\psi_r^r}{t_r}$$

$$R_r = \frac{R_r^r}{t_r^2}$$

$$\tag{6.7}$$

$$L_r = \frac{L_r^r}{t_r^2}$$

For the current through the main inductance L_m (the magnetizing current $\mathbf{i}_\mu$) can be written:

$$\mathbf{i}_\mu = \mathbf{i}_s + \mathbf{i}_r \tag{6.8}$$

The reference to the stator side is primarily relevant for the treatment of the squirrel-cage IM ($\mathbf{u}_r = 0$) which shall be also the object of the further derivations. Because of the interchangeability of both approaches this does not represent any essential restriction of the generality.

If the flux-linkage in (6.1), (6.2) is replaced by (6.3), (6.4), the equations of the stator and rotor voltage can be changed into the form:

$$\mathbf{u}_s = R_s \mathbf{i}_s + L_{s\sigma} \frac{d\mathbf{i}_s}{dt} + L_m \frac{d\mathbf{i}_\mu}{dt} \tag{6.9}$$

$$0 = R_r \mathbf{i}_r + L_{r\sigma} \frac{d\mathbf{i}_r}{dt} + L_m \frac{d\mathbf{i}_\mu}{dt} - j\omega\psi_r \tag{6.10}$$

The mesh equations (6.9) und (6.10) describe the so called *T equivalent circuit* shown in the figure 6.2a. After the transition into the Laplace domain the following voltage equations will be obtained for the stationary operation ($s \rightarrow j\omega_s$):

$$\mathbf{u}_s = R_s \mathbf{i}_s + j\omega_s \left(L_{s\sigma} \mathbf{i}_s + L_m \mathbf{i}_\mu \right) \tag{6.11}$$

$$0 = \frac{R_r}{s} \mathbf{i}_r + j\omega_s \left(L_{r\sigma} \mathbf{i}_r + L_m \mathbf{i}_\mu \right) \qquad (6.12)$$

with the slip $s = (\omega_s - \omega)/\omega_s$, represented in the figure 6.2b.

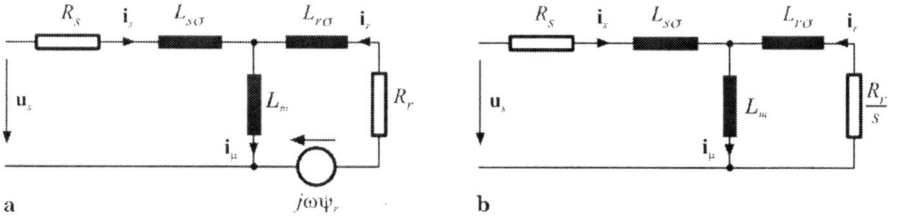

Fig. 6.2   T equivalent circuit of the induction machine: (**a**) non-stationary, (**b**) stationary

With R_s, L_m, $L_{s\sigma}$, $L_{r\sigma}$ and R_r the T equivalent circuit contains five parameters. The stator impedance, determinable by measuring stator quantities, contains on the other hand powers of the stator frequency from zero to three and is defined by four parameters (cf. the chapter 6.4.3). Therefore the T equivalent circuit is over-determined and not completely identifiable by measuring the stator quantities. For this reason $L_{r\sigma} = L_{s\sigma} = L_\sigma$ is often assumed. However, for many tasks it is advisable to change to an equivalent circuit with a reduced parameter number.

The two following representations achieve this by transformation of the leakage inductances into the stator or rotor mesh and by introduction of a total leakage inductance. At the same time this implies a redefinition of the cross or magnetizing current and of the main inductance with the consequence for these quantities losing their physical equivalent. As long as all parameters can be assumed constant and linear, this fact is of minor importance, though. Both new equivalent circuits are derived under the premise that in the case of the squirrel-cage IM no transformation of the stator quantities, measurable at the terminals, takes place.

6.1.1.2 Inverse Γ equivalent circuit

A modified equivalent circuit with the total leakage inductance in the stator mesh can be obtained by the introduction of a new cross current $\mathbf{i}_m$:

$$\mathbf{i}_m = \frac{\psi_r}{L_m} = \mathbf{i}_s + \frac{L_r}{L_m} \mathbf{i}_r \qquad (6.13)$$

After some transformations to eliminate the current $\mathbf{i}_\mu$ in equations (6.9), (6.10), and after the introduction of the leakage factor:

$$\sigma = 1 - \frac{L_m^2}{L_s L_r} \tag{6.14}$$

new voltage equations

$$\mathbf{u}_s = R_s \mathbf{i}_s + \sigma L_s \frac{d\mathbf{i}_s}{dt} + (1-\sigma)L_s \frac{d\mathbf{i}_m}{dt} \tag{6.15}$$

$$0 = \frac{L_m^2}{L_r^2} R_r \left(\frac{L_r}{L_m} \mathbf{i}_r \right) + (1-\sigma)L_s \frac{d\mathbf{i}_m}{dt} - j\omega(1-\sigma)L_s \mathbf{i}_m \tag{6.16}$$

which can be represented by the so called inverse-Γ-equivalent circuit (figure 6.3) are obtained. The newly introduced cross current $\mathbf{i}_m$ is according to (6.13) identical with the rotor ampere-turns. This explains why this equivalent circuit is particularly suitable for the treatment of rotor flux orientated control methods. For stationary operation a representation (figure 6.3b) which is equivalent to the figure 6.2b is here possible as well.

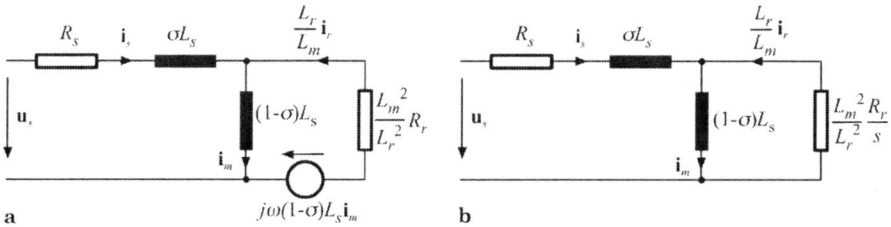

Fig. 6.3   Inverse-Γ-equivalent circuit for the induction machine: (**a**) non-stationary, (**b**) stationary

6.1.1.3 Γ equivalent circuit

To transform the leakage inductance *into the rotor side* a new cross current:

$$\mathbf{i}_{ms} = \frac{\psi_s}{L_s} = \mathbf{i}_s + \frac{L_m}{L_s} \mathbf{i}_r \tag{6.17}$$

is introduced analogously to the inverse Γ equivalent circuit. After substitution of $\mathbf{i}_\mu$ the equations of the Γ equivalent circuit represented in the non-stationary and stationary form in figure 6.4 will be obtained:

$$\mathbf{u}_s = R_s \mathbf{i}_s + L_s \frac{d\mathbf{i}_{ms}}{dt} \tag{6.18}$$

$$0 = \frac{L_s^2}{L_m^2} R_r \left(\frac{L_m}{L_s} \mathbf{i}_r \right) + \frac{\sigma L_s}{1-\sigma} \left(\frac{L_m}{L_s} \frac{d\mathbf{i}_r}{dt} \right) + L_s \frac{d\mathbf{i}_{ms}}{dt} - j\omega \frac{L_s}{L_m} \psi_r \tag{6.19}$$

Also at this place equation (6.14) is valid for the leakage factor σ. The rotor quantities appear, analog to the inverse Γ equivalent circuit, in transformed form.

As recognizable in the figure, the stator inductance now becomes the cross or magnetization inductance, and the stator flux linkage assumes the role of the main flux linkage. Therefore the Γ equivalent circuit is particularly suitable for the treatment of stator flux orientated control methods.

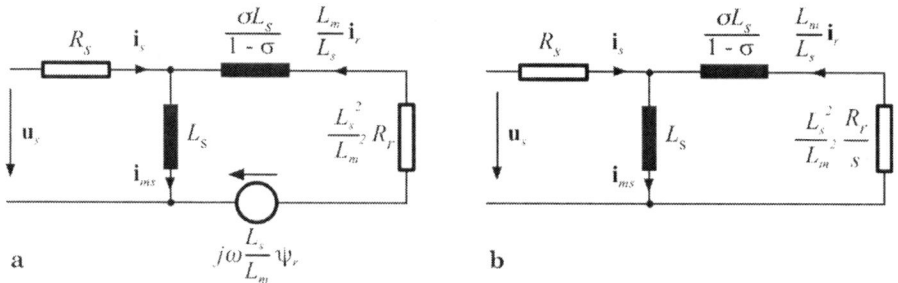

Fig. 6.4 Γ equivalent circuit of the induction machine: **(a)** non-stationary, **(b)** stationary

6.1.2 Equivalent circuits of the PMSM

Due to the permanent magnet excited pole flux the relations here are very simple. To derive a common equivalent circuit for both longitudinal and traverse axes it will usually be accepted that the same inductance is valid for both. The following equation holds for the stator voltage:

$$\mathbf{u}_s = R_s \mathbf{i}_s + L_s \frac{d\mathbf{i}_s}{dt} + j\omega \psi_p \tag{6.20}$$

With (6.20) the equivalent circuit represented in the figure 6.5 is obtained.

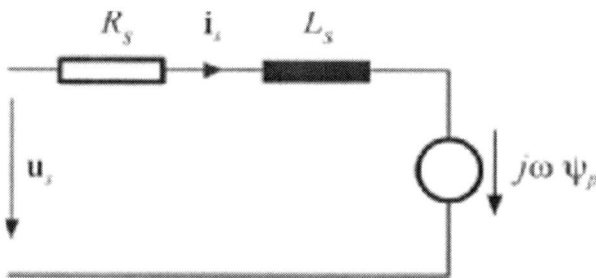

Fig. 6.5   Equivalent circuit of the PMSM

6.2 Modelling of the nonlinearities of the IM

For many control tasks the assumption of constant and state independent machine parameters represents a too rough approximation which leads to considerable deviations between model and reality at the examination of non-stationary operations. Therefore, the embedding of nonlinearities which are significant for different operating states into machine models and equivalent circuits shall be discussed in the following sections. Following the physical conditions, *magnetic saturation, current displacement and iron losses* are discussed in separate approaches and models. Symmetrical conditions and sinusoidal winding distribution are still presupposed.

In mathematical sense nonlinear relations are indicated by the fact that the superposition principle is not valid. Therefore an isolated treatment of the nonlinearities is, strictly speaking, not permitted. With respect to an engineer-like analysis however, it is fundamentally important to find easily comprehensible and utilizable approaches also for nonlinear relations. In the case of the 3-phase AC machines it is advantageous that the most important nonlinearities are describable as state dependent parameters. Since different parameters are affected, or the variable parameters depend on different state variables, a separate treatment is justified additionally.

6.2.1 Iron losses

Losses in the iron appear in the form of eddy-current losses and hysteresis losses. Because the rotor frequency remains small compared with the stator frequency unless at very small speeds, the rotor iron losses generally can be neglected compared to the stator side ones. The hysteresis losses are produced by the flux reversal energy consumed due to the sinusoidal with time varying iron magnetization. They are therefore proportional to the area of the hysteresis loop $(\sim |\psi_\mu|^2)$ and to the number of flux reversals per time unit $(\sim \omega_s)$ [Lunze 1978], [Philippow 1980]. The eddy-current losses are proportional to the square of the voltage $(\sim (\omega_s |\psi_\mu|)^2)$ induced in the iron and the effective electrical conductivity of the iron core lamellae. They significantly increase in converter fed motors because of the harmonic components in current and voltage.

Modeling is made difficult because the effects of eddy-currents and hysteresis and from sinusoidal magnetization are overlapping in a not exactly determinable way, and generally different magnetic conditions occur in yoke and teeth. The hysteresis losses depend on the effective permeability and therefore on the instantaneous flux amplitude. They disappear as soon as the area of the magnetic saturation is left (the upper field weakening area).

The following, strongly idealized model following [Murata 1990] takes into account hysteresis and eddy-current parts by respectively constant factors k_{hy} and k_w. Through consideration of the slip frequency ω_r even operating states in which the slip frequency will have a significant magnitude compared to the stator frequency ω_s are included:

$$p_{v,fe} = \frac{3}{2}\left[k_{hy}\left(\omega_s + \omega_r\right) + k_w\left(\omega_s^2 + \omega_r^2\right)\right]\left|\psi_\mu\right|^2 \tag{6.21}$$

with:

$$\psi_\mu = L_m i_\mu \tag{6.22}$$

After separating the stator frequency and the slip $s = \left(\omega_s - \omega\right)/\omega_s$, and with the general equation for the iron losses

$$p_{v,fe} = \frac{3}{2}\frac{\left(\omega_s\left|\psi_\mu\right|\right)^2}{R_{fe}} \tag{6.23}$$

an iron loss resistance R_{fe} as concentrated component describing the iron losses can be introduced:

$$R_{fe} = \frac{1}{k_w\left(1+s^2\right) + \dfrac{k_{hy}}{\omega_s}\left(1+s\right)} \tag{6.24}$$

Because better usability in some circumstances the iron loss conductance $G_{fe} = 1/R_{fe}$ is also used. A measured R_{fe} characteristics is represented exemplarily in the figure 6.6. The curves are the result of no-load measurements at an inverter-fed and external driven motor so that the influence of the friction losses is eliminated.

The iron loss power is dominated by the hysteresis losses rising nearly linearly in the basic speed range. With field weakening setting in, at first a strong drop can be observed because of the flux reduction. The eddy-current losses dominate in the upper field-weakening area. In addition, the inverter dependent eddy-current losses decrease strongly at the maximum voltage (= less high-frequency voltage harmonics) so that different factors k_w are used in constant flux area and constant voltage range. The corresponding diagrams calculated by least-square approximation and the model approach (6.24) are drawn in the figure 6.6 (dotted lines). It turns out that this simple approach with the above-mentioned modification describes the actual behavior quite well.

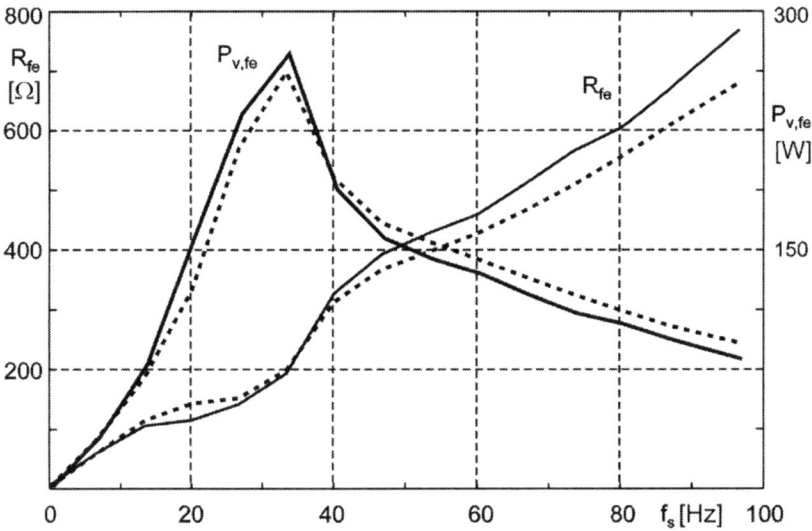

Fig. 6.6   Iron losses and iron loss resistance

Further analysis of the R_{fe}-diagram in the figure 6.6 suggests, however, the possibility of using a yet more simplified model which only contains a linear relation between loss resistance and stator frequency:

$$R_{fe} = R_{feN} \frac{\omega_s}{\omega_{sN}} \tag{6.25}$$

For this model only one parameter, the loss resistance at nominal frequency, must be determined by measurement.

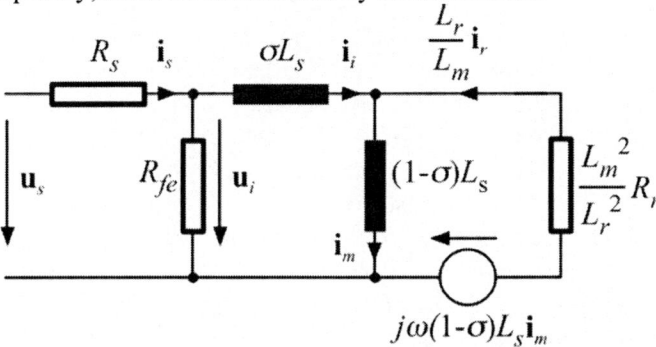

Fig. 6.7   Extended inverse Γ equivalent circuit with iron losses

A comfortable inclusion in the equation system of the IM will be obtained, if (as shown in the figure 6.7) the iron loss resistance in the inverse Γ equivalent circuit can be represented by a parallel resistance in the stator circuit [Schäfer 1989]. The necessary supplementation of the equation system is immediately recognisable from the figure 6.7. The actual input voltage of the machine will now be formed by the inner voltage $\mathbf{u}_i$ which drives the inner current $\mathbf{i}_i$. Following equations (6.15) and (6.16) the modified system is obtained to:

$$\mathbf{u}_i = \sigma L_s \frac{d\,\mathbf{i}_i}{dt} + (1-\sigma)L_s \frac{d\,\mathbf{i}_m}{dt} \tag{6.26}$$

$$0 = (1 - j\omega T_r)\mathbf{i}_m + T_r \frac{d\,\mathbf{i}_m}{dt} - \mathbf{i}_i \tag{6.27}$$

$$\mathbf{u}_i = \mathbf{u}_s - R_s \mathbf{i}_s \tag{6.28}$$

$$\mathbf{i}_i = \mathbf{i}_s - \frac{\mathbf{u}_i}{R_{fe}} \tag{6.29}$$

6.2.2 Current and field displacement

With regard to current and field displacement effects it must be distinguished between effects caused by the fundamental of the current on one hand and by inverter dependent current harmonics on the other hand. The principle physical mechanism is the same in both cases. The current displacement leads to a frequency dependent increase of the resistance values, and the field displacement to a reduction of the leakage inductances. As a consequence the current harmonics produce higher losses. Because the harmonic spectrum of fast switching inverters with sine modulation is orders of magnitude above the fundamental wave, its significance for control related parameters remains small. The consequences of the fundamental dependent current displacement, however, must be investigated for the modeling of the machine.

In stator windings of induction machines, fundamental wave dependent current displacement effects can usually be neglected since they are intentionally suppressed by a number of constructive measures. An exception would merely be the big machines with accordingly large winding diameters at high frequencies. For the bars of the rotor squirrel-cage such a neglection is not possible from the outset because of the large bar heights and diameters, except for rotors with intentionally current displacement free construction. In the normal (stationary) operation current displacement effects do not play a considerable role, however, because of the low rotor frequency (slip). This turns different in special non-stationary operation modes with high slip frequencies, where the current displacement is used with purpose to increase the resistance, or if the input quantities are controlled differently to the normal operation. In a field-orientated control system however, also at start-up extreme slip values will not appear due to the current being controlled with defined amplitude and slip.

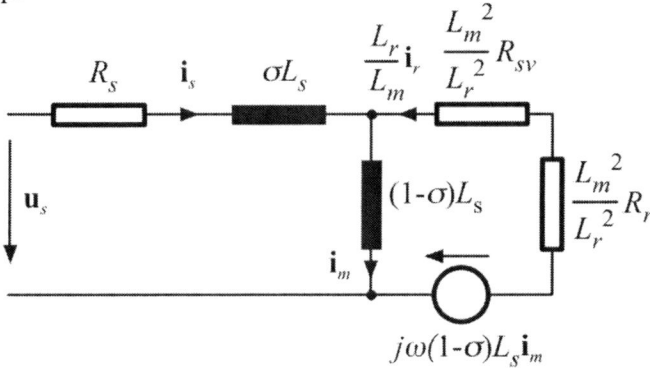

Fig. 6.8   Extended equivalent circuit with current displacement in rotor

A consideration of the resistance changes in the machine model is possible by an additional resistance R_{sv} being inserted in series to the rotor resistance in the rotor circuit (figure 6.8). The size of this resistance is a generally very unhandy function of material constants, construction data and the rotor frequency. The following equation can be learned from [Vogt 1986] for a rectangular deep-bar rotor with the height h_l:

$$k_r = \frac{R_r + R_{sv}}{R_r} = \beta \frac{\sinh 2\beta + \sin 2\beta}{\cosh 2\beta - \cos 2\beta} \tag{6.30}$$

There β is a normalized height of the conductor and is calculated by:

$$\beta = h_l \sqrt{\frac{1}{2}\omega_r \mu_0 \kappa} \tag{6.31}$$

μ_0 Absolute permeability

κ Conductivity of rotor bars

Because the height of rotor bars is often designed over-critically (rotor with current displacement), h_l assumes values of up to about 70 mm. Following [Vogt 1986] this corresponds to a machine with a rated power of about 2 MW. Therefore it is not possible to come up with a uniform approximation for k_r for the complete interesting parameter range of β. The following variants for approximation approaches can be derived:

For $\beta > 2$ there will be $\sinh\beta \gg \sin\beta$, $\cosh\beta \gg \cos\beta$ and $\sinh\beta \approx \cosh\beta$, and therefore holds:

$$k_r \approx \beta \tag{6.32}$$

For $\beta \leq 2$, k_r can be obtained by series expansion of the transcendental function with a maximum relative fault of 0.036 to:

$$k_r \approx \frac{1 + \dfrac{2}{15}\beta^4}{1 + \dfrac{2}{45}\beta^4} \qquad (6.33)$$

A next approximation is possible for $\beta \leq 1$ by partial division of (6.33):

$$k_r \approx 1 + \frac{4}{45}\beta^4 \qquad (6.34)$$

Besides the increase of the electrical resistance by current displacement, fast changes of the flux will cause a field displacement recognized by the reduction of the leakage inductance. Also here, noticeable effects appear only in rotors with current displacement because of the larger conductor height. A reduction factor k can be given in analogy to (6.30). Following [Vogt 1986] this factor can be written for square deep-bar rotors with the fictitious series inductance $L_{\sigma v}$ to:

$$k_x = \frac{L_\sigma - L_{\sigma v}}{L_\sigma} = \frac{3(\sinh 2\beta - \sin 2\beta)}{2\beta(\cosh 2\beta - \cos 2\beta)} \qquad (6.35)$$

As above the following approximations can be derived.
- For $\beta > 2$:

$$k_x \approx \frac{3}{2\beta} \qquad (6.36)$$

- For $1 < \beta \leq 2$:

$$k_x \approx \frac{1 + \dfrac{2}{105}\beta^4}{1 + \dfrac{2}{45}\beta^4} \qquad (6.37)$$

- For $\beta \leq 1$:

$$k_x \approx 1 - \frac{8}{315}\beta^4 \qquad (6.38)$$

The equations (6.30) and (6.35) are represented in figure 6.9. For the better classification the normalized bar height β is additionally referred to the rotor frequency at different absolute bar heights. The computation was made for copper bars because these also have greater values β due to their greater conductivity compared with aluminum at the same frequency.

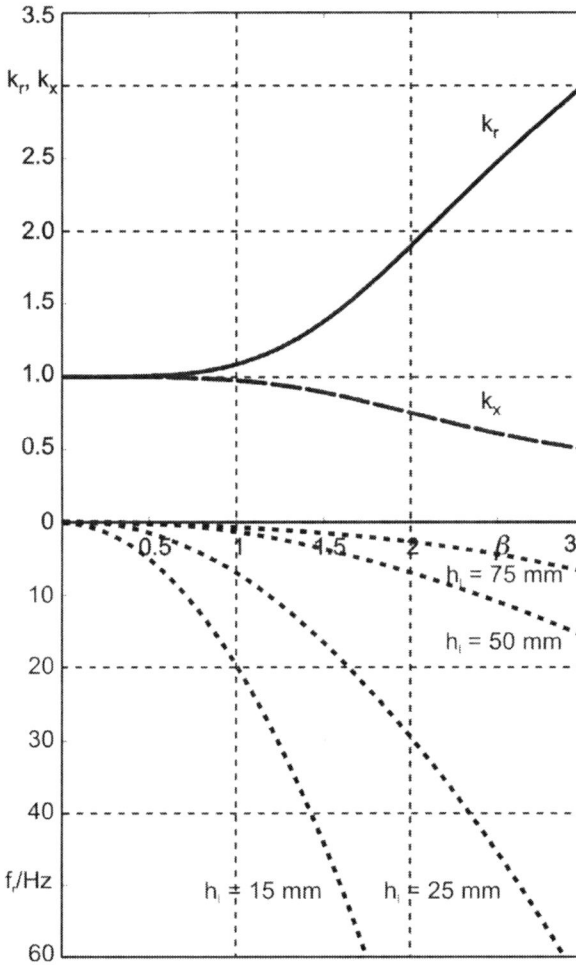

Fig. 6.9   Electrical resistance increase (___) and leakage inductance reduction (.....) due to current displacement in the rotor, parameter: bar height and rotor frequency

For very big machines in the megawatt range β already reaches great values at rotor frequencies below 10 Hz, and k_r and k_x also become significant. Such machines have a small nominal slip of typical below one hertz, and will be less overloaded in dynamic operation, though. In low power drive systems, only three to four times the nominal slip will be applied in dynamic operation using *field-orientated control*. Thus β will not exceed values of 1.5 through the whole power range, and for motors in the medium and low power range there are values of $\beta = 1$ to be expected at maximum. Because the rotor leakage inductance only shares about one half of the total leakage inductance, a special consideration of the inductance reduction can be abandoned in the model for field-orientated control. A consideration of the electrical resistance increase in the model for the field-orientated control is required only for machines above some hundred kilowatts rating.

The existence of the flux weakening and resistance increase must be taken into account though at the estimation of parameter variations or for example to define suitable excitation signals for the parameter identification especially at higher frequencies (cf. section 6.4).

The structure of the equations (6.30) and (6.35) is essentially correct also for usual bar cross-sections which differ from the rectangle form, though with other coefficients. Approximately the same relations hold for square bars with $d = h_l$ and for rods (diameter d). For wedge bars the value k_r increases in the extreme case (ratio of the trapezium front sides of 1:10) at $\beta = 2$ by 50%. k_x assumes more favourable values [Vogt 1986]. Thus the above statements remain also valid for these bar forms.

6.2.3 Magnetic saturation

At first the magnetic saturation has the consequence that the value of the inductances is a nonlinear function of the amplitude of the actual flux linkage. In addition, a general analysis of the saturated induction machine must take into account that the spatial distribution of the saturation depends on the current direction of the accompanying flux vector. This has the consequence that in the right-angled coordinate system the inductances assigned to the coordinate axes assume different values in the dynamic case, and mutual couplings appear [Vas 1990]. These depend on the sine of the angle between the main flux vector and the reference axis (real axis) of the used coordinate system.

The main field saturation has essential significance for the dynamic behavior of the machine, primarily in the field weakening and at great torques. Its correct or reasonable approximated consideration shall be examined in the following. At first the leakage inductances are considered as constant.

For a representation as generally as possible the machine equations are represented in the following in a right-angled coordinate system circulating with the angular velocity ω_k. The main flux linkage

$$\psi_\mu = L_m i_\mu \qquad (6.39)$$

is introduced into the general voltage equations of the induction machine (cf. section 3.2). With (6.39) the following voltage equations will be obtained:

$$\mathbf{u}_s = R_s \mathbf{i}_s + L_{s\sigma} \frac{d\,\mathbf{i}_s}{dt} + \frac{d\,\mathbf{\psi}_\mu}{dt} + j\omega_k (L_{s\sigma} \mathbf{i}_s + L_m \mathbf{i}_\mu) \tag{6.40}$$

$$\mathbf{u}_r = R_r (\mathbf{i}_\mu - \mathbf{i}_s) + L_{r\sigma} \frac{d(\mathbf{i}_\mu - \mathbf{i}_s)}{dt} + \frac{d\,\mathbf{\psi}_\mu}{dt}$$
$$+ j(\omega_k - \omega)(L_r \mathbf{i}_\mu - L_{r\sigma} \mathbf{i}_s) \tag{6.41}$$

The equations (6.40) and (6.41) are here still represented in an arbitrary orientated coordinate system and contain no restrictions regarding their validity at main field saturation. Following [Vas 1990] the next equation holds for the derivative of the main flux:

$$\frac{d\,\mathbf{\psi}_\mu}{dt} = \begin{pmatrix} M_x & M_{xy} \\ M_{xy} & M_y \end{pmatrix} \frac{d\,\mathbf{i}_\mu}{dt} \tag{6.42}$$

with: $M_x = L_m' \cos^2 \mu + L_m \sin^2 \mu$

$M_y = L_m' \sin^2 \mu + L_m \cos^2 \mu$

$M_{xy} = (L_m' - L_m) \sin \mu \cos \mu$

Thereby $L_m(|\mathbf{i}_\mu|)$ is the static and $L_m'(|\mathbf{i}_\mu|) = \dfrac{d|\mathbf{\psi}_\mu|}{d|\mathbf{i}_\mu|} = L_m + \dfrac{dL_m}{d|\mathbf{i}_\mu|}|\mathbf{i}_\mu|$ the

differential main inductance, μ is the angle between the magnetizing current vector $\mathbf{i}_\mu$ and the real axis of the coordinate system. For the non-saturated machine $L_m = L_m'$ and $\dfrac{d\,\mathbf{\psi}_\mu}{dt} = L_m \dfrac{d\,\mathbf{i}_\mu}{dt}$ hold.

For the controller realization and also for a simulation of the saturated machine the correct representation of the flux derivation following (6.42) is quite unhandy. For the controller design primarily the rotor equation is important because the estimation equation of the rotor flux, required for field-orientated control, is derived from it. Therefore it would be desirable to maintain the rotor flux oriented description.

A first simplification (without validity restriction or loss of precision) arises with representation of (6.40) and (6.41) in a coordinate system related to the main flux. If the axis of the main flux vector and the real axis (x-axis) are identical, it will be $\mu = 0$, and the next equation can be obtained for the flux derivative:

$$\frac{d\,\mathbf{\psi}_\mu}{dt} = \begin{pmatrix} L_m'(|\mathbf{i}_\mu|) & 0 \\ 0 & L_m(|\mathbf{i}_\mu|) \end{pmatrix} \frac{d\,\mathbf{i}_\mu}{dt} \tag{6.43}$$

In addition, there are $\psi_{\mu x} = |\mathbf{\psi}_\mu|$ and $\psi_{\mu y} = 0$.

Since obviously the flux derivative is the most problematic part of the mathematical model, it seems reasonable to look for a form of presentation which gets along without its explicit calculation. In addition, no derivative of a state variable must be connected with the main inductance. Such a model was developed in [Levi 1994]. The rotor current is replaced by the flux linkage in the rotor voltage equation:

$$\mathbf{u}_r = \dot{\boldsymbol{\psi}}_r + j\left(\omega_k - \omega\right)\boldsymbol{\psi}_r + \frac{R_r}{L_{r\sigma}}\left(\boldsymbol{\psi}_r - \boldsymbol{\psi}_\mu\right) \tag{6.44}$$

The mutual flux can be calculated from:

$$\boldsymbol{\psi}_\mu = \boldsymbol{\psi}_r - L_{r\sigma}\left(\mathbf{i}_\mu - \mathbf{i}_s\right) \tag{6.45}$$

with:
$$\mathbf{i}_\mu = \frac{\boldsymbol{\psi}_\mu}{L_m\left(|\boldsymbol{\psi}_\mu|\right)}$$

It shall be remarked that this model also does not contain any restrictions regarding the saturation and is neutrally formulated with respect to coordinates. The calculation of a differential inductance is not required. As opposed to (6.41), the equation (6.45) contains with $\boldsymbol{\psi}_\mu = f(L_m(|\boldsymbol{\psi}_\mu|))$ an algebraic loop which can cause oscillations and limit cycles depending on the sampling period or on the integration steps in the realization.

A third model can be obtained after trying to introduce the saturation into the rotor equation immediately without further substitutions. This means, though, that the saturation is coupled to the rotor flux instead to the main flux, what does not correspond to the physical conditions correctly. The error can, however, be acceptable for many applications because the leakage flux on the rotor side will always remain small compared to the main flux. In any case this variant has the advantage to provide the simplest and clearest model. A possibility for its derivation immediately arises from the equivalent circuit in figure 6.3a. After substitution of the rotor current and division by the leakage factor $(1-\sigma)$, which shall be assumed as saturation invariant, the following equation for the rotor mesh can be obtained after transition into the coordinate system circulating with ω_k:

$$\mathbf{u}_r = R_r\left(\mathbf{i}_m - \mathbf{i}_s\right) + \frac{d(L_s\mathbf{i}_m)}{dt} + j\left(\omega_k - \omega\right)L_s\mathbf{i}_m \tag{6.46}$$

After dissolving the derivative, the rotor equation can finally be written in a detailed notation:

$$\mathbf{u}_r = \mathbf{i}_m - \mathbf{i}_s + \frac{L'_m(|\mathbf{i}_m|) + L_\sigma}{R_r} \frac{d\mathbf{i}_m}{dt}$$

$$+ j(\omega_k - \omega)\frac{L_m(|\mathbf{i}_m|) + L_\sigma}{R_r}\mathbf{i}_m \qquad (6.47)$$

In the same way the stator voltage equation can be obtained for an assumed constant leakage inductance:

$$\mathbf{u}_s = R_s\mathbf{i}_s + \sigma L_s \frac{d\mathbf{i}_s}{dt} + j\omega_k\sigma L_s\mathbf{i}_s$$

$$+ (1-\sigma)\left(L'_m(|\mathbf{i}_m|) + L_\sigma\right)\frac{d\mathbf{i}_m}{dt} \qquad (6.48)$$

$$+ j\omega_k(1-\sigma)\left(L_m(|\mathbf{i}_m|) + L_\sigma\right)\mathbf{i}_m$$

The modified system matrix of the continuous state-space representation finally can be derived from (6.47) and (6.48). It reads in complex notation with the abbreviations $L_{s\mu} = L_m\left(|\mathbf{i}_m|\right) + L_\sigma$ and $L'_{s\mu} = L'_m\left(|\mathbf{i}_m|\right) + L_\sigma$:

$$\mathbf{A} = \begin{pmatrix} -\dfrac{1}{T_\sigma} - j\omega_k & \dfrac{1-\sigma}{\sigma L_s}\left(R_r - j\omega L_{s\mu}\right) \\[2ex] \dfrac{R_r}{L'_{s\mu}} & -\dfrac{R_r}{L'_{s\mu}} - j(\omega_k - \omega)\dfrac{L_{s\mu}}{L'_{s\mu}} \end{pmatrix} \qquad (6.49)$$

with: $\qquad T_\sigma = \dfrac{\sigma L_s}{R_s + (1-\sigma)R_r}$

At first the model (6.47) yields the rotor-side magnetizing current $\mathbf{i}_m$. The rotor flux linkage can be calculated from $\psi_r = L_m\mathbf{i}_m$. For this calculation the main inductance has to be used depending on the magnetizing current $\mathbf{i}_\mu$. With (6.8) and (6.4) its amplitude can be derived in the rotor flux oriented coordinate system ($i_m = i_{md}$, $i_{mq} = 0$):

$$|\mathbf{i}_\mu| = \sqrt{\left(\frac{L_{r\sigma}}{L_r}i_{sd} + \frac{L_m}{L_r}i_{md}\right)^2 + \left(\frac{L_{r\sigma}}{L_r}i_{sq}\right)^2} \qquad (6.50)$$

For the stationary case the following result arises:

$$|\mathbf{i}_\mu| = \sqrt{i_{md}^2 + \left(\frac{L_{r\sigma}}{L_r}i_{sq}\right)^2} \qquad (6.51)$$

The knowledge of the magnetization characteristic, either in the form $\psi_\mu = f(\mathbf{i}_\mu)$ or its inverse form $\mathbf{i}_\mu = g(\psi_\mu)$, is required for all saturation dependent models. Thereby the use of a closed representation is advisable, because this can more simply be implemented in the model and simultaneously allows the calculation of the differential inductance without difficulties from the measurement of the stationary characteristic. In the literature different approaches can be found, which are based on exponential or power functions and differ from each other in the number of contained parameters. Polynomial approaches are also used. Exponential and power functions have the advantage to provide good models even at strong saturation. Two power functions, which are built on each other, shall be examined and compared in the following. They are based on a approach introduced by [de Jong 1980] and further developed by [Klaes 1992].

The main inductance is understood as a parallel circuit of a constant air-gap inductance L_0 (it corresponds to the main inductance in the linear range) and a saturation dependent part which is a power function of the obtained main flux or magnetizing current:

$$L_m(\gamma) = \frac{1}{\dfrac{1}{L_0} + \dfrac{\gamma^s}{L_{sat}}}; \gamma = \frac{i_\mu}{i_1} \qquad (6.52)$$

With measured values $L_0 = L_m(0), L_1 = L_m(1), L_2 = L_m(\gamma_2)$ the remaining parameters are obtained as follows:

$$L_{sat} = \frac{1}{\dfrac{1}{L_1} - \dfrac{1}{L_0}}; s = \frac{\ln\left(\dfrac{1/L_2 - 1/L_0}{1/L_1 - 1/L_0}\right)}{\ln \gamma_2} \qquad (6.53)$$

An extended approach takes into account, that the main inductance converges towards a fixed final value and not towards zero for large flux amplitudes, as it would result from the estimation function (6.52). This is considered by the addition of a limit inductance L_∞ into (6.52):

$$L_m(\gamma) = \frac{1}{\dfrac{1}{L_0 - L_\infty} + \dfrac{\gamma^s}{L_{sat}}} + L_\infty \qquad (6.54)$$

For the calculation of the coefficients however, no explicit solution can be derived. With the additional measurement $L_3 = L_m(\gamma_3)$ the saturation parameter s arises from the iterative solution of the following equation:

$$\left(\frac{\gamma_1}{\gamma_2}\right)^s - \frac{(L_2 - L_3)(L_1 - L_0) + (L_2 - L_1)(L_0 - L_3)\left(\dfrac{\gamma_1}{\gamma_3}\right)^s}{(L_3 - L_1)(L_0 - L_2)} = 0 \qquad (6.55)$$

For the remaining parameters the next equations hold:

$$L_\infty = \frac{L_3(L_1 - L_0) + L_1(L_0 - L_3)\left(\dfrac{\gamma_1}{\gamma_3}\right)^s}{L_1 - L_0 + (L_0 - L_3)\left(\dfrac{\gamma_1}{\gamma_3}\right)^s}$$ (6.56)

$$\frac{1}{L_{sat}} = \frac{1}{\gamma_2^s - \gamma_3^s}\left(\frac{1}{L_2 - L_\infty} - \frac{1}{L_3 - L_\infty}\right)$$ (6.57)

The estimates are represented together with the characteristic for the differential main inductance for an 11kW motor in figure 6.10. The differential main inductance can be calculated for the three-parameter approach from:

$$L_m'(\gamma) = \frac{d\psi_\mu}{d i_\mu} = L_m(\gamma)\left(1 - s\gamma^s \frac{L_m(\gamma)}{L_{sat}}\right)$$ (6.58)

and for the four-parameter approach from:

$$L_m'(\gamma) = L_m(\gamma) - s\gamma^s \frac{(L_m(\gamma) - L_\infty)^2}{L_{sat}}$$ (6.59)

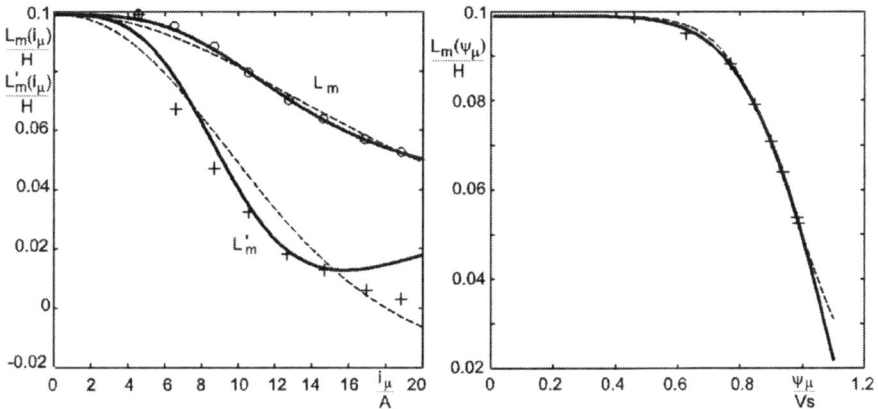

Fig. 6.10 Saturation calculation functions $L_m\left(i_\mu\right)$, $L_m'\left(i_\mu\right)$, $L_m\left(\psi_\mu\right)$ for 11kW motor. (----): three-parameter model; (——): four-parameter model

The measurements of the static inductance are visibly better approximated by the four-parameter approach. Apart from the area of extreme saturation, which practically plays not a role, this also applies to the differential inductance. It must be taken into account that the "measured values" of the differential inductance only were won by linear approximation from the static measurements, though.

The saturation of the leakage paths is the reason that the leakage inductances generally are functions of the current flowing through them. A general relation like for the main inductance however, can not be derived because it depends strongly on constructive influences according to the composition of the leakage field (slot leakage, tooth head leakage, winding head leakage, helical leakage). At the same time it is possible that almost no current dependency of the leakage inductance can be found for many motors in the complete current range. For this reason, a separate consideration in the model is abandoned. If required, a feed-forward adaptation $L_\sigma = f(|\mathbf{i}_s|)$ must be implemented.

6.2.4 Transient parameters

When discussing the current displacement effects it already became obvious that inductances and resistances of the induction machine generally have to be considered frequency dependent. Equivalent circuits can be developed with concentrated parameters which, however, have to be specified according to the operating point of interest and to the operation frequency. In an inverter-fed drive the highest frequencies practically appearing are determined by the switching slopes of the inverter. These have an effect on the effective leakage inductance of the motor which determines together with the voltage amplitude the current rise time. The effective leakage inductance is to be expected considerably smaller than the stationary leakage inductance σL_s and will be called in the following as the transient leakage inductance $\sigma L_s'$. At the same time it is the only parameter which must especially be defined in the practical controller design for transient operations.

6.3 Parameter estimation from name plate data

Lacking detailed and often not obtainable motor data sheets, the name plate of the motor represents the first and only information source for conclusions on the electrical parameters. For standard drives without high dynamic demands on the motor usage, usually it fully suffices to calculate the motor parameters from the name plate data. However, deviations from 50% to 100% depending on the parameter in question have to be taken into account, because:

- the manufacturer's information may be partly unreliable, and the actual motor parameters are subject to spreads,
- the name plate data refer to a certain working point (the nominal working point),
- not all parameters of the equivalent circuit can be directly set into a physical relation to the usual name plate data.

The procedure becomes impossible at the use of special machines with values differing from standard machines considerably. Understandably, the calculation of the inductance saturation characteristic has to be excluded.

The usual name plate data are:
- Nominal power P_N [kW]
- (Line-to-line) nominal voltage U_N [V]
- Nominal current I_N [A]
- Nominal frequency f_N [Hz]
- Nominal speed n_N [rpm]
- (Nominal) power factor $\cos\varphi$

Because the last information is not available in many cases, calculation equations are derived in the following without and with $\cos\varphi$. In the case PMSM, the following data are usually given by the name plate:
- (Line-to-line) nominal voltage U_N [V / 1000 rpm]
- Nominal current I_N [A]
- Nominal frequency f_N [Hz]
- Maximum speed n_{max} [rpm]
- Nominal torque m_N [Nm]

6.3.1 Calculation for IM with power factor cosφ

The method starts out from the equation of the IM in the stationary operation (cf. [Quang 1996]).

$$\mathbf{u}_s = R_s\mathbf{i}_s + jw_s\sigma L_s\mathbf{i}_s + jw_s(1-\sigma)L_s\frac{\psi_r}{L_m}$$
$$= R_s\mathbf{i}_s + jw_s\sigma L_s\mathbf{i}_s + \mathbf{e}_g \tag{6.60}$$

The parameters are approximately calculated for the nominal working point in the following steps:

1. *Calculation of the field-forming current component i_{sd}:*

   (1) Nominal power of the motor: $\quad P_N = 3U_{Phase}I_{Phase}\cos\varphi$

   (2) Amplitude of the nominal current: $\quad \hat{\imath}_N = \sqrt{2}\,I_N = \sqrt{\hat{\imath}_{sdN}^2 + \hat{\imath}_{sqN}^2}$

   (3) Impedance of one phase: $\quad Z_N = U_{Phase}/I_{Phase}$

(4) Approximate rotor resistance: $R_r \approx s Z_N$

(5) Nominal power of the motor: $P_N \approx 3 \left(\dfrac{R_r}{s} \right) \hat{i}_{sqN}^2$

(6) From (4), (5) is obtained: $\hat{i}_{sqN}^2 \approx \dfrac{P_N}{3 Z_N}$

(7) Inserting (1) into (6): $\hat{i}_{sqN}^2 \approx \dfrac{U_{Phase} I_{Phase} \cos \varphi}{Z_N}$

(8) Inserting (7) into (2): $\hat{i}_{sdN} \approx \sqrt{\hat{i}_N^2 - \dfrac{U_{Phase} I_{Phase} \cos \varphi}{Z_N}}$

The following approximate formula can be derived from (8):

$$\hat{i}_{sdN} \approx \sqrt{2} I_N \sqrt{1 - \cos \varphi} \qquad (6.61)$$

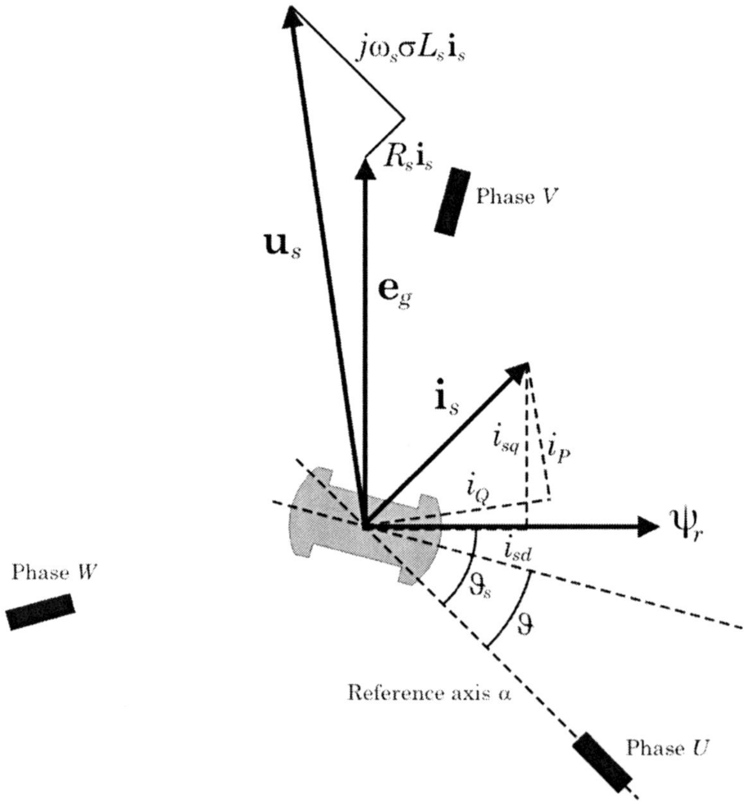

Fig. 6.11 Vector diagram of the IM in the stationary operation

In the step (5) the losses in the stator resistance were neglected without great loss of precision for the calculation of the power P_N.

2. *Calculation of the torque-forming current component i_{sq}*

$$\hat{I}_{sqN} \approx \sqrt{2I_N^2 - \hat{I}_{sdN}^2} \tag{6.62}$$

3. *Calculation of ω_r*

$$\omega_{rN} = 2\pi\left(f_N - \frac{z_p\, n_N}{60}\right) \tag{6.63}$$

4. *Calculation of the rotor time constant T_r*

$$T_r = \frac{\hat{I}_{sqN}}{\omega_{rN}\,\hat{I}_{sdN}} \tag{6.64}$$

5. *Calculation of the leakage reactance $X_\sigma = \omega_s \sigma L_s$*

The voltage drop over the stator resistance is neglected, which is justified for the nominal working point, compared to the voltage drop over the leakage inductance in the vector diagram in the figure 6.11. The simplified vector diagram of the figure 6.12 can be obtained then.

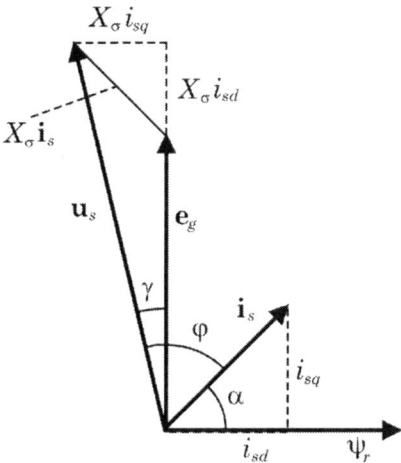

Fig. 6.12 Simplified vector diagram for calculation of X_σ

Using the figure 6.12 the following calculation steps can be used:

(1) Amplitude of the nominal phase voltage: $\hat{U}_N = \dfrac{\sqrt{2}\,U_N}{\sqrt{3}}$

(2) Relations between α, $\hat{I}_{sdN}$ and $\hat{I}_{sqN}$: $\sin\alpha = \dfrac{\hat{I}_{sqN}}{\hat{I}_N}; \cos\alpha = \dfrac{\hat{I}_{sdN}}{\hat{I}_N}$

(3) Relation between α, γ and φ: $\gamma = \varphi - \left(90^0 - \alpha\right)$

(4) Calculation of $\sin\gamma$:
$$\sin\gamma = \sin\left[\varphi - \left(90^0 - \alpha\right)\right]$$
$$= \sin\varphi\,\sin\alpha - \cos\varphi\,\cos\alpha$$

(5) Inserting (2) into (4):

$$\sin \gamma = \sin \varphi \frac{\hat{I}_{sqN}}{\hat{I}_N} - \cos \varphi \frac{\hat{I}_{sdN}}{\hat{I}_N}$$

(6) From fig. 6.12 it will be obtained:

$$X_\sigma \approx \sin \gamma \frac{\hat{U}_N}{\hat{I}_{sqN}}$$

(7) Inserting (5) into (6):

$$X_\sigma \approx \left(\sin \varphi - \cos \varphi \frac{\hat{I}_{sdN}}{\hat{I}_{sqN}} \right) \frac{\hat{U}_N}{\hat{I}_N}$$

With these results the formula for the calculation of X_σ is obtained:

$$X_\sigma \approx \left(\sin \varphi - \cos \varphi \frac{\hat{I}_{sdN}}{\hat{I}_{sqN}} \right) \frac{U_N}{\sqrt{3} I_N} \qquad (6.65)$$

6. *Calculation of the main reactance X_h:*

The main reactance $X_h = \omega_s (1 - \sigma) L_s \approx \omega_s L_s$ is the reactance of the EMF e_g. In the case $i_{sq} = 0$, i.e. no-load, the calculation equation is approximately obtained from the figure 6.12:

$$X_h \approx \frac{\hat{U}_N}{\hat{I}_{sdN}} - X_\sigma = \frac{\sqrt{2} U_N}{\sqrt{3} \hat{I}_{sdN}} - X_\sigma \qquad (6.66)$$

7. *Calculation of the stator resistance R_s:*

(1) It is assumed approximately:

$$R_s \approx R_r$$

(2) Calculation of the EMF amplitude:

$$\hat{e}_g = X_h \hat{I}_{sdN} \approx \frac{R_r}{\omega_{rN} / (2\pi f_N)} \hat{I}_{sqN}$$

The definite formula then looks as follows:

$$R_s \approx R_r \approx \frac{\omega_{rN}}{2\pi f_N} \frac{\hat{I}_{sdN}}{\hat{I}_{sqN}} X_h \qquad (6.67)$$

8. *Calculation of the total leakage factor σ:*

$$\sigma \approx \frac{X_\sigma}{X_h} \qquad (6.68)$$

9. *Calculation of the stator-side time constant T_s:*

$$T_s = \frac{L_s}{R_s} \approx \frac{X_h}{2\pi f_N R_s} \qquad (6.69)$$

The given calculation with using $\cos\varphi$ was tested successfully in the practice and is not subject to any restriction regarding motor power.

6.3.2 Calculation for IM without power factor $\cos\varphi$

Reference model is the inverse Γ equivalent circuit of the IM. All formulae are valid for motors with a nominal power of greater than 0.7kW.

The total leakage reactance determines fundamentally the short circuit behaviour of the motor or the current amplitude at nominal frequency at turn-on to the stiff grid. For standard motors the turn-on current maximum is 4 to 7 times the nominal current. Empirical values show that the most correct values for the leakage inductance σL_s will be obtained, if the 5 to 6-fold nominal current is used:

$$\sigma L_s \approx \frac{U_N}{5.5 I_N \omega_N \sqrt{3}} \tag{6.70}$$

For the transient leakage inductance a value of about $0.8 \sigma L_s$ can be started with.

The stator reactance is responsible for the current consumption of the no-load machine. This depends for comparable power ratings strongly on the magnetic utilization of the machine, thus on the nominal working point regarding the magnetic saturation, and therefore it can be subject to considerable variations for different manufacturers. Without using the power factor we can start out from the approximate rule that the nominal no-load current I_0 is about half of the nominal current at small power (until 7.5kW) and tends above this power towards to a good third of the nominal current. The following formula represents this empirical value:

$$I_0 \approx \frac{I_N + 1.9A}{2.6} \tag{6.71}$$

$$L_s = \frac{U_N}{I_0 \omega_N \sqrt{3}} \tag{6.72}$$

No physical relations can be given for the calculation of the stator resistance from the nominal data. We are here completely dependent on empirical values with the unavoidable uncertainties. The following formula provides usable results:

$$R_s \approx \frac{0.02 \, U_N}{I_N - 2.0 \, A} \tag{6.73}$$

The rotor resistance provides the physically best access. The stationary slip equation in field-orientated coordinates

$$\omega_r = \frac{I_{sq}}{I_{sd} T_r}$$

can be re-written for the nominal working point and $I_{sd} \approx I_0$ and solved to R_r:

$$R_r \approx \frac{2\pi\left(f_N - \frac{z_p n_N}{60}\right) L_s I_0}{\sqrt{I_N^2 - I_0^2}} \qquad (6.74)$$

The stator inductance and the no-load current can be taken from equations (6.71) and (6.72).

6.3.3 Parameter estimation from name plate of PMSM

Starting point for this is the stator voltage equation (cf. [Quang 1996]) in the stationary operation.

$$\begin{aligned}
\mathbf{u}_s &= R_s\,\mathbf{i}_s + j\omega_s L_s\,\mathbf{i}_s + j\omega_s\,\psi_p \\
&= R_s\,\mathbf{i}_s + j\omega_s L_s\,\mathbf{i}_s + \mathbf{e}_g
\end{aligned} \qquad (6.75)$$

At the nominal working point and in stationary operation the stator current $\mathbf{i}_s$ only contains the torque-forming component. This fact is represented by the stationary vector diagram in the figure 6.13.

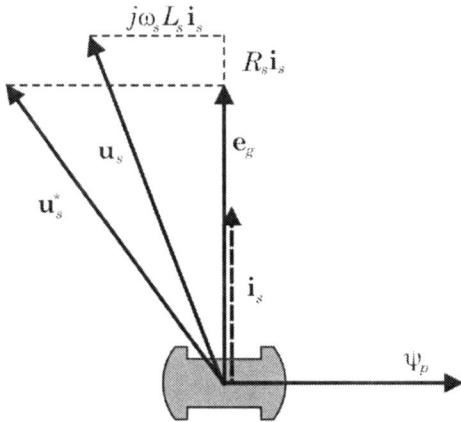

Fig. 6.13 Simplified vector diagram of the PMSM in stationary operation

1. *Calculation of pole flux Ψ_p:*

   (1) Calculation of torque-forming current: $\quad \hat{i}_{sqN} = \hat{i}_N = \sqrt{2}\,I_N$

   (2) Nominal torque using equation (3.63): $\quad m_N = \frac{3}{2} z_p\,\psi_p\,\hat{i}_{sqN}$

   (3) Inserting (1) into (2) it will be obtained: $\quad \psi_p = \frac{2}{3}\frac{m_N}{\sqrt{2}\,z_p\,I_N}$

2. *Calculation of stator inductance L_s:* The voltage drop over R_s is neglected. The next steps follow from figure 6.13:

(1) Amplitude of nominal voltage:
$$\hat{U}_N = n_N \sqrt{2}\left(\frac{U_N}{\sqrt{3}}\right)$$

(2) Amplitude of EMF:
$$\hat{\mathbf{e}}_g = 2\pi f_N \psi_p$$

(3) After substituting ψ_p the stator inductance L_s is given to:

$$L_s \approx \frac{\sqrt{\hat{U}_N^2 - \hat{\mathbf{e}}_g^2}}{2\pi f_N \hat{I}_{sqN}} \qquad (6.76)$$

6.4 Automatic parameter estimation for IM in standstill

6.4.1 Pre-considerations

For the complete description of the IM four parameters are required with a constant parameter model. If the inverse Γ equivalent circuit (cf. fig. 6.3) is chosen as the reference model, the four parameters are then the stator resistance R_s, the rotor resistance R_r, the total leakage inductance σL_s and the stator inductance L_s. The constant parameter model in its precision does not suffice for the synthesis of advanced algorithms, however. At least the inclusion of the saturation characteristics of the inductances is required. Because of the different saturation functions for main and leakage paths a division of the model inductance parameters into leakage inductance σL_s or L_σ and main inductance L_m can be made.

For a current controlled drive the slip is limited also in non-stationary states to values which not yet necessitate a consideration of the current displacement in the rotor for the modelling. Harmonic caused current displacement effects also shall be neglected for the modelling in accordance with the presumptions made (inverter-fed operation at high switching frequency). An exception for the consideration of frequency dependencies is the leakage inductance. Depending on the excitation frequency it has to be distinguished between different inductance values. This means in particular that a transient leakage inductance $\sigma L_s'$ for the current controller design and a stationary (fundamental wave) leakage inductance σL_s for the stator-frequent operation have to be estimated.

The consideration of the iron losses is not avoidable (cf. chapter 7 and 8) for some special tasks. Their identification is practically only possible with the no-load test in a classical way, however, and shall not be discussed more in-depth.

Furthermore, from practical considerations for a useful incorporation of the off-line adaptation into the technological regime of an inverter-fed drive some conditions, which fundamentally narrow down the choice of possible methods, have to be formulated:

1. If possible, no demands or prerequisites on the part of the identification algorithms should be made to technological conditions of the drive. This is the case if the identification runs at standstill and does not need a speed feedback.

2. The safety of the methods and their transferability onto different drive configurations increase if algorithms which run in the closed current control loop are used.

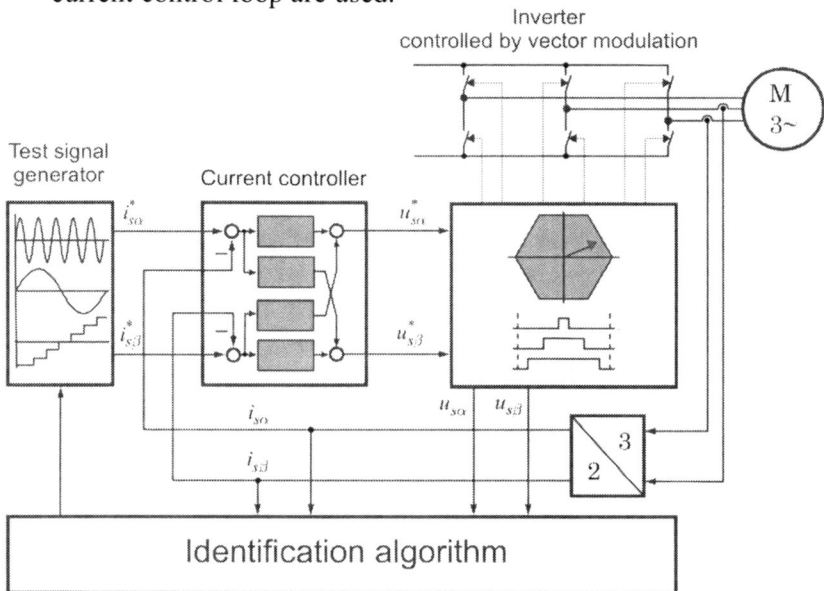

Fig. 6.14 Principle structure of the off-line parameter identification

Regardless that the frequency dependencies are not considered in the model except for the exception mentioned above, the choice of the identification methods has to take into account that such dependencies exist. Thus the test signal frequencies used by the identification should, on one hand, be located in the same range as the frequencies at which the models are operated later. On the other hand the test frequencies have to be selected for current displacement effects not invalidating the identified parameters. For this reason methods with *predefined appropriately selected excitation frequencies* will be preferred for the concrete identification methods in the following sections. The parameter estimation is essentially accomplished by evaluation of the *frequency responses* of current and voltage. The identification shall be implemented without voltage measuring sensors, and the voltage has to be estimated from the control signals of the inverter.

For the decoupled identification of the parameters, a further criterion for the choice of the excitation frequencies results from the consideration, if possible, not to influence the identification of one parameter by inaccurate other parameters. This suggests to optimize the excitation frequencies by evaluation of sensitivity functions.

The test signals for the parameter identification are produced by frequency inverters. These have a non-linear current-voltage characteristic because of the effects of blanking time, switching delays and voltage drops over the semiconductor switch primarily at small voltages. Just in this voltage area the parameter estimation takes place because of $\omega = 0$. Therefore, the current-voltage characteristic of the inverter must be considered in the model, and also identified for a generally usable identification algorithm. Because of the abandonment of voltage measuring sensors this measure is also imperative for an adequately exact voltage feedback.

The corresponding principle structure of the off-line parameter identification is shown in the figure 6.14.

6.4.2 Current-voltage characteristics of the inverter, stator resistance and transient leakage inductance

As indicated already, a great importance for the precision of the parameter identification for inverter feeding and abandonment of special measuring sensors relies on the voltage capturing. Blanking times and non-linear current-dependent inverter voltage drops have to be considered as error sources which have an effect in particular at small voltages and around current zero crossings. The suppression of their effects on the parameter identification is taken care of in two ways: Firstly by an appropriate choice of the excitation signals, and secondly by embedding the inverter characteristics into the motor model.

Suitable excitation signals are discussed in the context of the individual identification methods specifically. The stator voltage equation is amended by an additional current-dependent term to consider the inverter voltage drops $u_z(i_s)$ in the motor model and looks in the stationary case with $\omega = \omega_s = 0$ as follows:

$$u_s(i_s) = u_z(i_s) + R_s i_s \qquad (6.77)$$

At first the measurement of the complete characteristic $u_s(i_s)$ is carried out point wise by impression of DC currents. It has qualitatively the appearance of the dotted curve in figure 6.15. Because of a possible unbalance of the motor an averaging of the measurements from single tests of the three phases is advisable. Assuming that the voltage increase at high currents is only determined by the linear portions of the voltage drop, the stator resistance can be calculated from the ascent of the current-voltage curve at high current:

$$R_s = \frac{u_{s1} - u_{s2}}{i_{s1} - i_{s2}} \tag{6.78}$$

The then known linear term is now eliminated from (6.77), and the non-linear characteristic remains. For the non-linear inverter voltage drop $u_z(i_s)$ different approaches with constant and/or exponential sections have been proposed in the literature (cf. [Baumann 1997], [Rasmussen 1995], [Ruff 1994]). To avoid the on-line evaluation of exponential functions, a piecewise-linear approximation also can be carried out. A characteristic which also is represented qualitatively in figure 6.15 (solid line) is obtained.

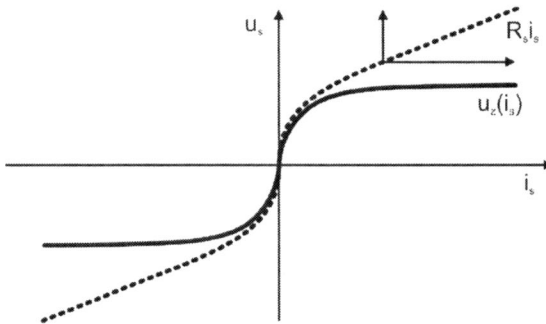

Fig. 6.15 Inverter current-voltage characteristic

The actual compensation is made by a sign and phase correct addition to the voltage reference values, similarly like described in section 2.3.3 for the protection time compensation. With $u_{zu} = u_z(i_{su})$, $u_{zv} = u_z(i_{sv})$ and $u_{zw} = u_z(i_{sw})$ the following voltage components are obtained in stator-fixed coordinates:

$$u_{z\alpha} = \frac{1}{4}\left(2u_{zu} - u_{zv} - u_{zw}\right)$$
$$u_{z\beta} = \frac{\sqrt{3}}{4}\left(u_{zv} - u_{zw}\right) \tag{6.79}$$

To measure the transient leakage inductance a short voltage impulse is applied to the stator winding, and the current gradient is measured. Since the time needed for this test pulse is very short, and the process is barely noticeable, this measurement can be carried out also outside a special identification run. The leakage inductance arises from:

$$\sigma L_s' = \frac{\hat{u}_s \Delta t}{\Delta i_s} \tag{6.80}$$

For an appropriate width of the voltage pulse and measurement over the complete current slope a good average inductance value will be obtained.

6.4.3 Identification of inductances and rotor resistance with frequency response methods

6.4.3.1 Basics and application for the identification of rotor resistance and leakage inductance

By impressing a sinusoidal current into the stator all desired motor parameters can theoretically be identified by measuring the waveforms of currents and voltages and subsequent frequency response analysis. However, before applying this method some preceding considerations are necessary which follow up the preliminary remarks and determine the most suitable environment.

The demand for an identification at standstill, and therefore the demand that no torque must be developed, can be fulfilled by a single-phase excitation.

The estimation of the stator impedance requires an exact acquisition of the current and voltage fundamental waves. The compensation of the inverter nonlinearities is decisive for the quality of the identification results because of the low voltage amplitudes at standstill (cf. section 6.4.2). Furthermore [Bünte 1995] worked out, that the remaining error only has an effect on the real part of the measured impedance, if the impressed current is sinusoidal. The latter is achieved if the identification is performed in the closed current control loop. Furthermore the current should, if possible, be free of zero crossings because the largest deviations from the sinusoidal form arise in the zero crossings.

A zero crossing free current can be produced by overlaying the sinus reference with a direct current component. This component is reasonably chosen close to the nominal magnetization. This corresponds to a direct current pre-magnetization, and a main field excitation alternating permanently around the working point is produced by the single-phase sinusoidal excitation. Therefore the derivation of the transfer function has to start out from equations of the saturated machine (6.40), (6.41) and (6.43). Because of $\omega = 0$ these equations are simplified to a great deal. In addition, the excitation only takes place in the α axis so that the dimension of the equation system is reduced to one. Under these prerequisites the following transfer function between stator voltage and stator current can be derived by elimination of i_μ (s = Laplace operator):

$$\frac{u_{s\alpha}}{i_{s\alpha}} = \frac{b_0 + b_1 s + b_2 s^2}{1 + a_1 s} \tag{6.81}$$

$$b_0 = R_s; b_1 = \left(L_m' + L_\sigma\right)\left(1 + \frac{R_s}{R_r}\right)$$

$$b_2 = \frac{L_\sigma}{R_r}\left(L_m' + L_\sigma\right); a_1 = \frac{L_m' + L_\sigma}{R_r}$$

In the steady-state operating condition ($s \to j\omega_e$, ω_e ... excitation frequency) the equation (6.81) can be written as a complex impedance:

$$\underline{Z}_s = \frac{\underline{u}_{s\alpha}}{\underline{i}_{s\alpha}} = \frac{b_0 + (a_1 b_1 - b_2)\,\omega_e^2}{1 + a_1^2 \omega_e^2} + j\omega_e \frac{b_1 - b_0 a_1 + a_1 b_2 \omega_e^2}{1 + a_1^2 \omega_e^2} \tag{6.82}$$

or

$$\underline{Z}_s = R_s + \frac{R_r\left(\omega_e L_m'\right)^2}{R_r^2 + \omega_e^2\left(L_m' + L_\sigma\right)^2} + j\omega_e\left(L_m' + L_\sigma - \frac{\left(\omega_e L_m'\right)^2\left(L_m' + L_\sigma\right)}{R_r^2 + \omega_e^2\left(L_m' + L_\sigma\right)^2}\right) \tag{6.83}$$

Under special conditions for the excitation frequency the formula (6.83) could further be simplified. For example, with $\sigma\left(\omega_e T_r\right)^2 \gg 1$ it can be written:

$$\underline{Z}_s \approx R_s + (1 - \sigma)R_r + j\omega_e \sigma L_s \tag{6.84}$$

This equation would be very comfortable for the calculation of the rotor resistance and leakage inductance. The excitation frequency should be within the range of at least 25 Hz, though. Here the current displacement effects in the rotor already have a considerable magnitude and markedly distort the estimated value of the rotor resistance. Under certain assumptions these effects could be taken into account by an additional approach. The safe way, if more than the leakage inductance shall be identified, consists, however, in the evaluation of the complete equation (6.81).

For the estimation of the four parameters of (6.81) current and voltage values have to be captured after achieving the steady-state operating condition over at least one period of the fundamental wave at two excitation frequencies ω_{e1} and ω_{e2}. Harmonics are conveniently suppressed by discrete Fourier transformation of the measurement values. Two complex resistance values are the result:

$$\underline{Z}_{s1}(\omega_{e1}) = c_1 + jd_1$$
$$\underline{Z}_{s2}(\omega_{e2}) = c_2 + jd_2$$

(6.85)

The coefficients of (6.81) can be calculated as follows:

$$a_1 = \frac{\omega_{e2}d_1 - \omega_{e1}d_2}{\omega_{e1}\omega_{e2}(c_2 - c_1)}$$

(6.86)

$$b_1 = \frac{d_2}{\omega_{e2}} + c_2 a_1$$

(6.87)

$$b_2 = a_1 b_1 + \frac{c_2\left(1 + a_1^2\omega_{e2}^2\right) - c_1\left(1 + a_1^2\omega_{e1}^2\right)}{\omega_{e1}^2 - \omega_{e2}^2}$$

(6.88)

$$b_0 = c_1\left(1 + a_1^2\omega_{e1}^2\right) - \omega_{e1}^2\left(a_1 b_1 - b_2\right)$$

(6.89)

Solving to the actual machine parameters is elementary. The obtained value for the differential main inductance L_m', however, is not immediately usable because only a small area of the hysteresis curve is passed through at every direct current working point and the gradient at this point does not or only at strong saturation coincide with the gradient of the actual magnetization characteristic. To identify the main inductance the frequency response method has to be adapted specifically (cf. section 6.4.3.3). The value for b_0 contains apart from the stator resistance the uncompensated inverter-caused voltage errors, and is therefore not representative as an estimate.

For the determination of the current dependency of the leakage inductance (saturation characteristic) a separate series of measurements is required because the magnitude of the current must be varied without pre-magnetization. Because of the zero crossing errors the received values differ a little from the leakage inductances found with DC offset. Because no general function for σL_s can be given due to the different leakage saturation behavior, a linear approximation between the test points or a polynomial approximation may be used.

6.4.3.2 Optimization of the excitation frequencies by sensitivity functions

Depending on the excitation frequency, changes of a motor parameter effect the frequency-dependent complex impedance $\underline{Z}_s$ in the equation (6.83) with different strength. This behavior can mathematically be described by the sensitivity function E(p) of the complex impedance $\underline{Z}_s$ regarding a parameter p. For the separate investigation on the influence on real and imaginary part of $\underline{Z}_s$, the sensitivity function is calculated one by one for real and imaginary part respectively:

$$E_I(s) = \frac{\partial \operatorname{Im}(\underline{Z}_s)}{\partial s} \frac{s}{\operatorname{Im}(\underline{Z}_s)}$$

$$E_R(s) = \frac{\partial \operatorname{Re}(\underline{Z}_s)}{\partial s} \frac{s}{\operatorname{Re}(\underline{Z}_s)}$$

(6.90)

After some transformations the following equations result:

$$E_I(L_m) = \left(1 - \frac{\omega_e^2 L_m^2 R_r^2 \left(3L_m' + 2L_\sigma\right) + \omega_e^4 L_m \left(L_m' + 2L_\sigma\right)\left(L_m' + L_\sigma\right)^2}{\left(R_r^2 + \omega_e^2 \left(L_m' + L_\sigma\right)^2\right)^2}\right) \frac{\omega_e L_m'}{\operatorname{Im}(\underline{Z}_s)}$$

(6.91)

$$E_I(L_\sigma) = \left(1 - \frac{\omega_e^2 L_m'^2 \left(R_r^2 - \left(L_m' + L_\sigma\right)^2\right)}{\left(R_r^2 + \omega_e^2 \left(L_m' + L_\sigma\right)^2\right)^2}\right) \frac{\omega_e L_\sigma}{\operatorname{Im}(\underline{Z}_s)}$$

(6.92)

$$E_I(R_r) = \frac{2R_r \omega_e^3 L_m'^2 \left(L_m' + L_\sigma\right)}{\left(R_r^2 + \omega_e^2 \left(L_m' + L_\sigma\right)^2\right)^2} \frac{R_r}{\operatorname{Im}(\underline{Z}_s)}$$

(6.93)

$$E_R(R_r) = \frac{\omega_e^2 L_m'^2 \left(\omega_e^2 \left(L_m' + L_\sigma\right)^2 - R_r^2\right)}{\left(R_r^2 + \omega_e^2 \left(L_m' + L_\sigma\right)^2\right)^2} \frac{R_r}{\operatorname{Re}(\underline{Z}_s)}$$

(6.94)

$$E_R(L_\sigma) = \frac{-2\omega_e^3 L_m'^2 \left(L_m' + L_\sigma\right) R_r}{\left(R_r^2 + \omega_e^2 \left(L_m' + L_\sigma\right)^2\right)^2} \frac{\omega_e L_\sigma}{\operatorname{Re}(\underline{Z}_s)}$$

(6.95)

For reasons which will be discussed in the next section, the sensitivity function of the real part regarding L_m is not of interest. For one example the sensitivity functions are represented in figure 6.16. The typical qualitative characteristic can be transferred and generalized to other motor power ratings.

Values between 2 and 12 Hz prove to be suitable excitation frequencies for the identification of rotor resistance and leakage inductance. The frequencies are still low enough to neglect current displacement effects, on the other side however, adequately high to achieve a decoupling of the main inductance. Optimal values to estimate the main inductance are in the area from 0.1 to 0.4Hz.

An exact pre-computation of optimal excitation frequencies with the help of the sensitivity functions, however, is not possible because these in turn contain the parameters to be identified. But an iterative optimization of the excitation frequencies within several identification runs is possible, threat no more than 2 - 3 iteration steps are generally required.

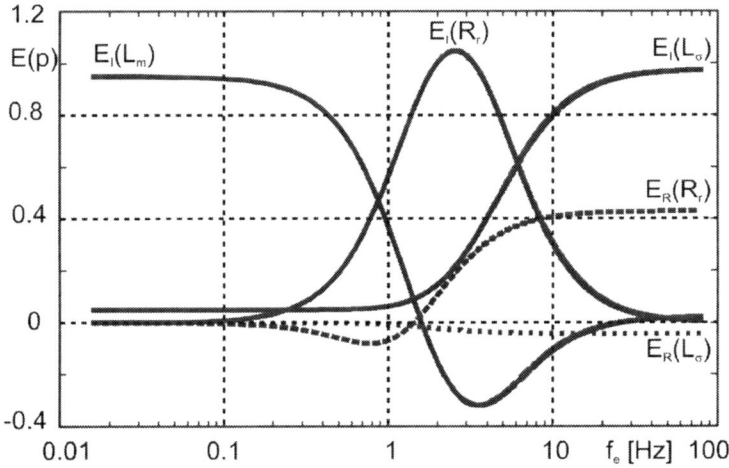

Fig. 6.16 Sensitivity functions of a 3kW motor

6.4.3.3 Peculiarities at estimation of main inductance and magnetization characteristic

Also the main inductance can be identified by single-phase sinusoidal excitation like leakage inductance and rotor resistance. The hysteresis problem mentioned in the previous section can be solved by working without direct current offset. Because of the necessary lower excitation frequencies, zero crossing errors have a less strong effect. Because the voltage measuring errors primarily distort the real part of the measured impedance, only the imaginary part of (6.83) is used for evaluation. Because the imaginary part is mainly determined by the phase shift between current and voltage, this phase shift must be measured with sufficient accuracy which in turn sets a lower limit of approximately 0.1 Hz for the excitation frequency. At this time R_r and L_σ are assumed as known parameters.

Solving the equation (6.83) yields for the main inductance:

$$L_m = \frac{R_r^2 + X_\sigma^2 + 2X_\sigma(X_\sigma - \mathrm{Im}(\underline{Z}_s)) - \sqrt{\left(R_r^2 - X_\sigma^2\right)^2 - 4\,\mathrm{Im}(\underline{Z}_s)R_r^2(\mathrm{Im}(\underline{Z}_s) - 2X_\sigma)}}{2\omega_e(\mathrm{Im}(\underline{Z}_s) - 2X_\sigma)}$$

$$(6.96)$$

Because the stator current is divided between inductance branch and rotor, the exact magnetization current has to be calculated:

$$i_\mu = i_s \sqrt{\frac{R_r^2 + X_\sigma^2}{R_r^2 + \omega_e^2 (L_m + L_\sigma)^2}} \qquad (6.97)$$

Different operating points on the magnetization characteristic are adjusted by different current amplitudes. It has to be taken into account that the identified main inductance is not identical with the effective main inductance at three-phase excitation. The reason is that the magnetizing current has constant amplitude at three-phase excitation, but changes sinusoidally at single-phase excitation. The amplitudes coincide for the two cases. The voltage at single-phase excitation is distorted due to saturation. *The amplitude of its fundamental wave* evaluated for the frequency response does therefore not represent the instantaneous maximum value of the magnetic field strength correctly. The described relations are qualitatively represented in the figure 6.17.

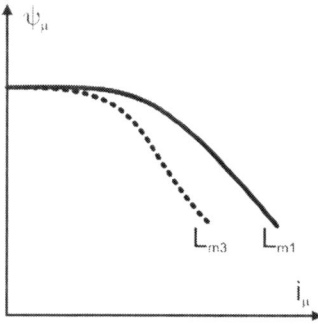

Fig. 6.17 Single-phase and three-phase main inductance

In [Klaes 1992] the difference between single-phase and three-phase inductance is compensated by a constant factor established heuristically which subsequently compresses the scale of the magnetizing current or flux axis. An interesting systematic solution was described in [Bünte 1995]. It assumes that single-phase and three-phase inductance curves $L_{m1}(\hat{I}_\mu)$ and $L_{m3}(i_\mu)$ can be described by polynomials of the n-th degree in the following form:

$$L_{m1}(\hat{I}_\mu) = \sum_{k=0}^{n} a_{1k} \hat{I}_\mu^k \; ; \; L_{m3}(i_\mu) = \sum_{k=0}^{n} a_{3k} i_\mu^k \qquad (6.98)$$

The single-phase main inductance L_{m1} is calculated for a sinusoidal magnetizing current $i_\mu(t) = \hat{I}_\mu \sin \omega_e t$ from the continuous Fourier coefficients of the fundamental of the magnetizing current, and the voltage drop over the main inductance u_μ from:

$$\omega_e L_{m1} \int_0^{\pi/\omega_e} i_\mu(t)\sin\omega_e t\, dt = \int_0^{\pi/\omega_e} u_\mu(t)\cos\omega_e t\, dt \qquad (6.99)$$

Furthermore the following equation applies for the voltage over the main inductance:

$$u_\mu(t) = \frac{d\psi_\mu}{dt} = \frac{dL_{m3}(i_\mu)}{dt}i_\mu + L_{m3}(i_\mu)\frac{di_\mu}{dt} \qquad (6.100)$$

After substituting, processing of integrals and comparison of coefficients the result is:

$$a_{1k} = b_k a_{3k} \quad \text{with} \quad b_k = \begin{cases} 2\displaystyle\prod_{l=0}^{k/2}\frac{2l+1}{2l+2} & \text{for } k \text{ even} \\[2ex] \dfrac{4}{\pi}\displaystyle\prod_{l=1}^{(k+1)/2}\frac{2l}{2l+1} & \text{else} \end{cases} \qquad (6.101)$$

For the above mentioned second method of the adjustment of the characteristic by coordinate axis compression, power-dependent compression factors:

$$i_{\mu 3} = c_k i_{\mu 1} \quad \text{mit} \quad c_k = \sqrt[k]{b_k} \qquad (6.102)$$

with values of $c_k = 0.85 \dots 0.88$ for a polynomial degree $n = 3$ can be derived. Thus this method also should provide a useable characteristic transformation. The polynomial approximation is obtained from the single measurements by applying least squares approximation (cf. section 12.3).

6.4.4 Identification of the stator inductance with direct current excitation

The basic concept of this method is derived from the fact that at impression of a direct current into the stator windings a part of the applied voltage is consumed by the stator resistance, the other part is used to build the stator flux. From the stator voltage equation:

$$u_s = R_s i_s + \frac{d\psi_s}{dt} \qquad (6.103)$$

and after integration for the stationary state it follows:

$$\int u_s\, dt = R_s \int i_s\, dt + L_s i_s \qquad (6.104)$$

Because the leakage inductance and the stator resistance are known from the previous measurement, the main inductance can be calculated from that theoretically without difficulty. Offset errors, stationary errors of the voltage measurement or an incorrectly estimated stator resistance can be eliminated, if the integral term on the right side of (6.104) is replaced by the stator voltage in the steady state condition ($t \rightarrow \infty$). In time-discrete notation the computation equation of L_s with the sampling period T, the time step k and the total integration time NT is:

$$L_s = \frac{T \sum\limits_{k=0}^{N} u_s(k) - u_s(\infty)NT}{i_s(\infty)} \qquad (6.105)$$

For the determination of the complete magnetization characteristic the identification is realized at different current levels. Because different single tests, particularly at small currents, partly show a considerable scattering of measurements, an averaging of the values from several tests is recommendable. The measurement should be carried out in all three windings to eliminate machine unbalances.

The figure 6.18 shows finally some measurement results. The values for the stator inductance L_s from alternating current and direct current methods delivered by the identification are plotted together with the results of the no-load test. The consequences of voltage measuring errors are most distinctive particularly at small currents and simultaneously small voltage amplitudes. A very good correspondence to the no-load characteristic is shown in the area of high saturation. Altogether, the precision of both methods can be considered as sufficient for the purposes of the self-tuning.

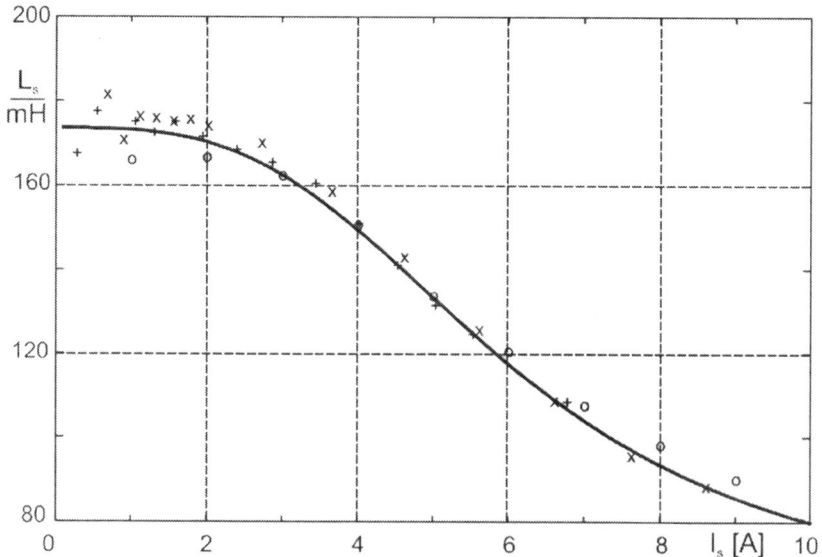

Fig. 6.18 L_s identification for a 5kW motor: no-load test (+), direct current method (o), alternating current method (×), solid line: four-parameter model from no-load measurements (regressed by polynomial of 3rd order)

6.5 References

Baumann T (1997) Selbsteinstellung von Asynchronantrieben. VDI-Fortschritt-Bericht Reihe 21 Nr. 230, VDI Verlag Düsseldorf

Bünte A, Grotstollen H (1995) Offline Parameter Identification of an Inverter-Fed Induction Motor at Standstill. EPE 1995 Sevilla, pp. 3.492 - 3.496

de Jong HCJ (1980) Saturation in Electrical Machines. Proc. of the Intern. Conf. on Electrical Machines, Athen, part 3, pp. 1545 – 1552

Klaes N (1992) Identifikationsverfahren für die betriebspunktabhängigen Parameter einer wechselrichtergespeisten Induktionsmaschine. VDI-Verlag Düsseldorf

Levi E (1994) Magnetic Saturation in Rotor-Flux-Oriented Induction Motor Drives: Operating Regimes, Consequences and Open-Loop Compensation. European Transactions on Electrical Power Engineering Vol. 4, No. 4, July/August, pp. 277 – 286

Lunze K (1978) Einführung in die Elektrotechnik. VEB Verlag Technik Berlin

Murata T, Tsuchiya T, Takeda I (1990) Quick Response and High Efficiency Control of the Induction Motor Based on Optimal Control Theory. 11. IFAC World Congress, Tallin, vol. 8, pp. 242 – 247

Philippow E (1980) Taschenbuch Elektrotechnik, Band 5: Elemente und Baugruppen der Elektroenergietechnik. VEB Verlag Technik Berlin

Quang NP (1996) Digital Controlled Three-Phase Drives. Education Publishing House Hanoi (Book in Vietnamese: Điều khiển tự động truyền động điện xoay chiều ba pha. Nhà Xuất bản Giáo dục Hà Nội)

Rasmussen H, Tonnes M, Knudsen M (1995) Inverter and Motor Model Adaptation at Standstill Using Reference Voltages and Measured Currents. Proceedings EPE 1995 Sevilla, pp. 1.367 - 1.372

Ruff M, Bünte A, Grotstollen H (1994) A New Self-Commissioning Scheme for an Asynchronous Motor Drive System. IEEE Industry Applications Society Annual Meeting Denver, pp. 616 – 623

Schäfer U (1989) Feldorientierte Regelung einer Asynchronmaschine mit Feldschwächung und Berücksichtigung der Eisensättigung und Erwärmung. Dissertation RWTH Aachen

Vas P (1990) Vector Control of AC Machines. Oxford University Press

Vogt K (1986) Elektrische Maschinen, Berechnung rotierender elektrischer Maschinen. VEB Verlag Technik Berlin

7 On-line adaptation of the rotor time constant for IM drives

A typical problem of the field-orientated control consists of the system having to evaluate the actual value of the rotor flux without flux sensors through a model from the measurable terminal quantities of the motor and the speed (cf. section 4.4). The often used current-speed model contains the rotor time constant of the motor as an essential parameter whose exact knowledge influences decisively the quality of the control. This fact and the working point dependence of this parameter motivate the introduction of special measures to primarily compensate the temperature dependence of the rotor resistance. To achieve this, *two approaches are in principle conceivable*: Either *the rotor flux model can be completed by an on-line adaptation method* which corrects the rotor resistance permanently, or *the rotor flux is estimated by an observer* which is insensitive against variations of the rotor resistance. The first approach is subject of this chapter.

In the first section the range and effects of temperature-dependent changes of the rotor resistance on other characteristic quantities are examined. A summary of published compensation methods follows. Thereafter *adaptation methods with a parametric error model* are discussed in greater detail. Such on-line adaptation methods use error models for the tracking of the parameters which in turn contain at least another two machine parameters. Their precision therefore also influences the quality of the field orientation. Thus these dependencies form a further main emphasis in the discussion besides *adaptation dynamics and problems of the adaptation in the non-stationary operation*.

7.1 Motivation

When using the i_s-ω flux model in field coordinates (cf. section 4.4) the amplitude and phase angle of the rotor flux linkage (model quantities indicated with ^) are:

$$\frac{d\hat{i}_{md}}{dt} = \frac{1}{T_r}\left(-\hat{i}_{md} + \hat{i}_{sd}\right) \tag{7.1}$$

$$\dot{\hat{\vartheta}}_s = \hat{\omega}_s = \omega + \frac{\hat{i}_{sq}}{T_r \hat{i}_{md}} \tag{7.2}$$

with: $\hat{i}_{md} = \dfrac{\hat{\psi}_{rd}}{L_m}$

Thus the rotor time constant T_r is obviously the decisive parameter for both dynamics and precision. Assuming an exact initial setting and the possibility of an exact modelling of the rotor inductance, the rotor resistance R_r remains as not predictably variable parameter. Considering the temperature coefficient and the possible change of the rotor temperature it can be shown that a resistance variation of about 50% has to be expected during operation. This undoubtedly causes a loss of quality in the system behaviour. The size of it and its tolerability or non-tolerability shall be examined in the following. The following criteria will be analyzed:

- Stationary torque and flux deviation (or difference).
- Linearity between torque and torque-producing quantity (the torque-forming current component).
- Dynamics of torque impression.

A faulty rotor time constant generates according to (7.2) a flux phase error and thus a phase difference between model current and motor current in the consequence:

$$\mathbf{i}_s = \hat{\mathbf{i}}_s e^{j\tilde{\vartheta}_s}, \tilde{\vartheta}_s = \hat{\vartheta}_s - \vartheta_s \tag{7.3}$$

After solving into components, the equation (7.3) can be written as follows:

$$\begin{aligned} i_{sd} &= \hat{i}_{sd}\cos\tilde{\vartheta}_s - \hat{i}_{sq}\sin\tilde{\vartheta}_s \\ i_{sq} &= \hat{i}_{sq}\cos\tilde{\vartheta}_s + \hat{i}_{sd}\sin\tilde{\vartheta}_s \end{aligned} \tag{7.4}$$

Because the speed is measured and the slip has to adjust to the existing load after dissipation of all transient processes, the slip values in the model and motor are identical in stationary operation, and therewith the next equation is valid according to (7.2):

$$\frac{i_{sq}}{i_{sd}\,T_r} = \frac{\hat{i}_{sq}}{\hat{i}_{sd}\,\hat{T}_r} \tag{7.5}$$

Using (7.4) and (7.5) the phase error $\tilde{\vartheta}_s$ can be calculated as follows:

$$\tan \hat{\vartheta}_s = \left(\frac{T_r}{\hat{T}_r} - 1\right) \frac{\hat{i}_{sd}\,\hat{i}_{sq}}{\hat{i}_{sd}^2 + \frac{T_r}{\hat{T}_r}\hat{i}_{sq}^2} \tag{7.6}$$

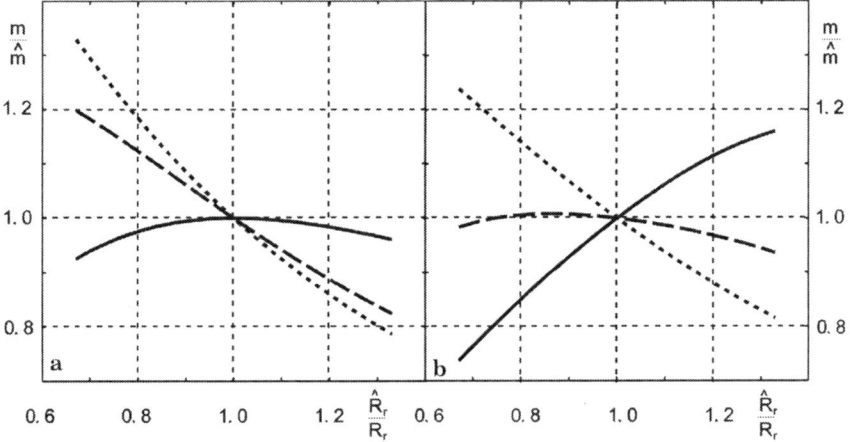

Fig. 7.1   Torque errors caused by inaccurate rotor resistance: a) without main field saturation; b) with main field saturation ($\underline{\quad}\ \hat{i}_{sq} = \hat{i}_{sd}$, ------ $\hat{i}_{sq} = 2\hat{i}_{sd}$, ······· $\hat{i}_{sq} = 3\hat{i}_{sd}$)

With the help of the stationary torque equation:

$$m_M = \frac{3}{2} z_p \frac{L_m^2}{L_r} i_{sd} i_{sq} \tag{7.7}$$

relations can now be derived for the stationary torque and flux amplitude deviation. After some intermediate steps the following formulae will be obtained:

$$\frac{m_M}{\hat{m}_M} = \frac{L_m^2}{L_r} \frac{\hat{L}_r}{\hat{L}_m^2} \frac{T_r}{\hat{T}_r} \frac{1+(\omega_r \hat{T}_r)^2}{1+(\omega_r T_r)^2} \tag{7.8}$$

$$\frac{\psi_r}{\hat{\psi}_r} = \frac{L_m}{\hat{L}_m} \sqrt{\frac{1+(\omega_r \hat{T}_r)^2}{1+(\omega_r T_r)^2}} \tag{7.9}$$

After inserting (7.4) and (7.6) into the torque equation (7.7), the torque characteristic $m_M(\hat{i}_{sq})$ will be obtained assuming constant $\hat{i}_{sd}$:

$$m_M(\hat{i}_{sq}) = \frac{3}{2} z_p \frac{L_m^2}{L_r} \frac{T_r}{\hat{T}_r} \hat{i}_{sd} \hat{i}_{sq} \frac{\hat{i}_{sd}^2 + \hat{i}_{sq}^2}{\hat{i}_{sd}^2 + \left(\frac{T_r}{\hat{T}_r}\right)^2 \hat{i}_{sq}^2} \tag{7.10}$$

The equation (7.8) is graphically represented with and without the consideration of the main field saturation in the figure 7.1 with the data of an 11kW standard motor. Without saturation the equation (7.8) does not contain any further machine parameters apart from the rotor time constant deviation and therefore describes a generally valid relation. With the consideration of the saturation there is also $\hat{L}_m \neq L_m$ for a wrong rotor time constant in the model because of $\hat{i}_\mu \neq i_\mu$. As shown in the figure, the saturation will have a weakening influence on the torque error at larger load. The reason is that the fraction in equation (7.8) and the terms before it describe contrary trends of the torque deviation. One of them predominates depending on the load, also an approximated compensation is possible, as in figure 7.1b for $\hat{i}_{sq} = 2\hat{i}_{sd}$.

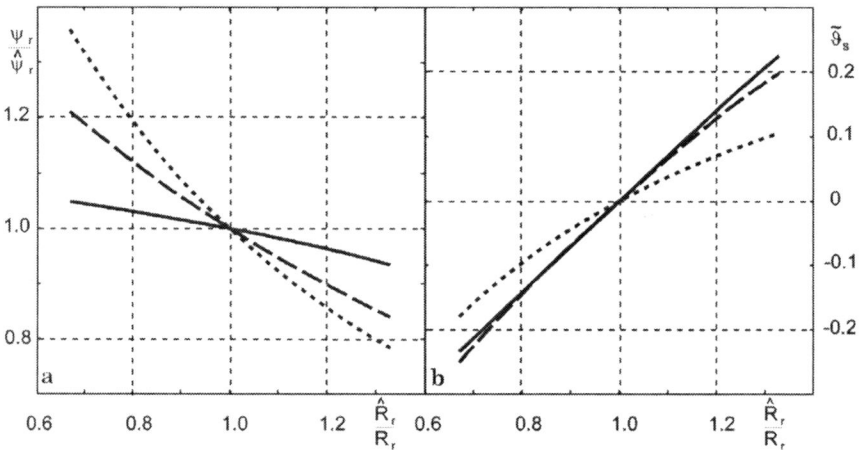

Fig. 7.2   a) Flux amplitude errors, b) Flux phase errors: by R_r deviation (with magnetic saturation); ——— $\hat{i}_{sq} = \hat{i}_{sd}$; ------ $\hat{i}_{sq} = 2\hat{i}_{sd}$; ······· $\hat{i}_{sq} = 3\hat{i}_{sd}$

The size of the torque deviation is approximately half as large as the model resistance error at nominal operation and therefore actually remarkable. Whether a too small or too big model resistance represents the more critical case can be recognized in connection with the flux deviation. The corresponding characteristics using (7.9) and (7.6) are shown in figure 7.2.

For a speed controlled drive the motor torque to be produced will correspond to the load torque in any case. If the rotor flux is weakened by a wrong orientation, a higher current must be applied to achieve the demanded torque which can possibly exceed (at full load) the maximum inverter current and then leads to premature breakdown or the drive not reaching its rated speed. According to figure 7.2a this is the case for a too big model resistance. With a too small model resistance, a flux increase will follow which at corresponding speed causes a premature approaching the voltage limit. It is possible that the reference speed can not be reached at nominal torque, and the error of the rotor time constant leads to a reduction of the available power. Because the drives are usually designed with a current reserve on the inverter side, but the voltage of the DC link cannot be increased beyond a certain limit, the second case (small model resistance) has to be classified as the more critical one.

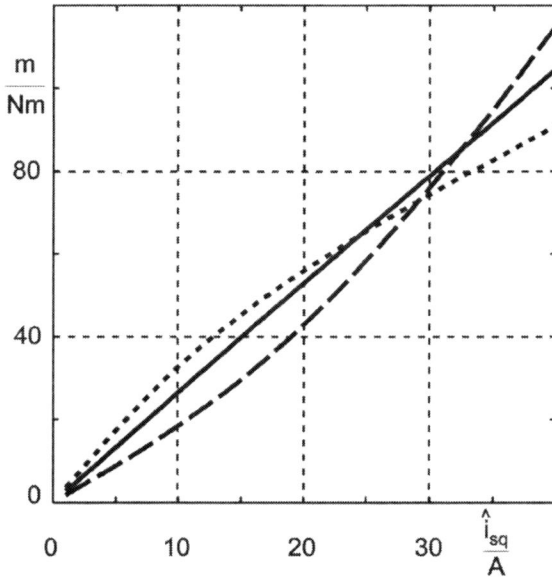

Fig. 7.3 $m_M(\hat{i}_{sq})$ characteristic: — $\hat{R}_r = R_r$, --- $\hat{R}_r = 0.66 R_r$, ···· $\hat{R}_r = 1.33 R_r$

The depicted area of the T_r deviation is approximately within the temperature-dependent limits which can be practically expected, if with regard to the initial settings of T_r the following two cases are considered: On one hand an initial setting on the cold machine, in which case an increase of the rotor resistance by 50% has to be taken into account during operation; and on the other hand an initial setting on the medium-warm machine with an operation dependent resistance change of $\pm 25\%$.

For the pictures 7.1 and 7.2 three different load cases were analyzed in which the largest load approximately corresponds to the rated torque.

For a speed-controlled drive without high dynamic and precision demands the appearing flux and torque errors are probably tolerable. A superimposed speed control will compensate stationary torque errors. With adjustment on the warm machine an unintentional flux increase and power reduction will be avoided. However, depending on technical conditions and demands on the drive quality and on the intended optimization goals the expected errors could become too large and therefore no more acceptable. Such cases can be:

- The current reserve of the inverter is so small (or there is no overrating at all) that a flux weakening caused by wrong orientation really leads to a prematurely reaching of the current limit.
- An exact control of the state variables at variable rotor flux and in the field weakening becomes impossible.
- Drives which require an exact torque impression cannot be designed without additional measures. This becomes clear also by the stationary characteristic $m_M(\hat{i}_{sq})$ in figure 7.3.
- The recommended operation to avoid power reduction with a reduced flux automatically leads to an increase of the slip, and thus to a worse efficiency.

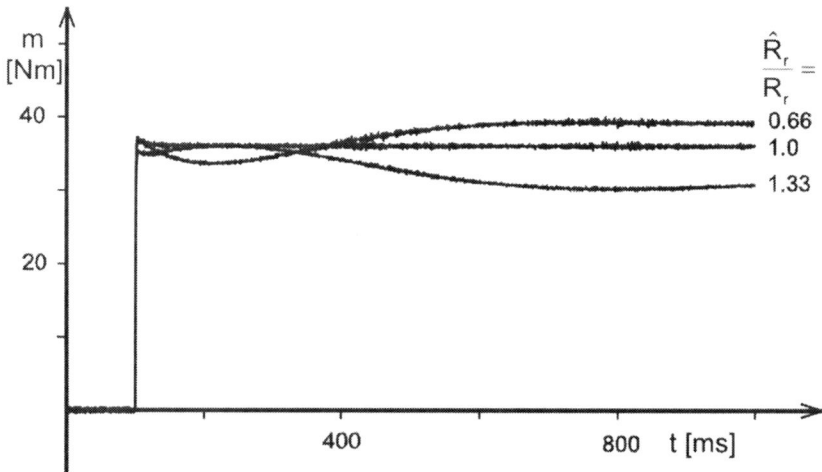

Fig. 7.4   Dynamic torque impression at faulty adjustment of T_r.

The figure 7.4 shows the influence of a wrongly adjusted rotor time constant on the dynamic torque impression in the constant flux area. Because the rotor flux remains constant in the first instance after the reference step, the actual torque responds non-delayed in the first place as in the case of correctly adjusted parameters. The following settling process is determined by the transient of the rotor flux and is finished if the rotor flux also has reached its new stationary state.

Thus the consequences of a wrong adjustment are less serious in the dynamic case in the basic speed range. *Altogether, the stationary torque and flux errors represent the more serious effects and ask for the search of compensation measures in high quality and highly utilized drives.*

A high-dynamic torque impression is not conceivable in the field weakening or at low voltage reserve without an exact machine model, though. These issues will be discussed in chapter 8 more thoroughly.

7.2 Classification of adaptation methods

Because of the significance of an online adaptation of the rotor time constant outlined in the previous section these problems are a standard topic of the pertinent technical literature with a mass of papers since the first publications about FOC. An overview is found in [Krishnan 1991] for example.

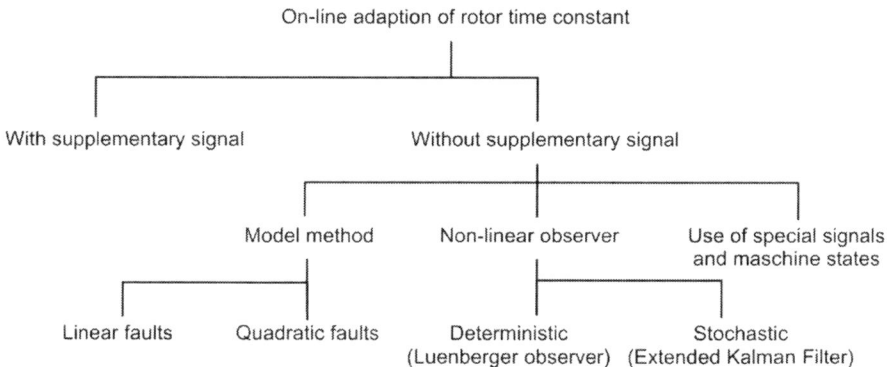

Fig. 7.5   Systematization of the methods for online adaptation of the rotor time constant

Because of the variety of different methods, only a certain group, namely the model methods with different kinds of error signals, shall be dealt with in detail subsequently. At first, a general survey will be worked out comprising a systematization and summaries of characteristic features to give to the reader a broader insight and the possibility to analyze the subject more deeply with the help of secondary literature. The figure 7.5 only shows a rough classification. In this picture only adaptation methods which work without physical changes on the motor (additional windings or the like) are included.

a) Methods using additional signal injection

The classic method in this group was published in [Gabriel 1982]. It uses the property of the rotor flux process to low-pass filter high-frequency disturbances in the flux forming current and therefore to keep the torque un-effected from such signals. But this is only valid in the case of exact field orientation, if the flux forming current does not directly contribute to torque production. If consequences of a high-frequency pseudo-noise signal injected on the flux forming current are provable in the torque (by correlation calculation), the field orientation is not exact and the noticed error can be used for the correction of the rotor time constant. A similar approach is used in [Nomura 1987]. A higher-frequency sine-wave signal, however, is fed into the *d* axis instead of the noise signal.

In the method described in [Chai 1992] with spectrum analysis the supplementary signal is added to the reference value of the rotor flux. At the same time, flux and torque forming current components are controlled in such way that no disturbance of the torque takes place. The suitable choice of the harmonic frequencies to be evaluated in relation to the stationary stator frequency enables an on-line emulation of the classical short circuit and no-load tests using simple algebraic equations for the parameter calculation following digital Fourier transformation of the measurement values. The method was used for the online adaptation of resistances and leakage inductances. Because of the harmonics produced additionally by the saturation, an estimation of the main inductance is not possible.

The method [Sng 1995] which was especially developed for extremely low speeds works similarly. A MRAS estimator for the rotor time constant is combined with an on-line estimation for resistances and leakage inductances which is excited by a high-frequency sine-wave signal.

A method which uses harmonics produced by the inverter as excitation was finally published in [Gorter 1994]. These harmonics are in the range of 300....600Hz. The rotor resistance, the leakage and main inductances are online-identified using the RLS method. The required linear machine model was derived by transition into rotor coordinates and use of a stator current - stator flux model.

b) Methods using models

The methods of this group work according to the model reference principle. A physical quantity of the motor is calculated by two different and independent models, and an error signal is derived from the output signals of both models. This error signal works as a driving quantity of an adjusting controller which corrects on-line the rotor resistance, rotor time constant or other parameters as well. Of course, the designed error signal must depend on the parameter to be estimated in a way which supports an unambiguous tracking. Input quantities of the models are measured terminal quantities of the machine, whereby a sub-model can immediately be identical with a measured quantity. In different ways stability considerations can be included, for example through an observer approach or by exploiting the theory of model reference adaptive systems (MRAS).

The methods described in the literature differ from each other primarily by the choice of the physical quantity used for the calculation of the model error. Furthermore it can be distinguished between the linear and quadratic error signals. Linear error signals are formed from stator current components [Pfaff 1989], [Reitz 1988] (in this publication all parameters are adapted by an adaptation law designed according to the Gauß-Newton method), stator voltage [Dittrich 1994], [Rowan 1991], motor EMF [Kaźmierkowski] or rotor flux [Ganji 1995]. Also the method described in [Fetz 1991], which works with a field-orientated open-loop current control and an adaptation signal derived from the output signal of a current by-pass controller implicitly uses the stator voltage components as reference values. Estimated and measured stator current trajectories are compared in [Holtz 1991] to calculate all machine parameters and rotor current components by using a gradient method.

Quadratic error signals can be derived from the amplitude [Rowan 1991] or the phase angle [Schumacher 1983] of the stator voltage or the motor EMF, from the air gap power [Dolal 1987], the active and/or reactive power [Dittrich 1994], [Koyama 1986], [Summer 1991], [Summer 1993], from the electrical torque [Lorenz 1990], [Rowan 1991], the magnitude of the stator flux [Krishnan 1986] or from especially designed signals [Vucosavić 1993], [Weidauer 1991].

c) Non-linear observers
These estimators also could be assigned to the methods with additional signal injection as far as extended Kalman filters (EKF) are used for the parameter estimation, because here the harmonics produced by the pulse-width modulated voltage are partly used as excitation signal [Zai 1987]. For the classification carried out at this place the observer approach shall play, however, the decisive role.

Compared with simple model methods, an observer approach offers the possibility of predicting the dynamic behaviour of the adaptive system in certain limits, and of targeted adapting the feedback matrix. Furthermore, certain properties like the robustness of the system, can be influenced by the suitable choice of the feedback matrix. The parameter adaptation is a by-product to the actual task of the observer, the flux estimation.

When state observers are used for parameter estimation at the same time, non-linear or extended observers [Zeitz 1979] arise. A complete observer for the electrical quantities of the induction machine with inclusion of one parameter would have the order of five. Because the currents usually are being measured, the order of a reduced observer (flux and one parameter) is cut down to three, and the realization expenditure is substantially more favourable. The observer error essentially corresponds to the stator voltage component error model mentioned above. Observers of reduced order with parameter adaptation are described in [Dittrich 1998], [Nilsen 1989], [Schrödl 1989].

As opposed to Luenberger observers, Kalman Filters (KF) or Extended Kalman Filters (EKF) take into account stochastic uncertainties of the system and measuring errors for a combined state and parameter estimation. As already mentioned they can also work with stochastic input signals. The realization effort is, however, considerable. Although the asynchronous machine represents a deterministic system, a number of papers have been published on the application of EKF [Atkinson 1991], [Loron 1993], [Pena 1993]. [Du 1993] is to mention as an interesting contribution on the topic of applying extended observers or EKF's.

d) Evaluation of special signals and machine states

All methods which work without injection of an additional signal and can not be assigned to other groups shall be assigned here. So [Vogt 1985] evaluates the speed oscillations caused by torque vibrations from an inaccurately adjusted rotor time constant. The method described in [Hung 1991] calculates the rotor flux and a correction signal for the rotor time constant from the third voltage harmonic caused by the magnetic saturation, and therefore independent of rotor parameters, this under the assumptions of an exact voltage measurement, operating the motor in the saturation and star-connection of the windings.

An essential weakness of the methods with additional signals is certainly the influence on the normal operation which can really have a disturbing effect, even if the torque remains undisturbed as indicated in [Chai 1992]. The adaptation can be carried out only in the stationary operation; a general proof of stability is barely possible. Furthermore great care is required to ensure that only answers to the excitation signals are actually evaluated and no harmonics and disturbances caused by other influences (saturation, mechanical oscillation). On the other side, an identification of the rotor parameters is also possible in no-load operation [Chai 1992] with appropriate design of the excitation signal.

This is fundamentally impossible for methods without additional signal. Furthermore it cannot be assumed that the signals appearing in the normal operation have an adequate information content suitable to carry out a multi-parameter identification (what is not intended in the context of this chapter, though). Error models for the identification of the rotor time constant always contain other machine parameters which decisively influence the precision of the adaptation. On the other hand an adaptation is conceivable and theoretically also possible in the dynamic operation. The system behaviour including stability can be designed and assessed in an uniform approach, at least with certain limitations (e.g. partial linearization).

7.3 Adaptation of the rotor resistance with model methods

In this section some approaches from the group of the model methods shall be discussed in more detail, whereat for design and stability analysis principles of the nonlinear observer theory will be applied. Linear and quadratic fault models (reactive power) are included. The online adaptation is focused on the compensation of temperature variations and therefore on the rotor resistance. The state variable dependent main inductance is adjusted in feed-forward mode. If the adaptation is implemented primarily for the optimization of the stationary operation, an immediate tracking of the rotor time constant as a whole is also conceivable and sufficient.

The observer is designed from a linearized process model based on a local approach of the system at small state errors. This approach is justified because a state observer is designed for the purpose to keep deviations minimal between observer and system state variables. Prerequisite is that the initial values of the observer states are chosen accordingly, i.e. close to the actual system states.

All fault models contain besides the rotor resistance at least two further machine parameters whose precision fundamentally influences the adaptation error and with that the precision and stability of the adjustment. For this reason corresponding sensitivity studies will occupy a relatively wide room in the following considerations.

In principle it is possible to implement the online adaptation like the flux model in arbitrary coordinate systems. But because the flux model was already established in the rotor flux orientated coordinate system, and thus the rotor flux is immediately available in this system, the adaptation methods are also designed in field-orientated coordinates.

7.3.1 Observer approach and system dynamics

As already indicated, the adaptation algorithm shall be designed using the theory of non-linear observers. This approach has the substantial advantage that the adaptation dynamics and stability can be examined in a uniform design procedure. The design is carried out for a linearized system in quasi-stationary operation. The essential design prerequisites are:

- The observer is designed exclusively for the rotor resistance, therefore being the only state quantity.
- For the analysis of the observer dynamics the steady-state condition with regard to the rotor flux vector is assumed:

$$i_m = i_{sd}, \hat{i}_m = \hat{i}_{sd}, \tilde{\vartheta}_s = \hat{\vartheta}_s - \vartheta_s = \text{const} \tag{7.11}$$

It has to be made sure for the functionality of this approach that the rotor resistance observer is assigned a sufficiently slow dynamics ensuring a dynamic decoupling to the remaining system. For the compensation of thermal resistance changes such a dynamics is completely sufficient.

With these prerequisites, for the system state and output equations can be written (for clarity the current time step k is written as an index in the following equations):

$$\begin{aligned} R_{r,k+1} &= R_{r,k} \\ \mathbf{y}_k &= \mathbf{h}(R_{r,k}, \mathbf{u}_k) \end{aligned} \tag{7.12}$$

Here $\mathbf{y}$ is the output vector, and $\mathbf{u}$ is the input vector of the system. The output equation represents the equation of the later error model. Therefore the distinction between input and output quantities has not to be understood in the strictly literal meaning. Both vectors are assumed multi-dimensional in the general case.

The observer for this system is formulated as extended Luenberger observer [Brodmann 1994], [Zeitz 1979]. It consists of a model of the system and a linear, but state variable and time dependent feedback of the output error (difference between model output and system output vector) to the observer state:

$$\begin{aligned} \hat{R}_{r,k+1} &= \hat{R}_{r,k} - \mathbf{k}(\hat{R}_{r,k}, \mathbf{u}_k)^{\mathsf{T}} \left(\hat{\mathbf{y}}_k - \mathbf{y}_k \right) \\ \hat{\mathbf{y}}_k &= \mathbf{h}(\hat{R}_{r,k}, \mathbf{u}_k) \end{aligned} \tag{7.13}$$

The equation (7.13) corresponds to a general adaptation approach with the adaptation fault

$$\varepsilon_k = -\mathbf{k}(\hat{R}_{r,k}, \mathbf{u}_k)^{\mathsf{T}} \left(\hat{\mathbf{y}}_k - \mathbf{y}_k \right) \tag{7.14}$$

in the structure of the figure 7.6. For the following design procedure two tasks remain:

- Specification of a fault model – subject of section 7.3.2.
- Design of the observer dynamics by determination of the weighting or feedback vector **k**.

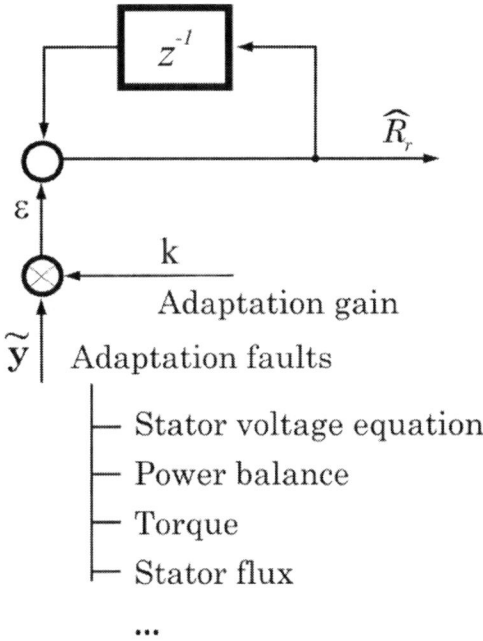

Fig. 7.6   Rotor resistance observer

The dynamic analysis is carried out using the difference equation of the observer error:

$$\tilde{R}_{r,k+1} = \hat{R}_{r,k+1} - R_{r,k+1} = \tilde{R}_{r,k} - \mathbf{k}(\hat{R}_{r,k}, \mathbf{u}_k)^{\mathsf{T}}\left(\hat{\mathbf{y}}_k - \mathbf{y}_k\right) \quad (7.15)$$

If suitable starting values which already are close to the system state are chosen for the observer state, being the rotor resistance, the design can be performed by a local analysis following the linearization around the model state. Then the equation (7.15) can be written as follows:

$$\tilde{R}_{r,k+1} = \left(1 - \mathbf{k}^{\mathsf{T}} \frac{\partial \mathbf{h}}{\partial R_r}\bigg|_{\hat{R}_{r,k},\mathbf{u}_k}\right) \tilde{R}_{r,k} \quad (7.16)$$

From this a relation for the linearized output error can be derived by comparison with (7.13):

$$\tilde{\mathbf{y}}_k = \frac{\partial \mathbf{h}}{\partial R_r}\bigg|_{\hat{R}_{r,k},\mathbf{u}_k} \tilde{R}_{r,k} \quad (7.17)$$

The actual design of the fault dynamics is carried out with help of the characteristic equation of the linearized fault system. This is in the z domain:

$$z - 1 + \mathbf{k}^\mathsf{T} \left. \frac{\partial \mathbf{h}}{\partial R_r} \right|_{\hat{R}_{r,k}, \mathbf{u}_k} = 0 \qquad (7.18)$$

or with (7.17):

$$z - 1 + \mathbf{k}^\mathsf{T} \frac{\tilde{\mathbf{y}}_k}{\tilde{R}_{r,k}} = 0 \qquad (7.19)$$

From this the coefficients of the feedback vector $\mathbf{k}$ can be calculated using the corresponding terms for the special error models. The most important design goals consist of achieving a time-invariant and system signal independent fault dynamics and of obtaining a stable transient response by specification of a constant eigenvalue z_1. In addition, to satisfy the demand for adequately slow adaptation dynamics (dynamic decoupling to the remaining system) z_1 should fulfil the equation:

$$0 < 1 - z_1 \le \frac{1}{2} \frac{T}{\hat{T}_r} \qquad (7.20)$$

For all simulation and test examples given subsequently the value of the upper limit was used respectively.

Because the observer is designed in field-orientated coordinates, the coordinate transformation is an integral component of the model. Therewith actual input quantities are currents and voltages in stator-fixed coordinates. If the observer equations are formulated and designed nevertheless in field-orientated coordinates, it has to be taken into account that the transformation angle ϑ_s or the phase error $\tilde{\vartheta}_s$ is a function of the state error $\tilde{R}_r$. All currents and voltages in field coordinates depend implicitly on $\tilde{R}_r$. Therefore the equation for the linearized output error has to be extended to the complete error difference:

$$\tilde{\mathbf{y}} = \left. \frac{\partial \mathbf{h}}{\partial R_r} \right|_{\hat{R}_r} \tilde{R}_r + \left. \frac{\partial \mathbf{h}}{\partial \vartheta_s} \right|_{\hat{\vartheta}_s} \tilde{\vartheta}_s \qquad (7.21)$$

Here $\tilde{\vartheta}_s$ is given by (7.6) for the interesting stationary case. For the relation between the phase error and the rotor resistance, it follows then in linearized form:

$$\tilde{\vartheta}_s = \left. \frac{\partial \tilde{\vartheta}_s}{\partial \tilde{R}_r} \right|_{R_r = \hat{R}_r} \tilde{R}_r = \frac{1}{\hat{R}_r} \frac{\hat{i}_m \hat{i}_{sq}}{\hat{i}_m^2 + \hat{i}_{sq}^2} \tilde{R}_r \qquad (7.22)$$

Using

$$\mathbf{i}_s^f = \mathbf{i}_s^s \, e^{-j\vartheta_s} \tag{7.23}$$

the next equation will be obtained for the derivative of the stator vector in field coordinates with respect to the phase angle:

$$\frac{\partial \mathbf{i}_s^f}{\partial \vartheta_s} = -j\mathbf{i}_s^s \, e^{-j\vartheta_s} = -j\mathbf{i}_s^f \tag{7.24}$$

Analogously the following equation is valid for the stator voltage:

$$\frac{\partial \mathbf{u}_s^f}{\partial \vartheta_s} = -j\mathbf{u}_s^s \, e^{-j\vartheta_s} = -j\mathbf{u}_s^f \tag{7.25}$$

7.3.2 Fault models

The stator voltage equations provide the first approach for the derivation of the output equation and in due course for the observer fault. In addition, the active and reactive power balance were chosen as an example for quadratic fault models. The reason for this special choice consists in the fact that characteristic model parameter constellations, which allow representative statements also for other methods, are produced in the context of these methods. These relations will be more exactly examined in the next section. At first the fault models shall be assembled, and the associated feedback coefficients shall be derived. The magnetic saturation remains so far unconsidered, and linear magnetic conditions or a constant rotor flux are assumed. If the saturation of the main inductance shall be taken into account for the adaptation in the field weakening area or for rotor flux transients, the corresponding relations from section 6.2.3 have to be used for the derivation of model equations.

7.3.2.1 Stator voltage models

In this case the system output equations can be derived immediately from the time-discrete stator voltage equation of the IM in field-orientated coordinates. To avoid the flux derivation it is started from the state equation with Euler discretization (cf. chapter 3):

$$\begin{aligned}
\mathbf{u}_{s,k} = R_s \mathbf{i}_{s,k} &+ \frac{\sigma L_s}{T}\left(\mathbf{i}_{s,k+1} - \mathbf{i}_{s,k}\right) + j\omega_s \sigma L_s \mathbf{i}_{s,k} \\
&-(1-\sigma)R_r\left(\mathbf{i}_{m,k} - \mathbf{i}_{s,k}\right) + j\omega(1-\sigma)L_s \mathbf{i}_{m,k}
\end{aligned} \tag{7.26}$$

Models which use voltage equations in d or q axis or as a combination of both components are practicable.

a) Stator voltage model in d axis

The output equation of the system is obtained by using the real component of (7.26):

$$y_{d,k+1} = \frac{1}{1-\sigma}\left[-u_{sd,k} + R_s i_{sd,k} + \frac{\sigma L_s}{T}\left(i_{sd,k+1} - i_{sd,k}\right)\right.$$
$$\left. - \sigma L_s \omega_s i_{sq,k}\right] - R_r\left(i_{m,k} - i_{sd,k}\right) = 0 \tag{7.27}$$

and the output error follows to:

$$\hat{y}_d - y_d = \frac{1}{1-\sigma}\left[-\hat{u}_{sd,k} + R_s \hat{i}_{sd,k} + \frac{\sigma L_s}{T}\left(\hat{i}_{sd,k+1} - \hat{i}_{sd,k}\right)\right.$$
$$\left. - \sigma L_s \omega_s \hat{i}_{sq,k}\right] - \hat{R}_{r,k}\left(\hat{i}_{m,k} - \hat{i}_{sd,k}\right) \tag{7.28}$$

To obtain the adaptation gain the linearized output error has to be calculated using (7.17) and (7.19). The calculation is carried out according to the assumption for the steady-state condition indicated by (7.11). The explicit indication of the current time step is omitted subsequently. Using (7.21), (7.24) and (7.25) it follows from (7.27):

$$\frac{\partial h}{\partial \vartheta_s} = \frac{1}{1-\sigma}\left(-u_{sq} + R_s i_{sq} + \sigma L_s \omega_s i_{sd}\right) = -\omega_s L_s i_m \tag{7.29}$$

Finally, the stationary linearized output error is found using (7.22):

$$\tilde{y}_d = -\omega_s \hat{T}_r \hat{i}_m \frac{\hat{i}_m \hat{i}_{sq}}{\hat{i}_m^2 + \hat{i}_{sq}^2} \tilde{R}_r \tag{7.30}$$

After substituting this expression into (7.19) the adaptation gain can now be calculated with the predefined eigenvalue z_1:

$$k_d = \frac{1 - z_1}{\omega_s \hat{T}_r \hat{i}_m \dfrac{\hat{i}_m \hat{i}_{sq}}{\hat{i}_m^2 + \hat{i}_{sq}^2}} \tag{7.31}$$

b) Stator voltage model in q axis

Analog to the *d* axis model the output equation immediately results from the imaginary component of (7.26),

$$y_{q,k+1} = \frac{1}{1-\sigma}\left[-u_{sq,k} + R_s i_{sq,k} + \frac{\sigma L_s}{T}\left(i_{sq,k+1} - i_{sq,k}\right)\right.$$

$$\left. + \sigma L_s \omega_s i_{sd,k}\right] + \omega_s L_s i_{m,k} = 0 \tag{7.32}$$

with the output error:

$$\hat{y}_q - y_q = \frac{1}{1-\sigma}\left[-\hat{u}_{sq,k} + R_s \hat{i}_{sq,k} + \frac{\sigma L_s}{T}\left(\hat{i}_{sq,k+1} - \hat{i}_{sq,k}\right)\right.$$

$$\left. + \sigma L_s \omega_s \hat{i}_{sd,k}\right] + \omega_s L_s \hat{i}_{m,k} \tag{7.33}$$

For the stationary linearized output error we obtain in the same way with:

$$\frac{\partial h}{\partial \vartheta_s} = \frac{1}{1-\sigma}\left(u_{sd} - R_s i_{sd} + \sigma L_s \omega_s i_{sq}\right)$$

$$+ \omega_s L_s i_{sq} = \omega_s L_s i_{sq} \tag{7.34}$$

the following equation:

$$\tilde{y}_q = \omega_s \hat{T}_r \hat{i}_{sq} \frac{\hat{i}_m \hat{i}_{sq}}{\hat{i}_m^2 + \hat{i}_{sq}^2} \tilde{R}_r \tag{7.35}$$

Accordingly the adaptation gain can be calculated as follows:

$$k_q = -\frac{1 - z_1}{\omega_s \hat{T}_r \hat{i}_{sq} \dfrac{\hat{i}_m \hat{i}_{sq}}{\hat{i}_m^2 + \hat{i}_{sq}^2}} \tag{7.36}$$

c) Voltage vector fault model

So far several approaches have already been suggested in the literature to combine both error components [Rowan 1991], [Schumacher 1983]. Thereat the amplitude or phase angle of the error vector was calculated. Furthermore it is conceivable to simply add both weighted error components derived above. However, this way shall not be followed here, because in spite of more information flowing into the adaptation the possibility to use the additional degree of freedom for dedicated weightings of the error components will be given away. The combination of the error components shall be aimed to define the dynamics with one weighting factor and to balance the contributions of both error components to the total error with a second factor. The addition of both components has to ensure that the sign of the total error (= adjustment direction of the rotor resistance) is only determined by the direction of the resistance deviation or the phase error, and not distorted by the combination of the error components. The following approach is chosen:

$$\varepsilon_{dq} = k_{d1}(\hat{y}_d - y_d) - k_{q1}(\hat{y}_q - y_q)\operatorname{sign}\hat{i}_{sq}$$
$$= k_{d1}\left[(\hat{y}_d - y_d) - k_{dq}(\hat{y}_q - y_q)\operatorname{sign}\hat{i}_{sq}\right] \qquad (7.37)$$
$$k_{d1} > 0, \quad k_{q1} > 0$$

The adaptation dynamics is determined by k_{d1}, and the error weighting by $k_{dq} = k_{q1}/k_{d1}$. By inserting (7.30) and (7.35) the linearized output error is obtained to:

$$\tilde{y}_{dq} = -\omega_s \hat{T}_r \left(k_{d1}\hat{i}_m + k_{q1}\left|\hat{i}_{sq}\right|\right)\frac{\hat{i}_m \hat{i}_{sq}}{\hat{i}_m^2 + \hat{i}_{sq}^2}\tilde{R}_r \qquad (7.38)$$

At positive $\hat{i}_m$ the bracket term is always positive, and thus the above-mentioned condition is fulfilled. The adaptation gain can be calculated as follows:

$$k_{d1} = \frac{1 - z_1}{\omega_s \hat{T}_r \left(\hat{i}_m + k_{dq}\left|\hat{i}_{sq}\right|\right)\dfrac{\hat{i}_m \hat{i}_{sq}}{\hat{i}_m^2 + \hat{i}_{sq}^2}} \qquad (7.39)$$

The derivation of a suitable value for k_{dq} is subject of section 7.3.3.

7.3.2.2 Power balance models

The first step at examination of an error model is to find out whether it is suitable for the rotor resistance adaptation at all. This is the case if the error signal proceeds steadily and there is an unambiguous connection between the variation of the rotor resistance and the sign of the model error. Particularly for error models of higher order these prerequisites are not obvious. With the method used here to analyze the adaptation problem with the help of the nonlinear state observer the corresponding proof can be adduced very comfortably.

Unlike to the previous models, for the power balance methods the system output vector is derived from the components of the complex power. Starting point is the equation of the complex power:

$$\mathbf{s} = \frac{3}{2}\mathbf{u}_s \mathbf{i}_s^* \tag{7.40}$$

Neglecting the factor 3/2 it follows in vector notation:

$$\mathbf{s} = \begin{pmatrix} p \\ q \end{pmatrix} = \begin{pmatrix} u_{sd}i_{sd} + u_{sq}i_{sq} \\ -u_{sd}i_{sq} + u_{sq}i_{sd} \end{pmatrix} \tag{7.41}$$

The output or model equations are obtained after inserting the voltage equations into (7.41). To simplify the representation they are written in the following in time-continuous form. As in the case of the stator voltage methods the dynamic analysis is carried out for steady-state condition regarding the rotor flux linkage and the stator current ($di_{sd}/dt = di_{sq}/dt = 0$).

a) Reactive power method
Following the described procedure the system output equation is:

$$y_b = u_{sd}i_{sq} - u_{sq}i_{sd} + \sigma L_s \left(\frac{di_{sq}}{dt} i_{sd} - \frac{di_{sd}}{dt} i_{sq} \right) +$$

$$+ \omega_s L_s \left[\sigma i_s^2 + (1-\sigma) i_m^2 \right] \tag{7.42}$$

$$+ (1-\sigma)\omega L_s i_m (i_{sd} - i_m) = 0$$

With:

$$\frac{\partial h}{\partial \vartheta_s} = 2(1-\sigma)\omega_s L_s i_m i_{sq} \tag{7.43}$$

the following expression is obtained for the stationary linearized error:

$$\tilde{y}_b = 2(1-\sigma)\omega_s \hat{T}_r \frac{\hat{i}_m^2 \hat{i}_{sq}^2}{\hat{i}_m^2 + \hat{i}_{sq}^2} \tilde{R}_r \tag{7.44}$$

The adaptation gain then will be:

$$k_b = -\frac{1 - z_1}{2\omega_s \hat{T}_r (1-\sigma)\dfrac{\hat{i}_m^2 \hat{i}_{sq}^2}{\hat{i}_m^2 + \hat{i}_{sq}^2}} \tag{7.45}$$

b) Active power method
The output equation arises as described:

$$y_w = -u_{sd}i_{sd} - u_{sq}i_{sq} + \sigma L_s \left(\frac{di_{sd}}{dt} i_{sd} + \frac{di_{sq}}{dt} i_{sq} \right) +$$

$$+ R_s i_s^2 + (1-\sigma)\omega_s L_s i_m i_{sq} \tag{7.46}$$

$$+ (1-\sigma)R_r i_{sd} (i_{sd} - i_m)$$

For the stationary linearized output error and for the adaptation gain it can be derived in the same way as above:

$$\frac{\partial h}{\partial \vartheta_s} = (1-\sigma)w_s L_s \left(i_{sq}^2 - i_m^2 \right) \tag{7.47}$$

$$\tilde{y}_w = (1-\sigma)w_s T_r \left(\hat{i}_{sq}^2 - \hat{i}_m^2 \right) \frac{\hat{i}_m \hat{i}_{sq}}{\hat{i}_m^2 + \hat{i}_{sq}^2} \tilde{R}_r \tag{7.48}$$

$$k_w = -\frac{1-z_1}{w_s \hat{T}_r (1-\sigma) \left(\hat{i}_{sq}^2 - \hat{i}_m^2 \right) \frac{\hat{i}_m \hat{i}_{sq}}{\hat{i}_m^2 + \hat{i}_{sq}^2}} \tag{7.49}$$

It shall be noted that obviously the active power method cannot produce any observer fault in the area $i_{sq} \approx i_m$, and therefore its usefulness is limited substantially. Thus further consideration is abandoned.

7.3.3 Parameter sensitivity

All error models contain machine model parameters which practically always show a certain deviation compared to the actual motor parameters. This means, however, that at use of such a faulty model the rotor resistance too cannot be estimated exactly. The knowledge of the relations between model parameter errors and a wrong adjustment of the rotor time constant resulting from these errors is therefore of essential importance for the choice and assessment of an adaptation method. Such an analysis shall be carried out now for the error models introduced in the previous section.

The vector of the error model parameters is referenced by $\hat{\mathbf{p}}$, and the vector of the machine parameters by $\mathbf{p}$. At first the adaptation is looked at in stationary operation. Stability of the overall system assumed, the adaptation algorithm will regulate the adaptation error to zero in every case:

$$\lim_{t \to \infty} \varepsilon(\hat{u}_s, \hat{i}_s, \hat{\mathbf{p}}) = 0 \tag{7.50}$$

For a speed and rotor flux controlled system, the system state and the model state are unambiguously determined by the motor torque m_M (= load torque) and by the set point $\hat{i}_m^*$ (= $\hat{i}_m$) of the rotor flux linkage. The current component $\hat{i}_{sq}$ of the model is adjusted by the speed controller according to the required torque. The connection between system and model currents is given by (7.4). A statement about the model parameter

dependent adjustment error of the rotor resistance can be obtained if an
analytical connection:

$$\widetilde{R}_r = f(m_M, \hat{i}_m, \widetilde{\mathbf{p}}) \tag{7.51}$$

can be derived. Unfortunately, this is not possible in explicit form,
though, so that a simulation study or an iterative calculation must be
undertaken. A possible approach for the iterative solution consists in
computing the resistance error $\widetilde{R}_r$ for a given $\widetilde{\mathbf{p}}$ for which the adaptation
error ε becomes zero. This way the searched dependencies (7.51) can be
determined point-wise.

In the first step the model current component $\hat{i}_{sq}$ shall be determined in
dependence of m_M and $\hat{i}_{sd}$. For this purpose (7.4) and (7.6) are inserted
into the torque equation (7.7). After some rewriting the following
expression arises:

$$m_M \left[\hat{i}_{sq}^2 + \left(\frac{\hat{T}_r}{T_r} \right)^2 \hat{i}_m^2 \right] = \frac{3}{2} z_p \frac{L_m^2}{L_r} \frac{\hat{T}_r}{T_r} \hat{i}_m \hat{i}_{sq} \left(\hat{i}_{sq}^2 + \hat{i}_m^2 \right) \tag{7.52}$$

which can be solved to $\hat{i}_{sq}$ iteratively. After that the motor currents can
be calculated by (7.4), and after inserting the result into the stator voltage
equation the motor voltages will be obtained. With that the model voltages
are finally found out also by (7.4) making all required quantities for the
calculation of the model error complete. Unstable areas of the overall
system are indicated by the fact that no solution $\varepsilon = 0$ for a given
parameter mistuning exists. In a dynamic system simulation this case can
be recognized by the fact that no stable stationary operating point will be
reached.

The results of the sensitivity calculations for the error models introduced
above are summarized in figure 7.7. They were obtained by using the data
of an 11kW standard motor. The voltage vector error method still remains
excluded in this figure because the second weighting factor to be
determined shall specifically used for the robustness improvement.
Although the validity of the results remains limited to the used motor,
characteristic trends can be recognized, which result from the structure of
the fault models and which also are transferable to other motors.

In the calculations the model parameters $\hat{R}_s$, $\hat{L}_m$ and $\sigma \hat{L}_s$ were varied,
and the motor parameters were held constant. For the discrimination of the
parameter influences only one parameter was changed at one time.
Therefore it is practically definitely possible that divergent results arise by
the overlapping of parameter errors. A parameter variation range of

-50%...+50% was examined for two characteristic rotational frequencies (1.6Hz and nominal speed) and two loads (nominal torque and half nominal torque). If a curve is not drawn over the full area, the overall control system reaches a stable operating point only in the marked area.

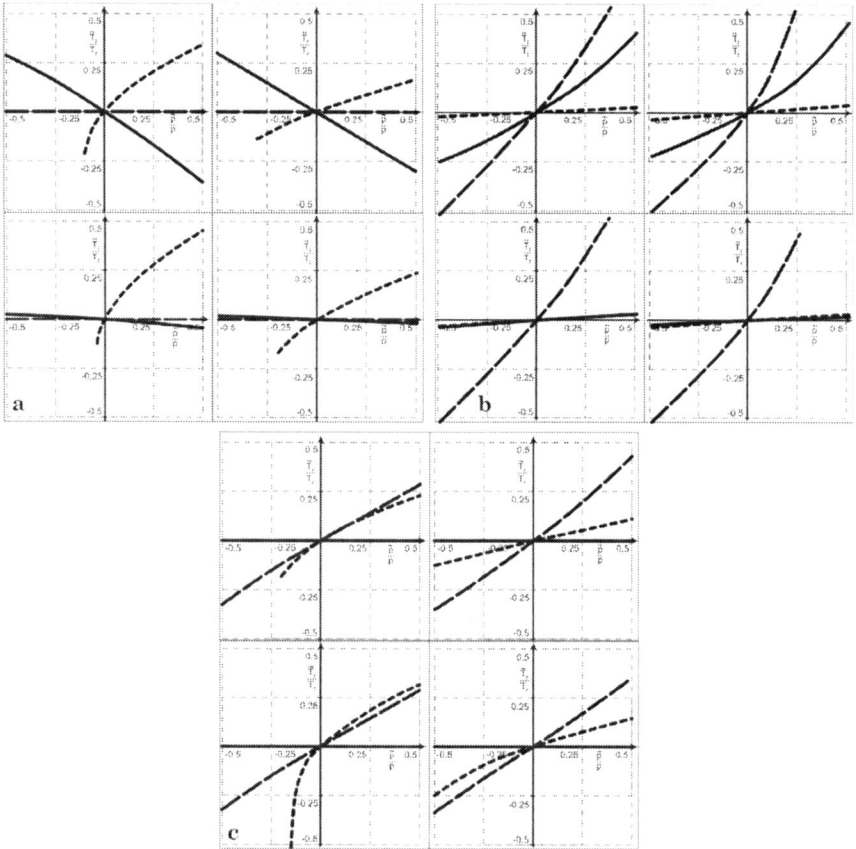

Fig. 7.7 (**a**) d axis voltage model, (**b**) q axis voltage model, (**c**) reactive power model: —— $\tilde{R}_s / R_s$, ---- $\tilde{L}_m / L_m$, ····· $\sigma\tilde{L}_s / \sigma L_s$, top: ω=10s^{-1}, bottom: ω=300s^{-1}, left: m=73 Nm, right: m=36 Nm

At first all parameter mistunings have an effect on the adjusted rotor time constant which thus represents a characteristic quantity representing the parameter errors. From this it can be concluded to flux amplitude and phase errors corresponding to section 7.1.

As a trend common to all measurements, it can be noticed that the sensitivity to the stator resistance drastically decreases at higher frequencies while the sensitivity to leakage and main inductance is only weakly or not frequency-dependent. This connection can be proved also arithmetically [Dittrich 1994]. In the models where the leakage inductance appears as a parameter (d axis voltage model and reactive power model) the influence of the leakage inductance is strongly load-dependent and a tendency towards a restriction of the stability area exists at increasing frequency and too small model values.

Among the examined methods the reactive power model proves to be the one with the most favorable robustness qualities despite the stability problems at leakage inductance errors. These can be avoided if the leakage inductance of the model is prevented from becoming smaller than the leakage inductance of the motor. The complete independence on the stator resistance must be highlighted as particularly positive.

At a more exact comparison of the results for the voltage models two facts can be noticed: The stator resistance sensitivity curves show a contrary trend, and at higher frequencies the main inductance or the leakage inductance determine the sensitivity characteristic, with their plots stretching from the third to the first quadrant.

From that it can be concluded, that it is possible to compensate the sensitivity to R_s almost completely if a combination of both fault models according to the voltage vector fault model established in the previous section is used. Only a certain balance of the sensitivity to the inductances will be reached, though, thereat the primary objective consists in extending the stability area. For the complete compensation of the R_s sensitivity the factor k_{dq} in (7.37) should be chosen to:

$$k_{dq} = \frac{\hat{i}_{sd}}{\hat{i}_{sq}} \tag{7.53}$$

The factor k_{dq} must approach this value at low stator frequencies. For weak load the weighting must be shifted to $(\hat{y}_d - y_d)$ because of the more favourable σL_s characteristics of the d-axis model, and at rising load and frequency a balanced weighting of both components should be obtained. One possibility to achieve this characteristic is provided by the following approach for the weighting factor k_{dq}:

$$k_{dq}(\hat{i}_s, \omega_s) = \frac{k_R \hat{i}_m + k_L |\hat{i}_{sq}|}{\hat{i}_m + |\hat{i}_{sq}| + k_L \hat{i}_m} \quad \text{and} \quad k_{dq} \le 1 \tag{7.54}$$

with $\quad k_R = 1 - 2\dfrac{|\omega_s|}{\omega_{sN}} \quad$ and $\quad k_R \ge 0$

$$k_L = \frac{1}{2} \frac{|\omega_s|}{\omega_{sN}} \frac{\left|\hat{i}_{sq}\right|}{\hat{i}_m} \frac{I_0}{I_N}$$

The factor k_R has the effect that k_{dq} will be shifted in the direction (7.53) at low stator frequencies because k_L also is small at low frequencies. At rising load and frequency a balanced weighting of both fault components is reached by k_L.

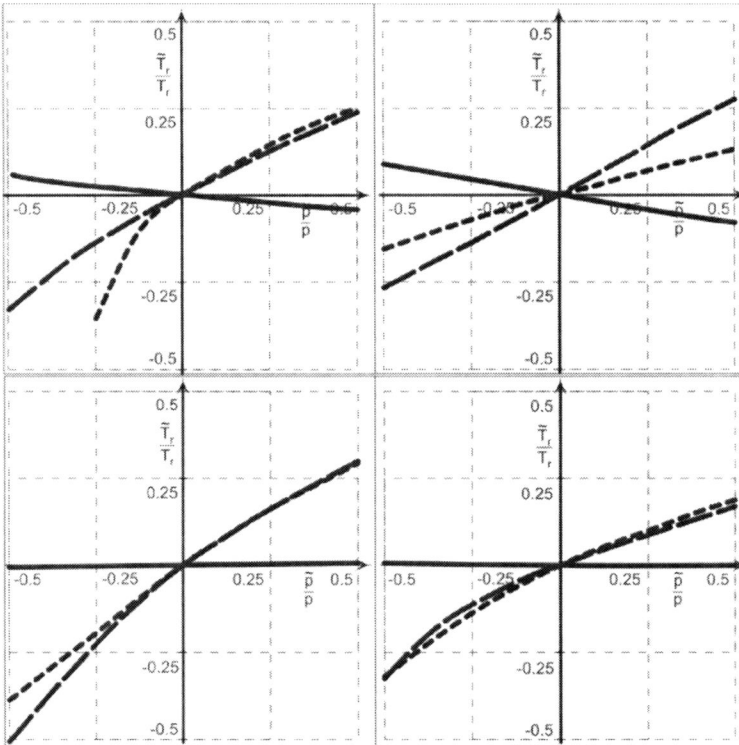

Fig. 7.8   Voltage vector fault model: —— $\tilde{R}_s / R_s$, - - - - $\tilde{L}_m / L_m$, ····· $\sigma\tilde{L}_s / \sigma L_s$; top: $\omega = 10\,\mathrm{s}^{-1}$, bottom: $\omega = 300\ \mathrm{s}^{-1}$, left: m = 73 Nm, right: m = 36 Nm

The obtained results are represented in the figure 7.8. The stator resistance sensitivity is reduced drastically compared to the individual methods, and the stability area extended significantly. The balance between leakage and main inductance sensitivity can be described as optimal in the upper frequency area. Also compared with the reactive power method, a reduction of the leakage inductance sensitivity at high frequency and strong load and of the main inductance sensitivity at weak load is established.

Altogether it is recognizable, that the influence of stator resistance inaccuracies can be suppressed almost completely. Despite compensation measures, the model errors of the inductances should not become greater than 10%, though.

7.3.4 Influence of the iron losses

The iron losses are a parameter which generally falls into the category "neglected or negligible quantity" at modeling for control design. This is generally justified because the iron loss resistance lying virtually parallel to the rotor resistance (cf. section 6.2.1) is about 1000 times greater than the rotor resistance, and the consequences of the neglection still remain acceptable for the normal operation of the field-orientated control. With inverter feeding and accordingly higher eddy current losses however, significant amplitude and phase errors of the rotor flux are already provable [Levi 1994]. Furthermore, for some operating states and control goals the perspective renders fundamentally different. One of these cases is the adaptation of the rotor time constant.

At first this can clearly be explained from the stationary equivalent circuit. The iron loss resistance is located quasi-parallel to the slip-dependent resistance R_r/s (section 6.2.1). Thus not R_r but the parallel connection of both resistances is estimated in reality. Particularly at small slip values near the no-load operation R_{fe} reaches the range of R_r/s and influences the estimation result significantly.

In order to approach the problem quantitatively the equation system introduced in section 6.2.1 is now pursued further. After some transformations the following stator and rotor voltage equations in the Laplace domain will be obtained assuming $R_s \ll R_{fe}$ (G_{fe} ... iron loss conductance):

$$\mathbf{u}_s = \frac{R_s \mathbf{i}_s + \sigma L_s \mathbf{i}_s \left(s + j\omega_s\right) + (1-\sigma)L_s \mathbf{i}_m \left(s + j\omega_s\right)}{1 + G_{fe}\sigma L_s \left(s + j\omega_s\right)} \tag{7.55}$$

$$0 = \left[1 + \left(s + j\omega_r\right)T_r\right]\mathbf{i}_m - \mathbf{i}_s + G_{fe}\mathbf{i}_s \tag{7.56}$$

The rotor voltage equation can be solved into real and imaginary components:

$$i_m = \frac{i_{sd} - G_{fe}u_{sd}}{1 + sT_r} \tag{7.57}$$

$$\omega_r = \frac{i_{sq} - G_{fe}u_{sq}}{T_r i_m} \tag{7.58}$$

The disorientating influence of the iron losses on the field orientation is owed primarily by the G_{fe} term in the slip equation. The R_{fe} model error produces an additional phase error, which also exists at no-load conditions and overlaps the phase errors produced by a rotor resistance model fault. Because all error models eventually derive their output error from the rotor flux phase error, the rotor resistance estimator may yield completely wrong results.

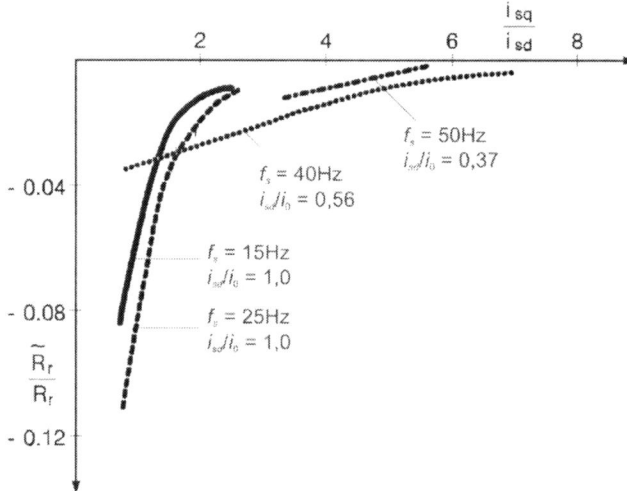

Fig. 7.9   Measured R_r estimation faults caused by iron losses

Some stationary measurements shall give a picture about the quantitative estimation error to be expected at different loads and stator frequencies. The results are shown in figure 7.9. For the R_r estimation the voltage vector error method is used. The ratio between torque and flux forming current components i_{sq}/i_{sd} serves as an equivalent for the motor load. It can clearly be recognized that the misadjustment is most critical at the upper limit of the constant flux area (greatest hysteresis losses). It diminishes considerably in the field weakening area, and also with increasing load. Altogether, it turns out that in stationary operation the estimation error can be safely kept below 4% if the adaptation is only allowed at current ratios of $i_{sq}/i_{sd} > 1,5$.

The conditions are more unfavorable in dynamic operation. The reason is that after starting a transient the error due to the rotor time constant difference is built-up delayedly, but the iron loss dependent error already exists in the no-load state. Therefore a restriction of the adaptation to appropriately great values of i_{sq} proves ineffective. Only the inclusion of the iron losses in the system equations according to the model in section 6.2.1 would make the additional error disappear completely.

7.3.5 Adaptation in the stationary and dynamic operation

At first the adaptation algorithms were developed for the stationary operation of the drive regarding current and rotor flux, and the parameter sensitivity examinations were also carried-out for this operation mode. Therefore it is interesting to examine, whether the methods can work stably also in dynamic operation and are able to adapt the rotor resistance. The discussion of the influence of the iron losses has already shown that different properties have to be expected in dynamic operation with regard to parameter sensitivity.

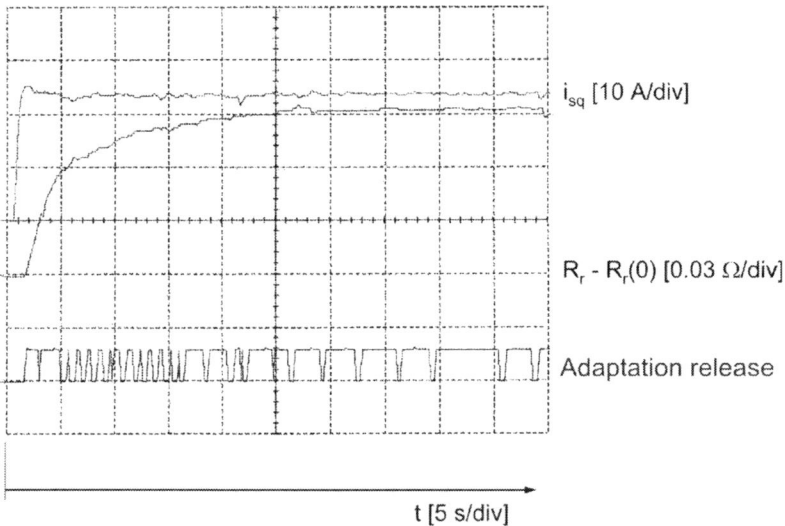

i_{sq} [10 A/div]

R_r - $R_r(0)$ [0.03 Ω/div]

Adaptation release

t [5 s/div]

Fig. 7.10 $\hat{R}_r$ adaptation cycles with voltage vector error method

At first the functionality of the adaptation shall, however, be illustrated by some examples during stationary operation. The oscillogram in the figure 7.10 shows a settling transient of the estimated rotor resistance after a load step. The initial error of the model rotor resistance is 30%. As already indicated in the previous section, the influences of the iron losses can be suppressed by switching-off the adaptation at insufficient torque. The detection of the stationary operation with adequate reliability is relatively easy by using a high pass filter for the torque forming current and the rotor flux. Together with the rotor resistance and the torque forming current the figure shows this adaptation release.

As pointed out above, the proportion between torque and flux forming currents is shifted by the rotor flux phase error at wrong model rotor resistance. In the case without adaptation a slow drift of the torque producing current in the model, caused by the warming-up of the machine, will be noticed at constant load until the thermal balance is reached. The effectiveness of the adaptation can thus be shown by the torque forming current keeping constant at constant load over long time. The oscillograms in the figure 7.11 show the corresponding plots during a warm-up process.

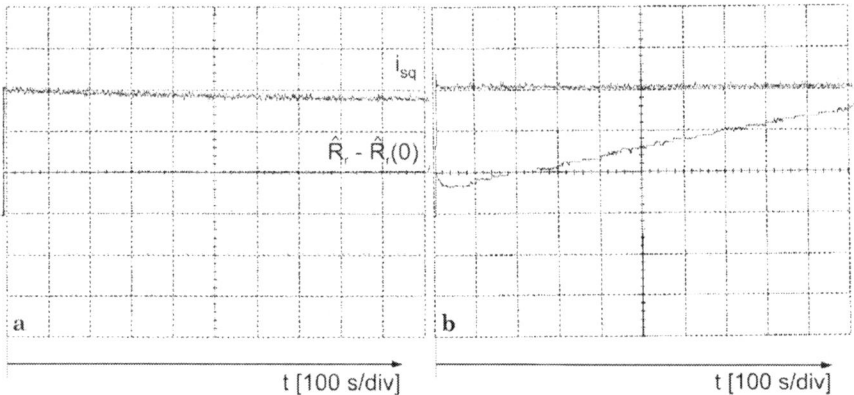

Fig. 7.11 $\hat{i}_{sq}$ and $\hat{R}_r$ at constant load, $\hat{i}_{sq}$ [10 A/div], $\hat{R}_r - \hat{R}_r(0)$ [0.03 Ω/div]: (**a**) without adaptation; (**b**) with adaptation

Similarly to the iron losses the parameter sensitivity to other model parameters also becomes more critical in the dynamic operation, and the demand to increase the precision of the error model increases. Like in the stationary operation, the influence of the leakage inductance is particularly strong and therefore shall be treated here with priority. Primarily this is caused by the following reasons:

- A dynamic speed change is connected to a short-time impression of a high torque forming current. The σL_s sensitivity assumes its most critical values just at high torque.
- Parameter errors in the error model have an immediate effect to the adaptation error. The model error of the rotor resistance, however, delayedly adds to the adaptation error through detuning of the phase angle $\tilde{\vartheta}_s$. Thus it is very probable at sufficiently short transients that the adaptation can only be activated (and detuned) by the model parameter errors.

For an appropriately exact adjustment of the error model the derived methods are definitely able to adapt the rotor resistance also in dynamic operation without an additional steady-state load torque. The figure 7.12 exemplarily shows such an adaptation process for the voltage vector error method recorded for longer time. The prerequisite is that the error equations, as done in section 7.3.2, are programmed using the dynamic machine equations.

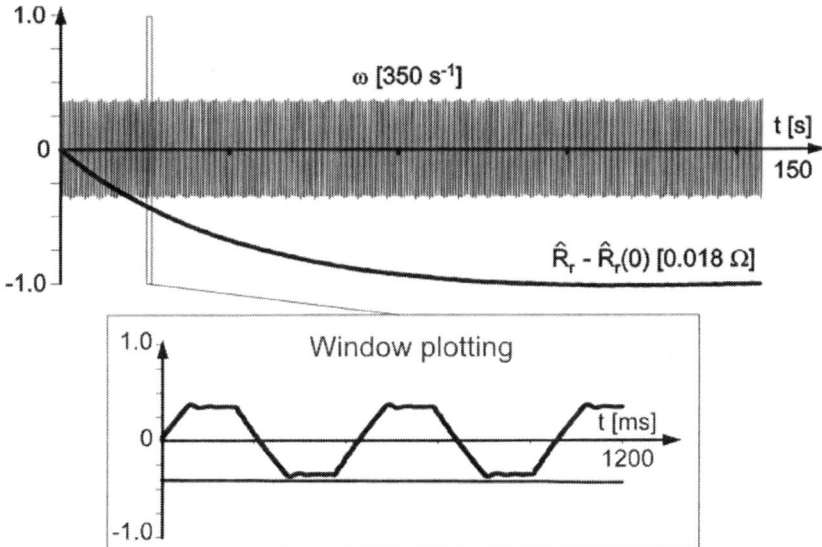

Fig. 7.12 Adaptation cycle with voltage vector error model, $\hat{R}_r(150\,s)=0.27\,\Omega$

The result of a comparison of different methods with regard to the sensitivity to leakage inductance changes and iron losses in dynamic operation is represented in the figure 7.13. The adaptation was excited by speed transients between 200 and 700 rpm. The error of the model value of the leakage inductance was +5% for all tests. The initial error of the model rotor time constant is zero.

The reference curve (curve 1) was taken with a simultaneous adaptation of rotor resistance and leakage inductance. The used method is not transferable to arbitrary operating states, though. It uses different properties of the model error contributions caused by the leakage inductance and the rotor resistance deviation in regenerative and motor operation.

The adaptation to the motor warm-up is already visible in the second part of the plots. As far as possible, the tests were recorded until achieving a stationary state of the adaptation.

Related to the final stationary value of curve 1 the following final deviations develop:

- Curve 2: -1.6 %
- Curve 3: -5.4 %
- Curve 4: -45 %

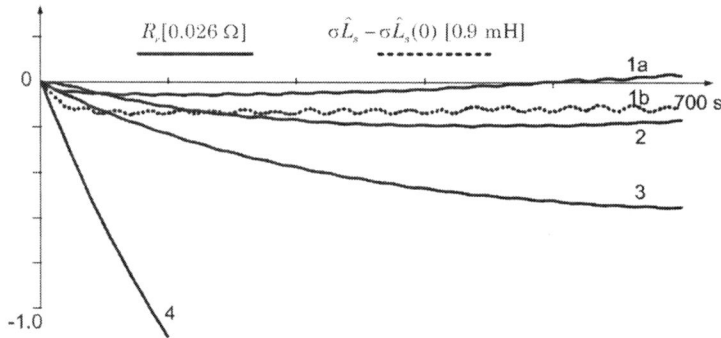

Fig. 7.13 $\hat{R}_r$ detuning caused by model error of leakage inductance and iron losses: curves (**1 – 3**) with compensation of iron losses, (**1**) combined adaptation of $\hat{R}_r$ and $\sigma\hat{L}_s$, voltage vector error, (**2**) reactive power method, (**3**) voltage vector error, (**4**) like curve (**3**), without compensation of iron losses

With that it is clear that an adaptation without consideration of the iron losses in dynamic operation must be considered impossible. The reactive power model proves to be the most suitable method for the dynamic operation here. The explanation can be found in that for this model the σL_s part of the flux phase error $\tilde{\vartheta}_s$, representing the rotor resistance deviation, is weighted approximately twice stronger than in the linear error model of the voltage vector method. This can be shown by deriving an error expression including all parameter errors according to (7.38) or (7.44) [Dittrich 1998]. Thus its share on the total error is increased with the consequence of a better suppression of parameter errors of the model.

7.4 References

Atkinson DJ, Acarnley PP, Finch JW (1991) Observers for Induction Motor State and Parameter Estimation. IEEE Transactions on Industry Applications, Vol. IA-27, pp. 1119 – 1127

Brodmann M (1994) Beobachterentwurf für nichtlineare zeitdiskrete Systeme. VDI-Fortschrittsberichte Nr. 416, VDI-Verlag Düsseldorf

Chai H, Acarnley PP (1992) Induction Motor Parameter Estimation Algorithm Using Spectral Analysis. IEE Proceedings-B, Vol. 139, No. 3, May, pp. 165 – 174

Dittrich JA (1994) Parameter Sensitivity of Procedures for Online Adaptation of the Rotor Time Constant of Induction Machines with Field-Oriented Control. IEE Proceedings-B, Vol. 141, November, pp. 353 – 359

Dittrich JA (1998) Anwendung fortgeschrittener Steuer- und Regelverfahren bei Asynchronantrieben. Habilitationsschrift, TU Dresden

Dolal D, Krishnan R (1987) Parameter Compensation of Indirect Vector Controlled Induction Motor Drive Using Estimated Airgap Power. Conference Record IEEE-IAS Annual Meeting, pp. 170 – 176

Du T, Brdys MA (1993) Implementation of Extended Luenberger Observers for Joint State and Parameter Estimation of PWM Induction Motor Drive. Proceedings EPE Brighton, pp. 4-439 - 4-444

Fetz J (1991) Parameter Adaptation for a Field Oriented Induction Machine Fed by a PWM-Inverter and Determination of the Fundamental Currents in the Range of Overmodulation. Proceedings EPE Firenze, pp. 138 – 144

Gabriel R (1982) Feldorientierte Regelung einer Asynchronmaschine mit einem Mikrorechner. Dissertation, TU Braunschweig

Ganji AA, Lataire P (1995) Rotor Time Constant Compensation of an Induction Motor in Indirect Vector Controlled Drives. EPE Sevilla, pp. 1.431 - 1.436

Gorter RJA, Veltman A, van den Bosch PPJ (1994) Parameter Estimation for Induction Motors, Using the Output-Error Identification Method. Proceedings of EPE Chapter Symposium Electric Drive Design and Applications, Lausanne, pp. 209 – 214

Holtz J, Thimm Th (1991) Identification of the Machine Parameters in a Vector-Controlled Induction Motor Drive. IEEE Transactions on Industry Applications, Vol. IA-27, No. 6, November/December, pp. 1111 – 1118

Hung TA, Lipo TA, Lorenz RD (1991) A Simple and Robust Adaptive Controller for Detuning Correction in Field Oriented Induction Machines. Wisconsin Electric Machines and Power Electronics Consortium, Research Report 91-17, May

Kazmierkowski MP, Sulkowski W (1986) Transistor Inverter Fed Induction Motor Drive with Vector Control System. Conference Record of the IEEE Industry Applications Society Annual Meeting, Part 1, pp. 162 – 168

Koyama M, Yano M, Kamiyama I, Yano S (1986) Microprocessor-Based Vector Control System for Induction Motor Drives with Rotor Time Constant Identification Function. IEEE Transactions on Industry Applications, Vol. IA-22, No. 3, May/June, pp. 453 – 459

Krishnan R, Pillay P (1986) Sensitivity Analysis and Comparision of Parameter Compensation Schemes in Vector Controlled Induction Motor Drives. Conference Record of the IEEE Industry Applications Society Annual Meeting, Part 1, pp. 155 – 161

Krishnan R, Bharadwaj AS (1991) A Review of Parameter Sensitivity and Adaptation in Indirect Vector Controlled Induction Motor Drive Systems. IEEE Transactions on Power Electronics, Vol. 6, No. 4, October, pp. 695 – 703

Levi E (1994) Detuned Operation of Field Oriented Induction Machines due to Iron Losses. Proceedings PCIM , Nürnberg, pp. 243 – 253

Lorenz RD, Lawson DB (1990) A Simplified Approach to Continuous Online Tuning of Field-Oriented Induction Machine Drives. IEEE Transactions on Industry Applications, Vol. IA-26, No. 3, May/June, pp. 420 – 424

Loron L (1993) Stator Parameters Influence on the Field-Oriented Control Tuning. Proceedings EPE Brighton, pp. 5-79 - 5-84

Nilsen R, Kaz;´mierkowski MP (1989) Reduced-Order Observer with Parameter Adaption for Fast Rotor Flux Estimation in Induction Machines. IEE Proceedings-B, Vol. 136, No. 1, January, pp. 35 – 43

Nomura, Tadashi, Masaguki, Nahamura (1987) A High Response Induction Motor Control System with Compensation for Secondary Resistance Variation. IEEE Power Electronics Specialists Conference Record, pp. 46 – 51

Pena RS, Asher GM (1993) Parameter Sensitivity Studies for Induction Motor Parameter Identification Using Extended Kalman Filters. Proceedings EPE Brighton, pp. 4-306 – 4-311

Pfaff G, Segerer H (1989) Resistance Corrected and Time Discrete Calculation of Rotor Flux in Induction Motors. Proceedings EPE Aachen, pp. 499 – 504

Reitz U (1988) Online-Berechnung der Parameter der Asychronmaschine bei schnell veränderlicher Belastung. Dissertation, TH Aachen

Rowan TM, Kerkman RJ, Leggate D (1991) A Simple Online Adaption for Indirect Field Orientation of an Induction Machine. IEEE Transactions on Industry Applications, Vol. IA-27, No. 4, July/August, pp. 720 – 727

Schrödl M (1989) Nachführung der Rotorzeitkonstanten von transient betriebenen Asynchronmaschinen mit Hilfe eines nichtlinearen Beobachterkonzepts. etzArchiv Bd. 11, H. 3, S. 83 - 88

Schumacher W, Leonhard, W (1983) Transistor-Fed AC Servo Drive with Microprocessor Control. International Power Electronics Conference Tokyo, March, pp. 1465 – 1476

Sng EKK, Liew AC (1995) On Line Tuning of Rotor Flux Observers for Field Oriented Drives Using Improved Stator Based Flux Estimator for Low Speeds. EPE Sevilla, pp. 1.437 – 1.442

Sumner M, Asher GM (1991) The Experimetal Investigation of Multi-Parameter Identification Methods for Cage Induction Motors. Proceedings EPE Firenze, pp. 389 – 394

Sumner M, Asher GM, Pena R (1993) The Experimetal Investigation of Rotor Time Constant Identification for Vector Controlled Induction Motor Drives During Transient Operating Conditions. Proceedings EPE Brighton, pp. 5-51 – 5-56

Vogt G (1985) Digitale Regelung von Asynchronmotoren für numerisch gesteuerte Fertigungseinrichtungen. Springer-Verlag

Vucosavic;´ SN, Stojic;´ MR (1993) Online Tuning of the Rotor Time Constant for Vector-Controlled Induction Motor in Position Control Applications. IEEE Transactions on Industrial Electronics, Vol. IE-40, No. 1, February, pp. 130 – 137

Weidauer J, Dittrich JA (1991) A New Adaptation Method for Induction Machines with Field-Oriented Control. Proceedings EPE Firenze

Zai L, Lipo T (1987) An Extended Kalman Filter Approach to Rotor Time Constant Measurement in PWM Induction Motor Drives. Conference Record IEEE-IAS Annual Meeting, pp. 176 – 183

Zeitz M (1979) Nichtlineare Beobachter. Regelungstechnik 27 (1979) 8, S. 241 - 249

8 Optimal control of state variables and set points for IM drives

8.1 Objective

At the design of drives an energetically optimal operation represents an essential point of view. Losses increase the energy requirement and produce heat which must be dissipated by additional measures and constructive efforts. Modern power electronic devices achieve efficiencies of 98%, motors of medium and high-power ratings of over 95% at the nominal working point. A different picture arises in the partial load area where the efficiency can decline considerably. Besides optimization possibilities in the hardware sector and the use of loss-optimized pulse pattern for inverter control, also the "soft" control faction is challenged to come forth with *approaches for an efficiency optimized operation.* In order to keep the analytical and realization effort within reasonable limits, only *stationary or quasi-stationary solutions* are examined.

Another question arises from the technically existing limitations of the hardware equipment with regard to currents and voltages. The control of the state variables should be designed for the drive or the motor to always being utilized as optimal as possible.

The method of the field orientation provides the tools to realize a decoupled control of rotor flux and torque by impressing torque and flux forming current components. In a speed controlled system the set point of the torque forming current is provided by the speed controller, in a torque controlled system it is an independent control quantity. Thus *the amplitude of the rotor flux or the ratio of both components, the slip frequency, remains as a degree of freedom for the optimization.*

As shown in the next sections, the exact knowledge of difficult measurable machine parameters is required for an effective optimization of the efficiency, or this optimization can only be implemented with reasonable effort by a dynamically slow control algorithm. For this reason a second optimization approach, the *torque optimal* control, becomes interesting. The optimization goal consists here to control machine and inverter in the best possible way *from the point of view of torque production* at the given limitations, i.e. the demanded torque has to be generated by the minimal current, or the maximum torque has to be provided at limited current or limited voltage. Such an optimization strategy also will deliver a good efficiency because of the current dependency of the ohmic losses although this does not represent the optimization goal in the first place.

8.2 Efficiency optimized control

At first it is required to perform an analysis of the controllable losses. Losses in the motor appear in the form of stator and rotor copper losses, iron losses and additional losses. Additional losses are produced in the iron and copper by the non-sinusoidal field distribution, and they can be calculated with a factor of approximately 0.3 proportionally to the copper losses [Murata 1990]. A quantitative expression for the copper losses can be derived from the active power equation:

$$p_w = \text{Re}\left\{\mathbf{u}_s \mathbf{i}_s^*\right\} = \frac{3}{2}\left(u_{sx}i_{sx} + u_{sy}i_{sy}\right)\,^1 \qquad (8.1)$$

After replacing the voltages with the help of the stator voltage equation (cf. chapter 3 and 6) this equation can be re-written in field-orientated coordinates for stationary operation in the following form:

$$p_w = \frac{3}{2}\left[R_s i_{sd}^2 + \left(R_s + (1-\sigma)R_r\right)i_{sq}^2 + (1-\sigma)\omega L_s i_{sd}i_{sq}\right] \qquad (8.2)$$

The copper losses including one part for the additional losses added with the factor k_z can be separated to:

$$p_{Cu} = \frac{3}{2}(1+k_z)\left\{R_s i_{sd}^2 + \left[R_s + (1-\sigma)R_r\right]i_{sq}^2\right\} \qquad (8.3)$$

According to the section 6.2.1, the iron losses can be calculated approximately to:

$$p_{Fe} = \frac{3}{2}\frac{\left(\omega_s \psi_\mu\right)^2}{R_{fe}} \qquad (8.4)$$

Using $R_{fe} \approx R_{feN}\,\omega_s/\omega_{sN}$ (cf. also section 6.2.1) and $\psi_\mu \approx L_m i_{sd}$ it can be finally written for the total losses:

---

[1] *xy* can be either *dq* or $\alpha\beta$

$$p_v = \frac{3}{2}\left\{\left[\left(1+k_z\right)R_s + \frac{\omega_{sN}L_m^2}{R_{feN}}\omega_s\right]i_{sd}^2 + \left(1+k_z\right)\left[R_s + \left(1-\sigma\right)R_r\right]i_{sq}^2\right\}$$

(8.5)

Therefore the total losses can be split into an i_{sd}-dependent (flux dependent) and a torque (or i_{sq}-) dependent part, in which the partition is defined by the parameters of the machine, and the flux dependent part is a function of the stator frequency:

$$p_v = a\left(\omega_s\right)i_{sd}^2 + bi_{sq}^2 = p_{v1}(i_{sd}) + p_{v2}(i_{sq})$$

(8.6)

With the side condition of a constant or given torque:

$$m_M \sim i_{sd}i_{sq} = \text{const}$$

the condition:

$$p_{v1}(i_{sd}) = p_{v2}(i_{sq})$$

(8.7)

follows for the minimal total losses, i.e. the flux dependent and the torque dependent loss components must have the same value.

A clear representation of the relationship between both loss parts can be derived if it is referred to the slip frequency. The corresponding equations are obtained similarly as above:

$$p_{Cu} = \frac{3}{2}(1+k_z)(1-\sigma)L_s i_{sd}i_{sq}\left[\left(1+\frac{R_s}{(1-\sigma)R_r}\right)\omega_r + \frac{R_s R_r}{L_m^2}\frac{1}{\omega_r}\right]$$

(8.8)

The factored out term is proportional to the torque and can be treated as a constant for the further calculation. For the iron losses the more exact two-parameter model from the section 6.2.1 is used now which leads to the following relation:

$$p_{Fe} = \frac{3}{2}(1-\sigma)L_s i_{sd}i_{sq}\left[R_r\left(k_{hy} + k_w\omega\right)\left(2 + \frac{\omega}{\omega_r}\right) + 2R_r k_w\omega_r\right]$$

(8.9)

The slip-dependent loss balance at two speeds with the nominal and half the nominal torque is represented corresponding to the equations (8.8) and (8.9) for an 11kW standard motor in the figure 8.1. Because of the main flux getting smaller by an increasing slip at the same torque, the iron losses behave inversely to the slip frequency. The copper losses drastically increase at small slip because of the magnetization current demand strongly increasing with higher saturation. They show a minimum and increase once more at the slip getting greater. The optimal point with respect to the total losses depends on the respective share of the iron losses. The operation at nominal speed represents the operating point with the greatest part of the iron losses in the total losses, because here the maximum stator frequency without field weakening is reached.

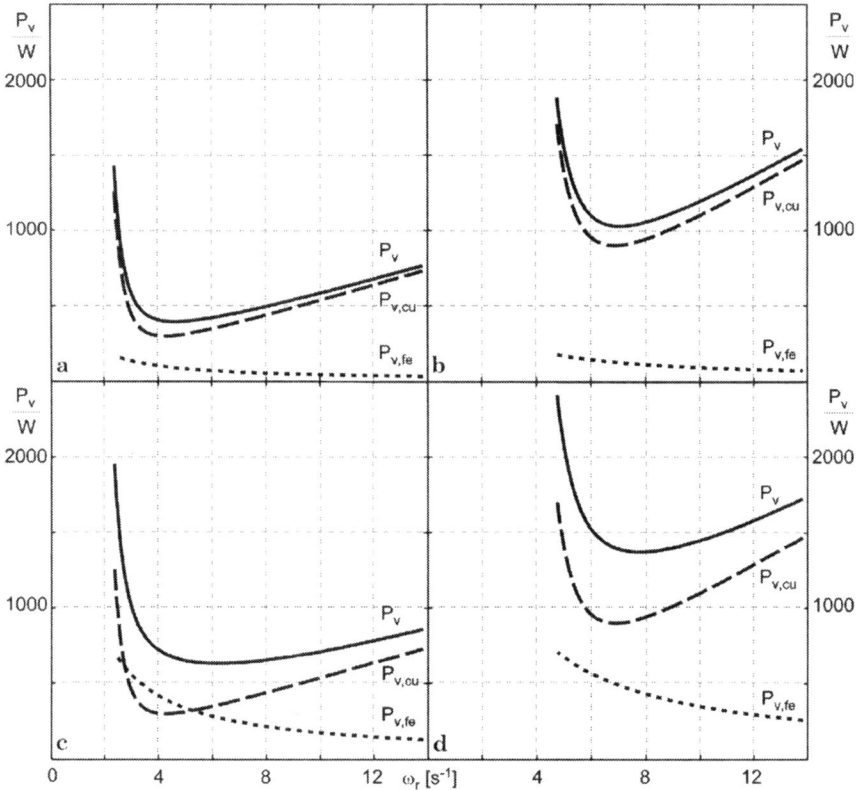

Fig. 8.1   Losses as a function of the rotor frequency: (**a**) $m=0.5m_N$, $n=500rpm$, (**b**) $m=m_N$, $n=500rpm$, (**c**) $m=0.5m_N$, $n=1500rpm$, (**d**) $m=m_N$, $n=1500rpm$

The explanations so far open up different possibilities for the practical realization of an optimal control strategy with respect to efficiency, in which the optimization goal is predefined by equation (8.7). Two variants shall be discussed in more detail.

a) Balancing of torque and flux dependent losses

The equation (8.7) shows the way for a direct control of the balance between the two parts. The method is schematically represented in the figure 8.2 (cf. [Rasmussen 1997]). The flux dependent losses can be directly controlled by the rotor flux without influencing the torque dependent losses. According to the equations (8.3) and (8.4) model values of the two loss parts are calculated. The difference of both forms the input quantity (control difference) for an I or PI controller which adjusts the equality of both parts with the rotor flux set point as a control variable.

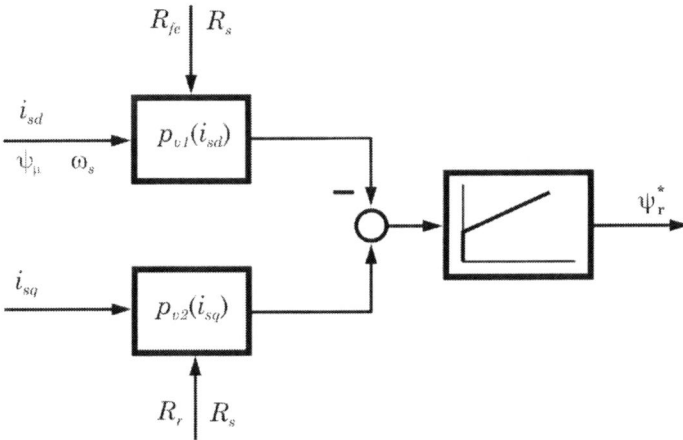

Fig. 8.2   Loss optimization by balancing of torque and flux dependent parts

A comparatively fast dynamics is achieved for adjusting the optimum, an exact compensation requires, however, an exact knowledge of the model parameters, which is difficult and requires some effort particularly with regard to the iron losses.

b) Loss compensation with search algorithm

The active power is calculated by equation (8.1), and by means of a search algorithm the rotor flux is modified until the minimum of the active power and with that the minimum of the losses is reached. Different search strategies are applicable with fixed or variable step, cf. e.g. [Moreno 1997]. A careful adjustment is required because of possible convergence problems. Such a method does not need any model parameters. Thus the power could also be measured at the input of the inverter, and the inverter losses could be included into the optimization. Caused by the searching method and the at first unknown "suitable" adjustment direction of the rotor flux, the method works slowly in this simple implementation, and is suitable exclusively for steady-state operation.

8.3 Stationary torque optimal set point generation

8.3.1 Basic speed range

Initially it shall be noted, that the relations discussed in this section are not exclusively limited to the basic speed range. They are valid everywhere where *no limitation of the stator voltage* becomes effective, thus also in the *field weakening area at low load*. The derivations start out, however, from the basic speed range initially because no voltage limitation occurs here in stationary operation also at the current limit.

Under this presumption maximum torque at given stator current amplitude will be achieved if the operating point is always on the maximum of the slip-torque-characteristic (cf. figure 8.3). This maximum corresponds graphically to the breakdown torque of the known torque-speed-characteristic.

The torque equation is with consideration of the main field saturation in stationary operation (cf. chapter 3):

$$m_M = \frac{3}{2} z_p \frac{L_m(i_\mu)^2}{L_m(i_\mu) + L_{r\sigma}} i_{sd} i_{sq} \tag{8.10}$$

It is easily comprehensible from (8.10) that the maximum torque will be reached at a given stator current with $i_{sd} = i_{sq}$ for constant inductances. Because of the magnetic saturation the calculation of the maximum point becomes, however, essentially more troublesome and requires the iterative solution of a nonlinear system of equations. Parts of this system are besides (8.10) the relation of the magnetization current amplitude (cf. section 6.2.3):

$$i_\mu = \sqrt{i_{sd}^2 + \left(\frac{L_{r\sigma}}{L_r} i_{sq}\right)^2} \tag{8.11}$$

the slip equation:

$$\omega_r = \frac{R_r}{L_m(i_\mu) + L_{r\sigma}} \frac{i_{sq}}{i_{sd}} \tag{8.12}$$

and the boundary condition:

$$i_s^2 = i_{sd}^2 + i_{sq}^2 \tag{8.13}$$

The figure 8.3 shows the calculated characteristics for an 11kW standard motor. The characteristic which would be obtained with constant main inductance (at the nominal working point) is drawn for comparison as dashed line.

As mentioned above, it would be the task of a torque optimal control to control the rotor flux in a way to keep the operating point always on the maximum of the torque-slip-characteristic depending on the demanded or available stator current. An online calculation of this point, however, practically has to be excluded because of the necessary iteration. Therefore it would be interesting to know, how the usual field-orientated operation with constant flux (nominal flux) fits into this analysis. The nominal value of the flux forming current would be calculated by:

$$I_{sdN} = \frac{U_N}{\sqrt{3}\omega_{sN} L_s (I_{sdN})} \qquad (8.14)$$

Equation (8.14) contains an iteration which would have to be solved in the practical implementation, but in the initialization phase and not in real time.

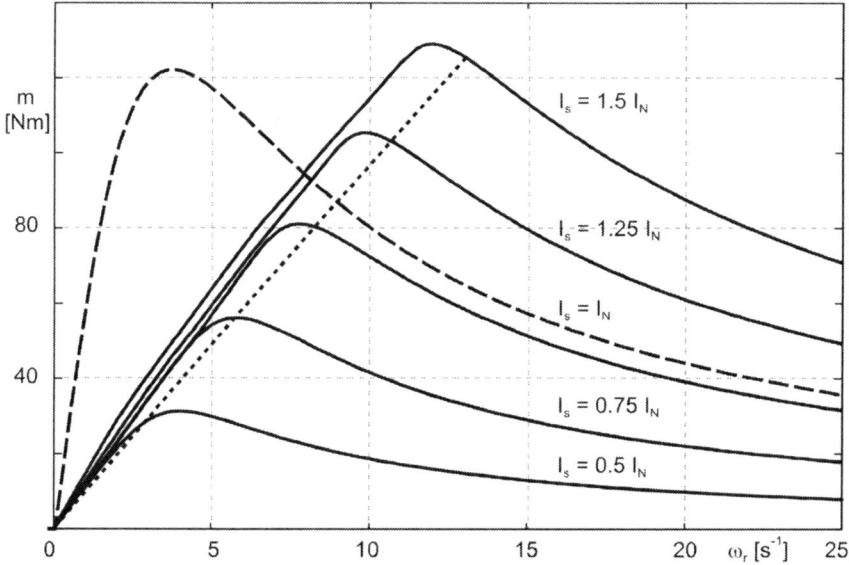

m
[Nm]

$I_s = 1.5\ I_N$

80

$I_s = 1.25\ I_N$

$I_s = I_N$

40

$I_s = 0.75\ I_N$

$I_s = 0.5\ I_N$

0   5   10   15   20 ω_r [s^{-1}]   25

Fig. 8.3   Slip – torque – characteristic as function of the stator current amplitude: ---- L_m = const; ..... $i_{sd} = i_{sdN}$

The corresponding curve is drawn as dotted line in figure 8.3 and shows, at least at higher currents, a surprisingly good approximation to the optimal value. Therefore the constant flux operation is obviously distinguished as a quasi torque optimal control strategy in the basic speed range at higher stator currents. This connection is understandable, because the motor is designed for the rated working point.

This is further illustrated in the figure 8.4. In figure a) the necessary flux forming current to achieve the exact torque maximum and additionally the control characteristics for constant flux and for $i_{sd} = i_{sq}$ are drawn. In the linear area the optimal characteristic coincides with the characteristic $i_{sd} = i_{sq}$ as expected. For higher stator currents the optimal characteristic deviates from the constant flux characteristic with a tolerance of approximately ±20%. The figure 8.4b shows the actual effects of the deviations on the torque. Obviously they are negligible in this case.

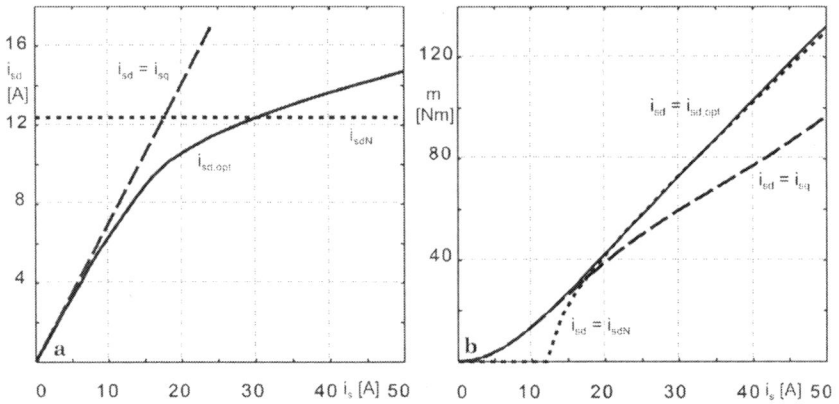

Fig. 8.4 (**a**) i_{sd} control characteristic, and (**b**) maximum torque as function of stator current and flux control

Under consideration of a minimal rotor flux which has to be kept to ensure a torque generation with an acceptable dynamics, the following simple control law can be used for the quasi torque optimal control:

$$\frac{\psi_{rd}^*}{L_m} = i_m^* = \begin{cases} i_{m,\min} & \text{for } i_{sq,f} < i_{m,\min} \\ i_{sq,f} & \text{for } i_{m,\min} \leq i_{sq,f} < i_{mN} \\ i_{mN} & \text{for } i_{sq,f} \geq i_{mN} \end{cases} \quad (8.15)$$

Here $i_{sq,f}$ is the low-pass filtered current i_{sq}. Because of this filtering and the anyway existing delay in the forming of the rotor flux, a dynamic decoupling is given between flux control and i_{sq} control. It has to be taken into account, however, that a variable flux inevitably leads to a deterioration of the torque dynamics. If a high torque dynamics represents the central optimization goal, the constant flux operation has to be maintained over the complete basic speed range. The torque dynamics then only depends on the dynamics of the current impression. This is illustrated by the figure 8.5 with some transients for constant flux operation and the described flux control algorithm. In addition, it is obvious that an approximately optimal operation mode with respect to efficiency in this area is not conceivable any more with fast flux tracking in the dynamic operation.

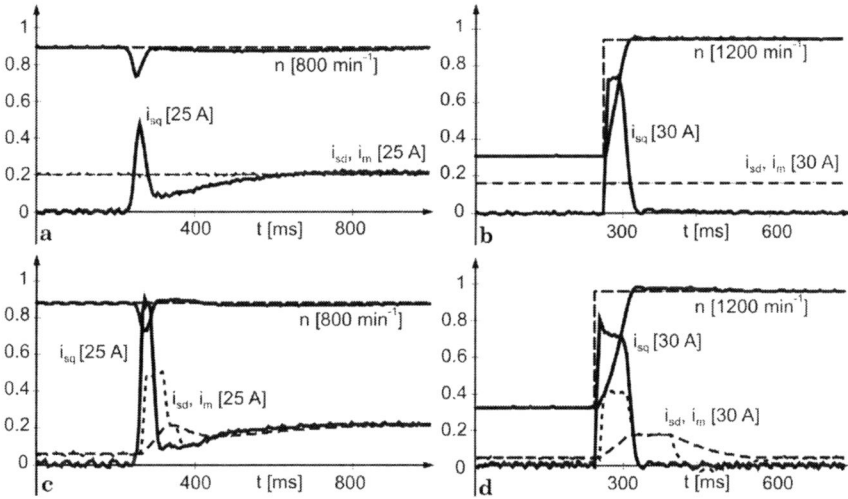

Fig. 8.5   Speed dynamics in the basic speed range: (**a**) Load step change with constant flux, (**b**) Set point step with constant flux, (**c**) Load step with torque optimal controlled flux, (**d**) Set point step with torque optimal controlled flux

8.3.2 Upper field weakening area

Different to the basic speed range, the limitation of the stator voltage represents a decisive additional influencing variable for the flux control in the field weakening. Two areas must be distinguished: The first area, in which the voltage limitation is the only deciding limiting variable, and a transition zone, in which both current and voltage limitation determine the character of the control characteristics. At first, only the voltage limitation shall be taken into account as a boundary condition, and the currents shall be assumed to develop freely.

Similar to the current-limited case, typical speed (slip) over torque characteristics can be calculated which contain the speed as a parameter. Because the calculations are only significant for the high field weakening area, the saturation can be neglected.

From the stator voltage equations in the field-orientated coordinate system (cf. chapter 3 and 6) the following equations will be obtained for steady-state operation with respect to the stator currents:

$$u_{sd} = R_s i_{sd} - \omega_s \sigma L_s i_{sq} \tag{8.16}$$

$$u_{sq} = R_s i_{sq} + \omega_s \sigma L_s i_{sd} + \omega_s (1 - \sigma) L_s i_m \tag{8.17}$$

with: $i_m = \dfrac{\psi_{rd}}{L_m}$

For constant rotor flux it follows from (8.17):

$$u_{sq} = R_s i_{sq} + \omega_s L_s i_{sd} \tag{8.18}$$

The system boundary condition is here:

$$u_{max}^2 = u_{sd}^2 + u_{sq}^2 \tag{8.19}$$

If the slip equation in field orientated coordinates (8.12) is inserted into (8.16) and (8.18) and the current components are eliminated, the following torque equation will be obtained:

$$m_M = \frac{3}{2} z_p \frac{L_m^2}{L_r} \frac{u_{max}^2}{R_s^2} \frac{\omega_r T_r}{\left(\omega_r T_r + \omega_s T_s\right)^2 + \left(1 - \omega_s \sigma T_s \omega_r T_r\right)^2} \tag{8.20}$$

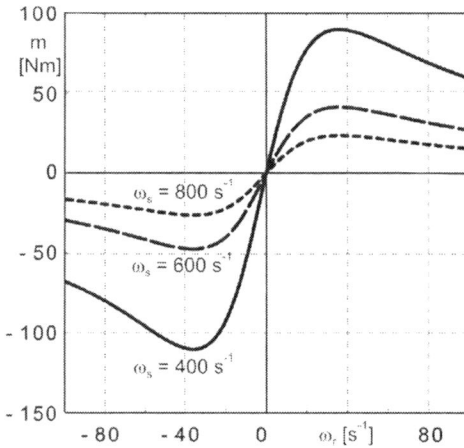

Fig. 8.6 Slip-torque characteristics at limited stator voltage; parameter: stator frequency

The equation (8.20) is represented in the figure 8.6 for the stator frequency as a parameter. This case corresponds to the natural behavior of the induction machine at frequency control. In contrast to the constant current case, differently high torque maxima, which are also characterized by differently high stator currents, appear for motor and regenerative operation. The slip frequency at the torque maximum or break-over point is calculated by (8.20):

$$\omega_{r,kipp} = \pm \frac{1}{T_r} \sqrt{\frac{1 + \left(\omega_s T_s\right)^2}{1 + \left(\sigma \omega_s T_s\right)^2}} \approx \frac{1}{\sigma T_r} \text{ for } \left(\sigma \omega_s T_s\right)^2 >> 1 \tag{8.21}$$

For speed controlled operation with the speed as a parameter the characteristics represented in figure 8.7 apply. In regenerative operation an absolute maximum at $\omega_s = 0$ appears. This means that the greatest regenerative braking torque can obviously be generated at a (negative) direct current feeding. The torque maximum itself, however, might be barely possible to be used, because it is connected to impractical high currents.

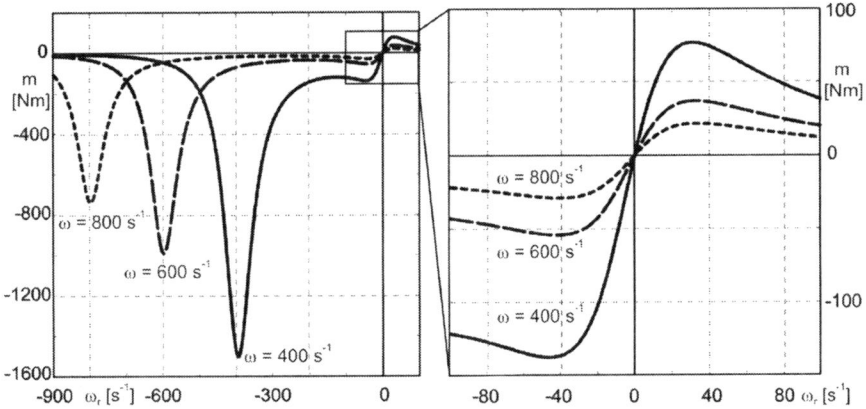

Fig. 8.7   Slip-torque characteristics at limited stator voltage; parameter: speed

Analog to the constant current operation, a torque optimal control law should adjust the currents for the maxima of the slip-torque characteristic. The solutions for the control characteristic will be obtained by an extreme value problem for the torque given by the equations (8.16), (8.18), (8.19) and the torque equation (8.10).

Because the solutions are interesting only for high stator frequencies the stator resistance can be neglected. This simplification actually makes the equation system accessible for a closed solution. Furthermore it is presupposed that because of the field weakening the main inductance can be regarded as constant.

At first, (8.16) and (8.18) have to be dissolved into components. With the mentioned simplification the following current equations are obtained (cf. equation (5.78)):

$$i_{sd} = \frac{u_{sq}}{\omega_s L_s} \tag{8.22}$$

$$i_{sq} = -\frac{u_{sd}}{\omega_s \sigma L_s} \tag{8.23}$$

Strictly speaking, the two equations also contain on the right side the current components i_{sd} and i_{sq} implicitly through the slip frequency, and would have to be solved further to equations with the speed as parameter. This would however make a closed solution impossible, whence the more transparent variant with the stator frequency as a parameter shall be kept. The arising error is tolerable because the share of the slip frequency in the stator frequency is small at high rotational speeds. An advantage of this consists of the fact that the same control characteristics are valid for both motor and regenerative operation because of the symmetrical position of the torque maxima with respect to the slip frequency (cf. figure 8.6).

After inserting (8.22) and (8.23) into the torque equation the following Lagrange function for the extreme value calculation can be formulated by inclusion of (8.19):

$$L(u_{sd},u_{sq},\lambda) = m(u_{sd},u_{sq}) + \lambda(u_{max}^2 - u_{sd}^2 - u_{sq}^2) \tag{8.24}$$

From the partial derivatives with respect to u_{sd} and u_{sq} the following equation system is obtained:

$$0 = k_m \left(i_{sq} \frac{\partial i_{sd}}{\partial u_{sd}} + i_{sd} \frac{\partial i_{sq}}{\partial u_{sd}} \right) - 2\lambda u_{sd} \tag{8.25}$$

$$0 = k_m \left(i_{sq} \frac{\partial i_{sd}}{\partial u_{sq}} + i_{sd} \frac{\partial i_{sq}}{\partial u_{sq}} \right) - 2\lambda u_{sq} \tag{8.26}$$

with: $k_m = \frac{3}{2} z_p \frac{L_m^2}{L_r}$

After solving to the current components it yields the control characteristics:

$$i_{sd,lim} = \frac{u_{max}}{\sqrt{2}\omega_s L_s} \tag{8.27}$$

$$i_{sq,lim} = \frac{u_{max}}{\sqrt{2}\omega_s \sigma L_s} \tag{8.28}$$

The figure 8.8 shows these characteristics for an 11kW motor together with the characteristics calculated by means of search method without the above mentioned approximation. They are identical for both motor and regenerative operation. The maximum inverter current i_{max} and the no-load current (the magnetizing current) i_0 are also included in the plots.

In order for the torque optimal control strategy to be effective, both current components must be able to develop freely. As shown in the diagrams, this depends fundamentally on the maximum inverter current, and for the sample drive this would be the case above approximately $\omega_s = 450$ s^{-1}. If the rotor flux is controlled below this frequency by the derived characteristic too, no torque optimal operation is achieved, because the inverter voltage cannot be utilized.

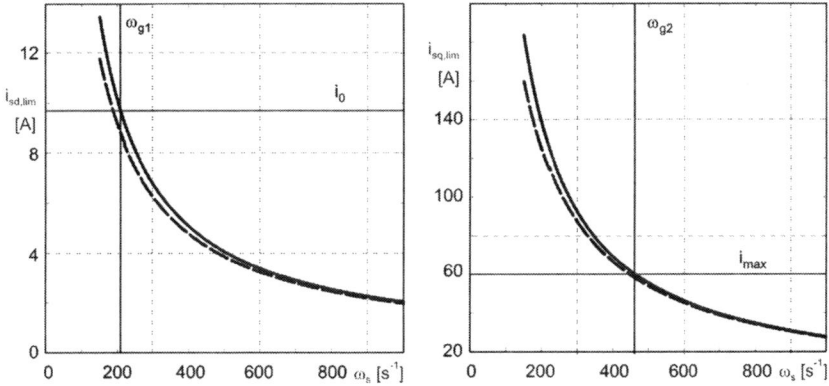

Fig. 8.8   Current limit characteristics for maximum torque at limited voltage: ---
exact curve with searching method, — approximate solutions (8.27) and (8.28)

Thus it is also shown that the flux with high probability at the cut-in
point of the torque optimal control characteristic may already be weakened
and the operating point having been shifted to the linear part of the
magnetization characteristic. Therefore the neglection of the saturation for
the derivation of the characteristics is justified.

8.3.3 Lower field weakening area

The lower field weakening area shall be understood as the zone between
the frequencies ω_{g1} and ω_{g2} (cf. figure 8.8). This area is indicated by the
following characteristics:

- Coming from lower frequencies, the stator voltage reaches its maximum
  value so that a flux reduction is needed to continue the frequency
  increase.
- An operation using the torque optimal control characteristics is not
  possible or expedient, because the torque forming current i_{sq}
  corresponding to these characteristics is either not needed or cannot be
  produced.

Without voltage limitation the drive would be controlled according to
the rules of the basic speed range, hence with $i_{sd} = i_{sdN}$. Thus it seems
reasonable to operate the control system as close as possible at this set
point in the lower field weakening area, meaning to operate the drive with
the maximum possible flux at the voltage limit. This rule is well known
and general practice.

For the implementation a voltage regulator is often used (figure 8.9).
The actual stator voltage feedback can be calculated from the (unlimited)
current controller output signal via low-pass filter to eliminate transient

parts. With that it is possible to keep the voltage always at the limiting level in stationary operation independent of motor parameters. This method has, however, also decisive disadvantages. It is attempted *to control a very fast variable quantity (stator voltage) by a slowly variable quantity (rotor flux)*. This demands for an artificial delay of the voltage dynamics and makes it impossible to react to variable operating states (load, acceleration) with an adequately fast change of the flux set point. For this reason a feed-forward controlled flux set point calculation shall be derived in the following.

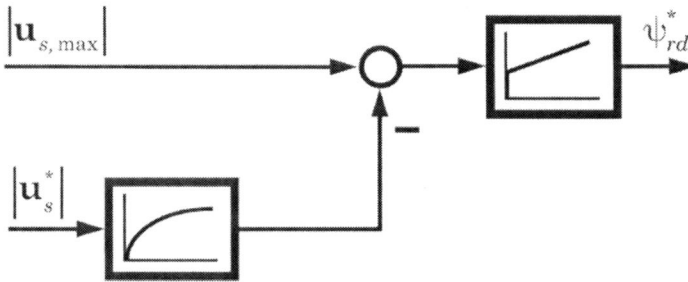

Fig. 8.9   Flux set point calculation using voltage controller

The stationary voltage equations (8.16) and (8.18) are again the starting point. The solution of the implicit relation current/stator frequency is refrained from because in this case a closed solution would require too far-reaching approximations. Both equations are squared and added up. The stator voltage amplitude is equated to the maximum voltage, and the equation is solved to $i_{sd} = i_m$. One obtains:

$$i_m^* = -\frac{(1-\sigma)R_s\omega_s L_s}{R_s^2 + (\omega_s L_s)^2} i_{sq} + \sqrt{\frac{u_{max}^2}{R_s^2 + (\omega_s L_s)^2} - \left[\frac{R_s^2 + \sigma(\omega_s L_s)^2}{R_s^2 + (\omega_s L_s)^2} i_{sq}\right]^2}$$

(8.29)

For $R_s^2 \ll \sigma(\omega_s L_s)^2$, which should be fulfilled in the interesting frequency area ($\omega_s > 300\ \text{s}^{-1}$), this relation can further be simplified. At the same time the current i_{sq} is replaced by its set point with regard to the practical implementation which enables a faster reaction to forthcoming i_{sq} changes:

$$i_m^* = -\frac{(1-\sigma)R_s}{\omega_s L_s} i_{sq}^* + \sqrt{\frac{u_{max}^2}{(\omega_s L_s)^2} - \left(\sigma i_{sq}^*\right)^2}$$

(8.30)

A pure flux set point control in an open-loop has of course the disadvantage of the parameter dependency which here would have the effect, that the available stator voltage would not be utilized in stationary operation, or the flux set point would be adjusted too high. Therefore it is useful to combine both methods, voltage controller and flux feed-forward control, in a suitable way. In this combination, the voltage controller has the task of keeping the voltage at the operating limit during stationary operation, and the open-loop set point control takes care that changes of the i_{sq} set point can be answered quickly with the corresponding flux change. The voltage controller should control a quantity corresponding to its input signal. The maximum voltage u_{max} in (8.30), which is regarded as variable now, would be such a suitable quantity.

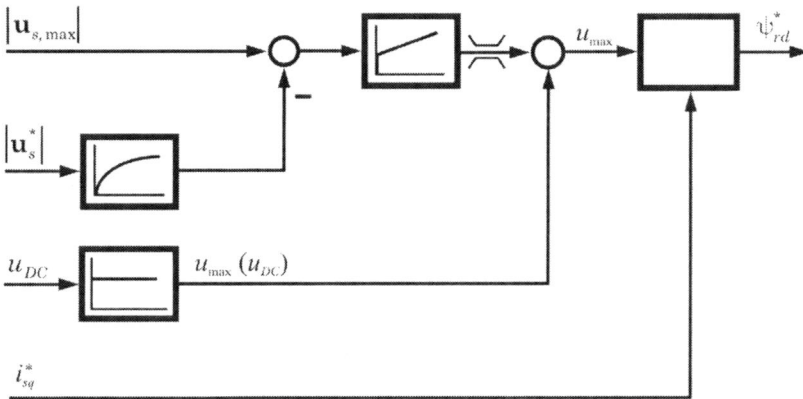

Fig. 8.10 Flux set point calculation using a combination of voltage (feedback) controller and set point control in open-loop

The figure 8.10 shows the correspondingly modified structure. The set point $u^*_{s,max}$ corresponds to the maximum output voltage of the current controller. The control variable of the voltage controller is added to the maximum stator voltage $u_{max}(u_{DC})$ [1] calculated from the DC link voltage. The sum of both forms the input quantity u_{max} of the control equation.

Because the control equation (8.30) was derived in exclusively algebraic way, the saturation does not influence the result. Operating point dependent parameters can be adapted on-line by an open-loop control. Remaining differences are stationarily compensated by the voltage controller.

---

[1] u_{DC} = DC link voltage

8.3.4 Common quasi-stationary control strategy

In the previous sections the theory for torque optimal control strategies has been outlined for basic speed, upper and lower field weakening area or limitation of current and/or voltage. For the implementation in a control system there are still the following additional tasks to be addressed:

- A common strategy which provides a continuous transition between the areas is to find.
- The developed strategy must be usable in a control structure, such as described in the section 1.2.

From the previous considerations the following conclusions can be summarized:

- The rules developed for the basic motor speed range can be generalized for all operating states in which no voltage limitation appears.
- The rules for the two field weakening areas show that the best utilization of the machine is always given at a maximum voltage output.
- The current limit characteristics in the upper field weakening area show such a relationship, that the limit characteristic of the torque forming current $i_{sq,lim}$ can be equated to the maximum value of the stator current with good approximation.

Therewith the following rules can verbally be formulated:

- *Rule* 1: The torque and flux forming current components are equated as long as either the flux forming current reaches its maximum value (nominal value) or the stator voltage goes into the limitation.
- *Rule* 2: If the torque forming current amplitude exceeds the flux forming current, either the flux forming current remains on its nominal value or is controlled to keep the drive always on the voltage limit.
- *Rule* 3: The stator current is limited to either its absolute maximum or the limit characteristic of the torque forming current in the upper field weakening area, depending on which of both quantities has the smaller value.

With the nominal value of the rotor flux linkage i_{mN} and the maximum inverter current i_{max} the rules can be summarized in equation form as follows:

$$i_m^* = \begin{cases} i_{m,\min} & \text{for } i_{sq,f} < i_{m,\min} \\ i_{sq,f} & \text{for } i_{m,\min} \le i_{sq,f} < i_{mN} \text{ and } i_{sq,f} < i_m^*(u_{\max},i_{sq}^*,\omega_s) \\ i_{mN} & \text{for } i_{sq,f} \ge i_{mN} \text{ and } i_{mN} < i_m^*(u_{\max},i_{sq}^*,\omega_s) \\ i_m^*(u_{\max},i_{sq}^*,\omega_s) & \text{otherwise} \end{cases}$$

$$(8.31)$$

$$i_{s,\max} = \begin{cases} i_{\max} & \text{for } i_{sq,\lim} > i_{\max} \\ i_{sq,\lim} & \text{otherwise} \end{cases} \tag{8.32}$$

The quantity $i_m^*(u_{\max}, i_{sq}^*, \omega_s)$ results from the equation (8.30). The rules or the control laws start out from the assumption that the set point i_{sq}^* exists as an independent input quantity. They will appear somewhat more complicated and contain components to be calculated possibly in iteration if the torque is immediately provided as a set point. In the configuration with superimposed speed control looked at here, the adjustment of the torque is subjected to the speed-feedback control loop.

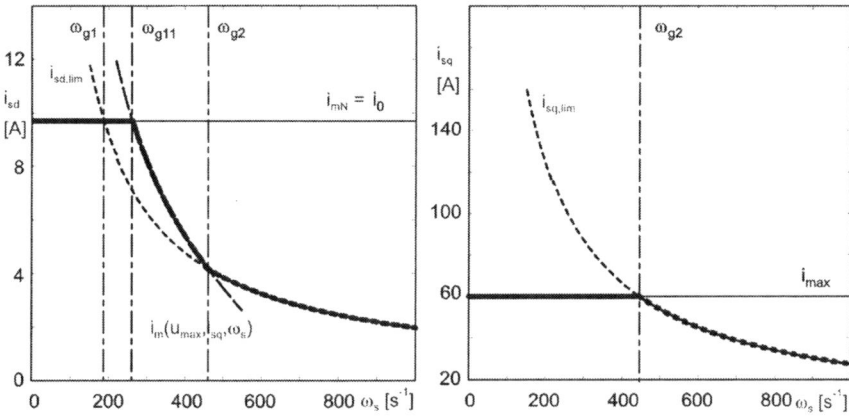

Fig. 8.11 Current limit characteristics for the torque optimal operation and maximum voltage output

The current limit characteristics are represented in expansion of figure 8.8 in figure 8.11 corresponding to the proposed algorithm for the complete speed range. The differences are recognizable clearly in the transition zone: The cut-in point of the field weakening is shifted to the frequency ω_{g11}, the maximum flux range is extended to substantially higher stator frequencies then defined by the torque optimal characteristic.

With this control, the area of the utilizable torque-speed range represented in the figure 8.12 for the first quadrant finally results. This area is delimited by the available current in the basic speed range, at high rotational speeds by the ceiling speed of the motor, and as discussed in the upper field weakening area by the available voltage, in the lower field weakening area by the maximum current and maximum voltage. The transition point between the upper and lower field weakening areas in turn is given by the frequency ω_{g2}.

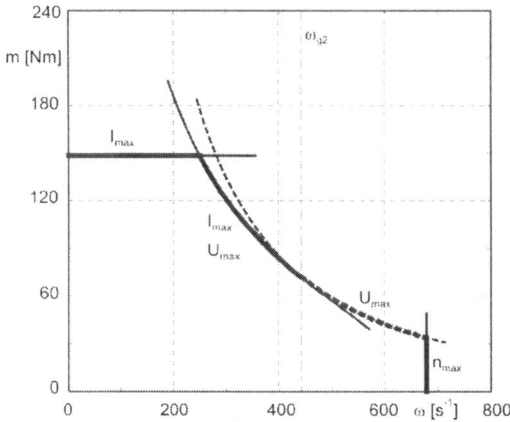

Fig. 8.12 Utilizable torque-speed range

The difference in the operating behaviour between torque optimal operation and conventional flux control can be shown best with the results of a practically realized control. For this purpose a speed reversal process is most suitable because quasi-stationary conditions (slowly variable currents and flux linkages) can be found here in wide areas. The comparison is made using a control characteristic of the form:

$$i_m^* = 0.9 i_0 \frac{\omega_N}{\omega} \frac{\sqrt{3} U_{s,max}}{U_N} \frac{L_{mN}}{L_m} \tag{8.33}$$

$$i_{s,max} = i_{max} \tag{8.34}$$

i.e. a flux characteristics, which is inversely proportional to the speed, and which adapts to changes of the DC link voltage and the main inductance. The plots of the most interesting quantities are shown in the figure 8.13. The differences in the reversal time are significant. An essential difference consists in the fact that *although the stator voltage with torque optimal control under load permanently resides at its limit during field weakening, the current controller predominantly, works in the linear area.* On the other hand *the controllability of the system is temporarily lost with the simple flux control.* It shall not remain unmentioned that similar results like those of the torque optimal control can be achieved also with a simple flux control in favorable parameter constellation and choice of the field weakening cut-in with respect to the reversal time. The temporary loss of the system controllability under loads is, however, hardly avoidable.

Furthermore it has to be noticed that a dynamically correct flux model (with saturation, cf. sections 6.2.3 and 4.4.1) is strongly necessary for the successful realization of the flux control algorithms.

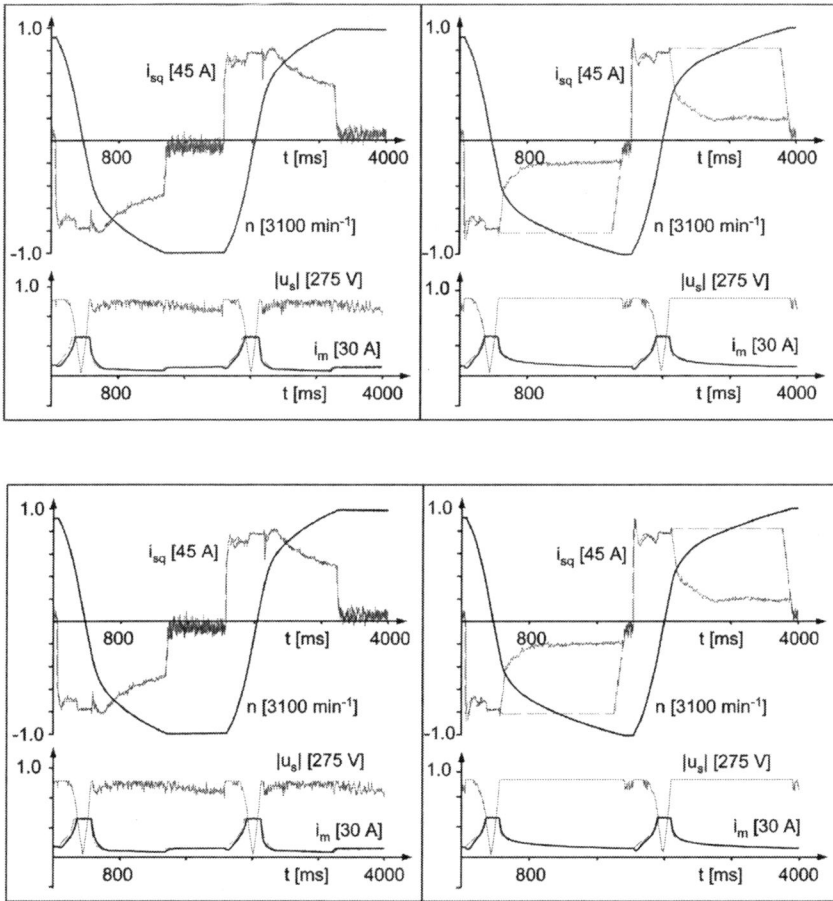

Fig. 8.13 Speed reversal processes +3100 rpm ↔ -3100 rpm: torque optimal control (**left**) and speed inverse flux control (**right**)

8.3.5 Torque dynamics at voltage limitation

As long as a sufficient voltage reserve is available, the torque dynamics is primarily a question of fast current impression. In the field weakening area this problem appears to be fundamentally more complex because of the missing voltage reserve. Regarding this the optimization goal consists here in reaching a rise time as short as possible also at the boundary condition of the limited voltage.

One of the outstanding features of the FOC consists in the possible high-dynamic impression of the torque because the torque rise time, constant rotor flux assumed, is identical with the rise time of the torque forming current. Prerequisite for a fast current impression is an adequate voltage reserve, though, so that a really fast torque impression is only possible in the basic speed range. According to the above derived rules, no voltage reserve would be available in the field weakening area at all. It is therefore necessary before an intended stepping-up of the torque to produce this voltage reserve by a (dynamic) flux reduction. The whole process should take place at unchanged maximum stator voltage for an optimal utilization of the machine.

After squaring and adding-up of (8.16) and (8.17) and neglection of R_s at the same time, the following equation is obtained:

$$u_s^2 = \left(w_s \sigma L_s i_{sq}\right)^2 + \left(w_s \sigma L_s i_{sd} + (1-\sigma)w_s L_s i_m\right)^2 \qquad (8.35)$$

From this relation it is recognizable that a dynamic control reserve can, considering the slow variability of i_m, be created by reducing the component i_{sd}. The derivation of the control law must, however, start out from the dynamically correct voltage equations. In the first instant the rotor flux linkage is constant changes compared to the stator current only very slowly, and therefore has influence on the calculation as a parameter and not as a variable. Regarding the searched-for current set point $i_{sd}^{\cdot*}$ quasi-stationary conditions can be assumed, and therefore the leakage time constant can be neglected. Furthermore it is assumed that the current transients are progressing approximately linearly and therefore the stator resistance is also negligible. The leakage inductance in the q axis must not be neglected however, because after the current i_{sd}^* to be calculated is reached, the decisive i_{sq} transient will unfold. Thus the initial equations are:

$$u_{sd} = -w_s \sigma L_s i_{sq} \qquad (8.36)$$

$$u_{sq} = \frac{\sigma L_s}{T_q} \Delta i_{sq} + w_s \sigma L_s i_{sd} + (1-\sigma)L_s w_s i_m \qquad (8.37)$$

The parameter T_q is the time needed by the i_{sq} transient, and represents a free design parameter in certain limits (see below). Δi_{sq} is the difference between the actual and the set point value. Both equations are squared, added-up and solved to i_{sd}:

$$i_{sd}^* = \frac{1}{\sigma}\left[-(1-\sigma)i_m - \frac{\sigma \Delta i_{sq}}{w_s T_q} + \sqrt{\left(\frac{u_{max}}{w_s L_s}\right)^2 - \left(\sigma i_{sq}\right)^2}\right] \qquad (8.38)$$

The radix expression in (8.38) shall be abbreviated to i_{m1}. The equation (8.38) is then:

$$i_{sd}^* = \frac{1}{\sigma}(i_{m1} - i_m) + i_m - \frac{\Delta i_{sq}}{\omega_s T_q} \qquad (8.39)$$

In the stationary state the last term disappears, and i_{sd}^* is equal to i_m. Therewith i_{m1} has also to be equal to i_m and can be understood as the stationary flux set point. The stationary set point for maximum voltage is provided normally by (8.30). Therefore also this value could be used instead of i_{m1} in (8.39) resulting in a sliding transition between dynamic and stationary flux control.

With respect to its structure, the equation (8.39) is comparable with an integral controller. The reciprocal leakage coefficient forms the controller gain, and the last term can be interpreted as an additional disturbance compensation. Nothing speaks against using this "controller" instead of the usual flux controller. *Caution is just necessary for machines with very small leakage factor and great sampling period of the flux control* because the controller gain then can accept inadmissible high values. In this case a PI flux controller may be used in stationary operation.

The determination of T_q must be carried out empirically. The influence of this parameter becomes clear by the simulation examples shown in the figure 8.14. A set point step of the torque forming current and thus of the torque was applied from zero to maximum for an acceleration process at about the double nominal speed. The sampling period T of the current control is 0.5 ms. The sampling period of the flux set point generator, flux and speed control is 2 ms. The optimum for T_q is found to approximately $T_q = 20T$ and valid for the complete field weakening area. Further reduction brings no benefit for the reduction of the rise time, just increases the overshooting of i_{sq}.

At the beginning of the transient a temporary drop of the torque forming current (and therewith the torque) appears. This is understandable because at first the voltage required for the impression of i_{sd} can only be gained at the expense of the voltage in q direction.

The rise times of the torque are in the area of 5....6 ms. It must not remain unmentioned that to obtain the shown transients besides the optimal set point generation a current controller is required which can impress the current components with the necessary dynamics and precision. The state-space dead-beat controller described in section 5.4 fulfills these prerequisites.

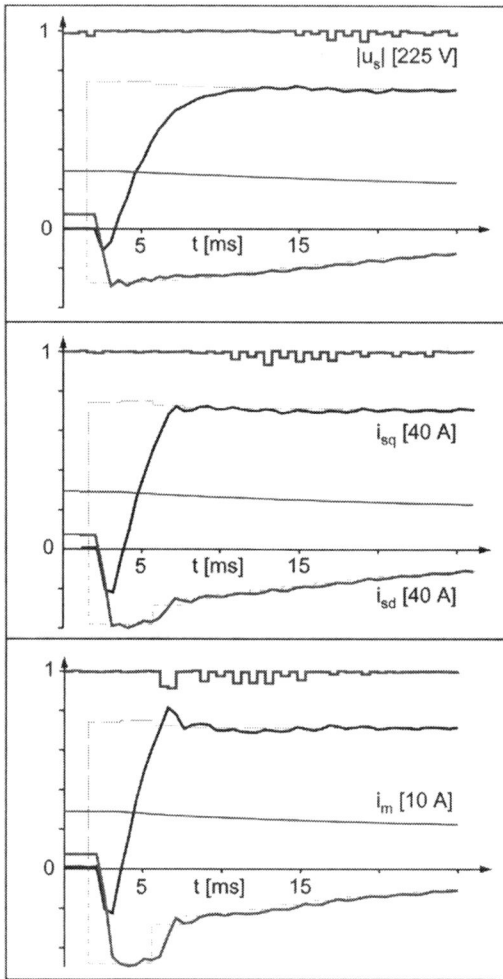

Fig. 8.14 Torque impression at maximum voltage: $T_q = \infty$ (**top**), $20T$ (**middle**), $10T$ (**bottom**), $\omega = 700s^{-1}$

Satisfactory results at the practical implementation fundamentally depend on the precision of the model parameters. A dynamically correct rotor flux model is required as an essential component. The figure 8.15 shows results from a sample drive. Because of the voltage maximum practically only one current component is actually controlled (in motor operation i_{sd}), all model inaccuracies and simplifications or unbalance and offset errors are mirrored in the second current component. The rise time of the torque forming current component confirms the values obtained by simulation.

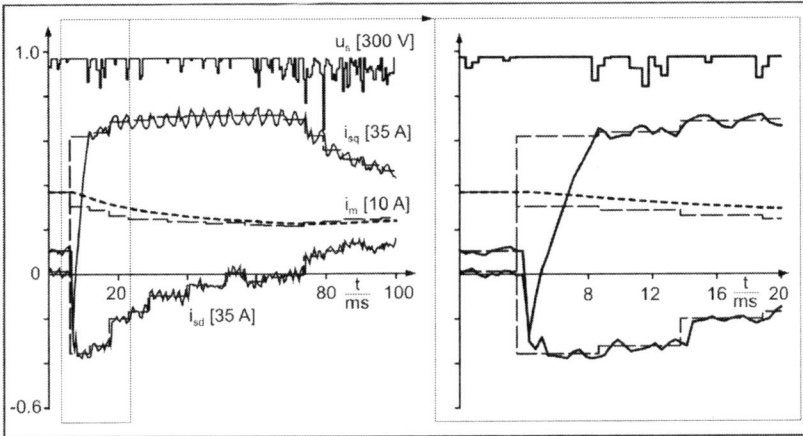

Fig. 8.15 Torque impression at voltage maximum, $\omega_s = 500$ s^{-1}

It has to be remarked that the described method manages with relatively moderate computation power. As shown by the figure 8.15, the current control works with a sampling time of 0.4 ms; the flux set point calculation, the flux and speed control are realized with a sampling time of 5 ms.

8.4 Comparison of the optimization strategies

Starting point of this chapter was the control of the state of the IM under the objective of energy optimal operation. The torque optimal control was introduced because of the advantages in the dynamic case and under the aspect of current and voltage limitation. In connection with this, it is interesting to compare the control methods with regard to the efficiency actually achieved.

The achieved efficiency for the different control methods is represented for the sample drive (11kW motor) in the figure 8.16. The curves for efficiency optimal and torque optimal control are approximately identical at loads above the half rated torque, because the total losses are dominated by the copper losses. The control with constant rotor flux also reaches the efficiency of the other methods at high torque, because the slip gets close to the optimal value (cf. section 8.3.1). Significant differences arise at medium and low torques, though. Here obviously the decisive possibilities for improvements by using optimized methods are to be found. For the examined example, the efficiency optimal control achieves visible differences compared to torque optimal control only at high speeds and small loads, because the share of the iron losses is here the biggest one. Also the values of the approximated torque optimal control described in the section 8.3.2 are only sparsely below those of the exact method. At higher loads a decrease of the efficiency can be found for all methods because of the increasing total current. This trend is more distinctive at smaller speeds, because the loss minimum shifts to lower slip values due to lower iron losses, and the main field saturation is reached earlier.

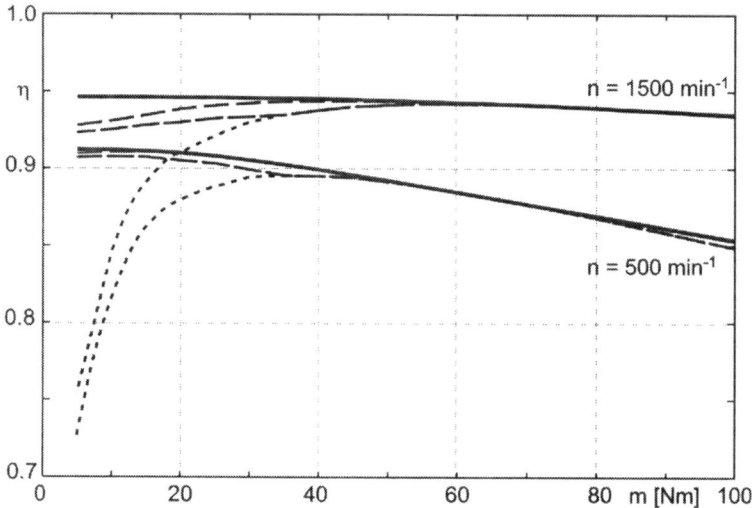

Fig. 8.16 Load-dependent efficiency for different control methods: ———— efficiency optimal control; - - - - - torque optimal control, exact calculation; ·········· constant rotor flux (= nominal flux); - - - - - (thin line) torque optimal control, approximated calculation

This behaviour can also be comprehended from the figure 8.17 in which the control characteristics for rotor flux and magnetizing current depending on the torque and speed are represented for the efficiency optimal and torque optimal control. Also here it can be recognized, that distinctive differences in the control characteristics only appear at high speeds and according share of the iron losses. The characteristics for torque optimal control are as expected independent of the speed, and insignificantly different from the efficiency optimal control at small speeds.

In the example, a rotor flux slightly higher than the nominal flux adjusts itself at high torques. The increase of motor voltage caused by this accompanying effect of the optimal control necessitates a voltage-dependent limitation of the rotor flux. Thus the optimum operation would be no longer practicable near the nominal operating point.

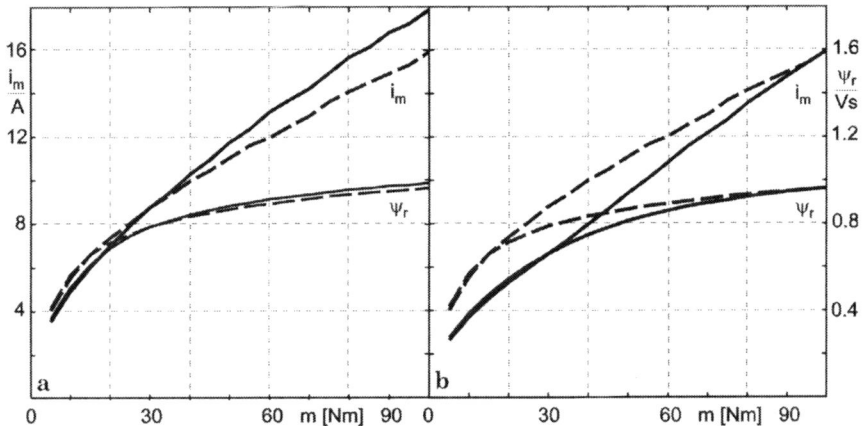

Fig. 8.17 i_m and ψ_r at torque optimal and efficiency optimal control: ———— efficiency optimal control; ·········· torque optimal control; **(a)** n = 500 min^{-1} **(b)** n = 1500 min^{-1}

In the field weakening area, the efficiency optimal operation will once more approach the torque optimal operation because of the strongly reduced iron losses. Thus noticeable efficiency improvements by an efficiency optimal control remain limited to the upper area of the basic speed range at low load. This statement is valid of course under the assumption that the ratio between copper and iron losses approximately corresponds to the one of the motor used for the calculations.

The following can be summarized to compare the control methods with respect to the efficiency behaviour:

- As opposed to constant flux operation the torque optimal control (optimization on minimal stator current at given torque) already delivers a considerable improvement particularly in the partial loads area.
- Further possibilities for efficiency improvement using a loss-optimal control strategy confine to the upper basic speed range at partial loads, and account to some per cent for usual ratios of iron and copper losses. The absolute loss reduction is even lower because visible effects are only obtained in the low load area.
- Further system boundary conditions like voltage and current limitation, which largely determine the optimization capabilities, come into play in the field weakening area. The iron losses decrease strongly with flux reduction.
- An exact optimization requires in any case iterative on-line calculations, because of the magnetic saturation, and measuring of the iron loss characteristic for loss-optimal operation. At least for the torque optimal

operation, however, simple control laws can be derived, which avoid these two problems and match the results of the exact control very closely.

Altogether, the torque optimal operation represents a very effective and recommendable control method also with respect to efficiency optimization. The efficiency optimal operation is generally recommendable if all optimization reserves have to be utilized (high motor power). It offers itself for stationary operation as an addition in form of an active power controller outlined in the section 8.2.

8.5 Rotor flux feedback control

The last section of this chapter deals with the rotor flux control in concentrated form because the calculated flux set points would be realized by a flux control loop according to the control structures discussed in chapter 1. Appropriately, the control is designed in the field-orientated coordinate system.

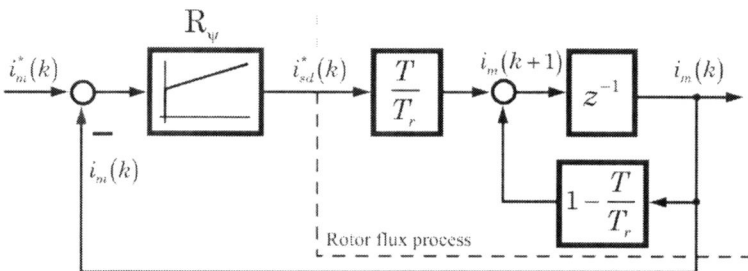

Fig. 8.18 Flux control loop with inner current loop

As described in the chapter 5, the inner-loop current control is optimized for finite adjustment times, and runs compared to the flux control with a considerably shorter sampling period of characteristically $T_i/T_\psi > 0.1$. Therefore the current impression can be regarded undelayed with respect to the flux control, and the actual controlled process is given by the rotor voltage equation in field-orientated coordinates (cf. chapter 3 and 6):

$$0 = i_m + T_r \frac{d i_m}{d t} - i_{sd} \tag{8.40}$$

It can be written in time-dicrete form:

$$i_m(k+1) = \left(1 - \frac{T}{T_r}\right) i_m(k) + \frac{T}{T_r} i_{sd}(k) \tag{8.41}$$

The current component i_{sd} forms the control variable. The control loop is represented in the figure 8.18. The flux controller is generally designed as a PI controller (digital magnitude optimum) and quasi-continuously optimized for transfer behaviour. If a PI controller is assumed with the equation:

$$R_\psi(z) = V_\psi \frac{1 - d_\psi z^{-1}}{1 - z^{-1}} \tag{8.42}$$

then the magnitude-optimal setting for the flux controller is:

$$V_\psi \approx \frac{1}{3\left(1 - e^{-T_\psi/T_r}\right)} \; ; \; d_\psi \approx e^{-T_\psi/T_r} \tag{8.43}$$

T_ψ = Sampling period of the flux controller; T_r = Rotor time constant

The figure 8.19 shows a transient process with magnetization and the following motor start-up up to the field weakening area, and illustrates the transient response of the flux at magnitude optimum.

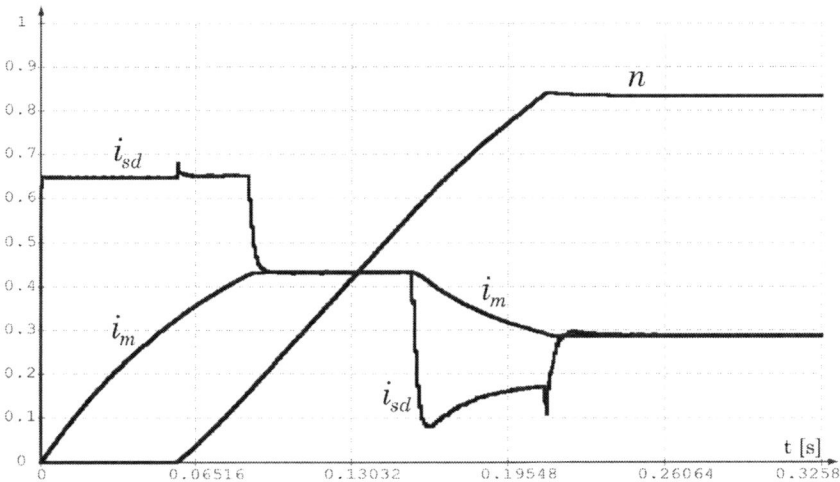

Fig. 8.19 Transient response of the flux at magnitude-optimal setting for the flux controller

A second approach for the flux controller, especially in connection with the optimization strategy discussed in this chapter, immediately arises from the control equation (8.39) for i_{sd} derived in the section 8.3.5 for the dynamically torque optimal current impression at the voltage limit. The equation is rewritten as follows:

$$i_{sd}^{*}(k)=i_{m}(k)+\frac{1}{\sigma}\left[i_{m}^{*}(k)-i_{m}(k)\right] \qquad (8.44)$$

The compensation term, dependent on the i_{sq} control error, was neglected because this term is only required for a fastest possible torque dynamics at the voltage limit. If the first term on the right side would not read $i_{m}(k)$ but $i_{sd}^{*}(k\text{-}1)$ (what is fulfilled in the stationary operation), the equation (8.44) would describe an I controller with the controller gain $V_{I}=1/\sigma$.

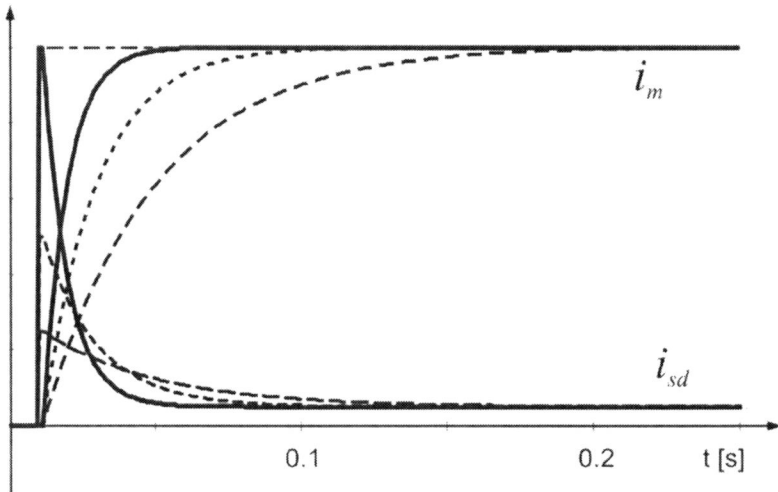

Fig. 8.20 Transient response of closed flux control loop with quasi-I controller

As shown in the figure 8.20, the closed control loop has in small signal operation approximately the transfer behaviour of a first-order delay function, in which the rise time is predefined by the controller gain. The unusual feature of the design consists in the fact that because of $V_{I}=1/\sigma$ the dynamics does not have to be predefined by "external" optimization rules, but is determined by the "point of view of the torque impression". For a very small leakage factor the controller gain can become too big, though, and cause unstable behaviour together with the amplitude quantization, hence a limit of $V_{I}=15...20$ should be set. On the other hand, if the dynamics of the controller is to be adapted to certain needs, nothing would speak against a deviation from the implicit setting $V_{I}=1/\sigma$.

8.6 References

Dittrich JA (1998) Anwendung fortgeschrittener Steuer- und Regelverfahren bei Asynchronantrieben. Habilitationsschrift, TU Dresden

Moreno-Eguílaz JM, Cipolla M, Peracaula J (1997) Induction Motor Drives Energy Optimization in Steady and Transient States: A New Approach. EPE Trondheim, pp. 3.705 - 3.710

Khater-Faeka MH, Lorenz RD, Novotny DW, Tang K (1991) Selection of Flux Level in Field-Oriented Induction Machine Controllers with Consideration of Magnetic Saturation Effects. IEEE Transactions on Industry Applications, Vol. IA-27, No. 4, July/August, pp. 720 – 726

Maischak D, Németh Csóka M (1994) Schnelle Drehmomentregelung im gesamten Drehzahlbereich eines hochausgenutzten Drehfeldantriebs. Archiv für Elektrotechnik 77, S. 289 – 301

Murata T, Tsuchiya T, Takeda I (1990) Quick Response and High Efficiency Control of the Induction Motor Based on Optimal Control Theory. 11. IFAC World Congress, Tallin, vol. 8, pp. 242 – 247

Rasmussen KS, Thogersen P (1997) Model Based Energy Optimizer for Vector Controlled Induction Motor Drives. EPE Trondheim, pp. 3.711 - 3.716

Vogt G (1985) Digitale Regelung von Asynchronmotoren für numerisch gesteuerte Fertigungseinrichtungen. Springer-Verlag

Wiesing J (1994) Betrieb der feldorientiert geregelten Asynchronmaschine im Bereich oberhalb der Nenndrehzahl. Dissertation, Uni. Paderborn

9 Nonlinear control structures with direct decoupling for three-phase AC drive systems

9.1 Existing problems at linear controlled drive systems

It is clearly recognizable that the 3-phase drive system engineering has reached a relatively mature stage of development (cf. chapters 1-8). The principle of the field orientated control also has largely asserted to be the most used method in commercial systems. The spectrum of solved questions extends from the control and observer structures over the problems of parameter identification (on-line, off-line) and adaptation to the self-tuning and the self-commissioning.

The most implemented structure (chapter 5 or [Quang 1999]) contains a 2-dimensional current controller for decoupling between the magnetization and the generation of torque as well as for undelayed impression of torque. Because of the decoupling the flux and speed control loops could be designed rather liberally. In these structures the current controller and the flux observer are always based on motor models linearized within a sampling period (cf. section 3.2.2).

The linearization is made under the assumption that the sampling time T is small enough for the stator frequency ω_s to be regarded constant within T. Because of this assumption the frequency ω_s is now a parameter in the system matrix, and *the bilinear model becomes a linear time-variant system* for which the known design methods of linear systems (cf. chapter 5) can be used.

Although the present concept was very successful, it is recognizable that:

- because of the nonlinear process model (the input quantity ω_s appears in the system matrix) in high-speed drives with synchronous modulation (cf. section 2.5.2) or in sensorless controlled systems (cf. section 4.3), or

- because of the nonlinear parameters (the main inductance is strongly dependent on the state variable i_m or ψ_r) with respect to the system stability in systems with parameter identification and adaptation,

some problems often appear, particularly if the system must work at the voltage limit (i.e. in the nonlinear mode) and therewith the condition $\omega_s =$ const is no longer fulfilled. If these problems remain unsolved, the drive quality will be affected considerably. In such cases at least a nonlinear design would be able to deliver better results.

Within the last approx. 15 years different new ways to design nonlinear controllers were shown ([Isidori 1995], [Krstíc 1995], [Wey 2001]) or even experimented in motor control ([Ortega 1998], [Bodson 1998], [Khorrami 2003], [Dawson 2004]), but they were mostly theoretical works. The practical developments were completely missing. Recently, some more thorough investigations ([Cuong 2003], [Ha 2003], [Duc 2004], [Nam 2004]) concerned with practical implementation of the methods had been forthcoming, particularly on the exact linearization method discussed in this book.

9.2 Nonlinear control structure for drive systems with IM

In the section 3.6.2 the nonlinear process model of the IM was already derived as a starting point to the controller design:

$$\begin{cases} \overset{\bullet}{\mathbf{x}} = \mathbf{f}(\mathbf{x}) + \mathbf{h}_1\,u_1 + \mathbf{h}_2\,u_2 + \mathbf{h}_3\,u_3 \\ \mathbf{y} = \mathbf{g}(\mathbf{x}) \end{cases} \tag{9.1}$$

$$\mathbf{f}(\mathbf{x}) = \begin{bmatrix} -d\,x_1 + c\,\psi'_{rd} \\ -d\,x_2 - c\,T_r\,\omega\,\psi'_{rd} \\ 0 \end{bmatrix}; \mathbf{h}_1 = \begin{bmatrix} a \\ 0 \\ 0 \end{bmatrix}; \mathbf{h}_1 = \begin{bmatrix} 0 \\ a \\ 0 \end{bmatrix}; \mathbf{h}_3 = \begin{bmatrix} x_2 \\ -x_1 \\ 1 \end{bmatrix} \tag{9.2}$$

$$y_1 = g_1(\mathbf{x}) = x_1;\, y_2 = g_2(\mathbf{x}) = x_2;\, y_3 = g_3(\mathbf{x}) = x_3$$

- Parameters: $a = 1/\sigma L_s\,;\, b = 1/\sigma T_s\,;\, c = (1-\sigma)/\sigma T_r\,;\, d = b + c$
- State variables: $x_1 = i_{sd}\,;\, x_2 = i_{sq}\,;\, x_3 = \vartheta_s$
- Input variables: $u_1 = u_{sd}\,;\, u_2 = u_{sq}\,;\, u_3 = \omega_s$
- Output variables: $y_1 = i_{sd}\,;\, y_2 = i_{sq}\,;\, y_3 = \vartheta_s$

9.2.1 Nonlinear controller design based on "exact linearization"

Using the idea in the section 3.6.1 the design can be realized in the following steps:

- Step 1: Calculation of the vector $\mathbf{r} = [r_1, r_2, \cdots, r_m]$ of the relative difference orders.
- Step 2: Calculation of the matrix $\mathbf{L(x)}$ using the formula (3.99) and check of its invertibility.

$$\mathbf{L(x)} = \begin{pmatrix} L_{h_1} L_f^{r_1-1} g_1(\mathbf{x}) & \cdots & L_{h_m} L_f^{r_1-1} g_1(\mathbf{x}) \\ \vdots & \ddots & \vdots \\ L_{h_1} L_f^{r_m-1} g_m(\mathbf{x}) & \cdots & L_{h_m} L_f^{r_m-1} g_m(\mathbf{x}) \end{pmatrix}; \det[\mathbf{L(x)}] \neq 0 \quad (9.3)$$

After that the fulfillment of the condition (3.95) is checked.

- Step 3: Realization of the coordinate transformation using (3.96).

$$\mathbf{z} = \begin{pmatrix} z_1 \\ \vdots \\ z_n \end{pmatrix} = \mathbf{m(x)} = \begin{pmatrix} m_1^1(\mathbf{x}) \\ \vdots \\ m_{r_1}^1(\mathbf{x}) \\ \vdots \\ m_1^m(\mathbf{x}) \\ \vdots \\ m_{r_m}^m(\mathbf{x}) \end{pmatrix} = \begin{pmatrix} g_1(\mathbf{x}) \\ \vdots \\ L_f^{r_1-1} g_1(\mathbf{x}) \\ \vdots \\ g_m(\mathbf{x}) \\ \vdots \\ L_f^{r_m-1} g_m(\mathbf{x}) \end{pmatrix} \quad (9.4)$$

- Step 4: Calculation of the state-feedback control law (3.98):

$$\mathbf{u} = \mathbf{a(x)} + \mathbf{L}^{-1}(\mathbf{x})\mathbf{w} = -\mathbf{L}^{-1}(\mathbf{x}) \begin{pmatrix} L_f^{r_1} g_1(\mathbf{x}) \\ \vdots \\ L_f^{r_m} g_m(\mathbf{x}) \end{pmatrix} + \mathbf{L}^{-1}(\mathbf{x})\mathbf{w} \quad (9.5)$$

Using the represented steps or the formulae (9.3) - (9.5) the design for drive systems with IM can now be proceeded as follows:

- Step 1: Calculation of the vector $\mathbf{r}$.
  a) Case $j = 1$:

$$L_{h_1} g_1(\mathbf{x}) = \frac{\partial g_1(\mathbf{x})}{\partial \mathbf{x}} \mathbf{h}_1 = \begin{bmatrix} 1 & 0 & 0 \end{bmatrix} \begin{bmatrix} a \\ 0 \\ 0 \end{bmatrix} = a \neq 0 \quad (9.6)a$$

$$L_{h_2} g_1(\mathbf{x}) = \frac{\partial g_1(\mathbf{x})}{\partial \mathbf{x}} \mathbf{h}_2 = \begin{bmatrix} 1 & 0 & 0 \end{bmatrix} \begin{bmatrix} 0 \\ a \\ 0 \end{bmatrix} = 0 \tag{9.6)b}$$

$$L_{h_3} g_1(\mathbf{x}) = \frac{\partial g_1(\mathbf{x})}{\partial \mathbf{x}} \mathbf{h}_3 = \begin{bmatrix} 1 & 0 & 0 \end{bmatrix} \begin{bmatrix} x_2 \\ -x_1 \\ 1 \end{bmatrix} = x_2 \neq 0 \tag{9.6)c}$$

Therewith $r_1 = 1$ follows from the equation (9.6).
b) Case $j = 2$:

$$L_{h_1} g_2(\mathbf{x}) = \frac{\partial g_2(\mathbf{x})}{\partial \mathbf{x}} \mathbf{h}_1 = \begin{bmatrix} 0 & 1 & 0 \end{bmatrix} \begin{bmatrix} a \\ 0 \\ 0 \end{bmatrix} = 0 \tag{9.7)a}$$

$$L_{h_2} g_2(\mathbf{x}) = \frac{\partial g_2(\mathbf{x})}{\partial \mathbf{x}} \mathbf{h}_2 = \begin{bmatrix} 0 & 1 & 0 \end{bmatrix} \begin{bmatrix} 0 \\ a \\ 0 \end{bmatrix} = a \neq 0 \tag{9.7)b}$$

$$L_{h_3} g_2(\mathbf{x}) = \frac{\partial g_2(\mathbf{x})}{\partial \mathbf{x}} \mathbf{h}_3 = \begin{bmatrix} 0 & 1 & 0 \end{bmatrix} \begin{bmatrix} x_2 \\ -x_1 \\ 1 \end{bmatrix} = -x_1 \neq 0 \tag{9.7)c}$$

From the equation (9.7), $r_2 = 1$ similarly follows.
c) Case $j = 3$:

$$L_{h_1} g_3(\mathbf{x}) = \frac{\partial g_3(\mathbf{x})}{\partial \mathbf{x}} \mathbf{h}_1 = \begin{bmatrix} 0 & 0 & 1 \end{bmatrix} \begin{bmatrix} a \\ 0 \\ 0 \end{bmatrix} = 0 \tag{9.8)a}$$

$$L_{h_2} g_3(\mathbf{x}) = \frac{\partial g_3(\mathbf{x})}{\partial \mathbf{x}} \mathbf{h}_2 = \begin{bmatrix} 0 & 0 & 1 \end{bmatrix} \begin{bmatrix} 0 \\ a \\ 0 \end{bmatrix} = 0 \tag{9.8)b}$$

$$L_{h_3} g_3(\mathbf{x}) = \frac{\partial g_3(\mathbf{x})}{\partial \mathbf{x}} \mathbf{h}_3 = \begin{bmatrix} 0 & 0 & 1 \end{bmatrix} \begin{bmatrix} x_2 \\ -x_1 \\ 1 \end{bmatrix} = 1 \neq 0 \tag{9.8)c}$$

From the equation (9.8), $r_3 = 1$ follows then.

- Step 2: Calculation of the matrix **L**.

$$\mathbf{L}(\mathbf{x}) = \begin{vmatrix} L_{h_1} g_1(\mathbf{x}) & L_{h_2} g_1(\mathbf{x}) & L_{h_3} g_1(\mathbf{x}) \\ L_{h_1} g_2(\mathbf{x}) & L_{h_2} g_2(\mathbf{x}) & L_{h_3} g_2(\mathbf{x}) \\ L_{h_1} g_3(\mathbf{x}) & L_{h_2} g_3(\mathbf{x}) & L_{h_3} g_3(\mathbf{x}) \end{vmatrix} = \begin{bmatrix} a & 0 & x_2 \\ 0 & a & -x_1 \\ 0 & 0 & 1 \end{bmatrix} \quad (9.9)$$

It is easy to see that $\det[\mathbf{L}(\mathbf{x})] = a^2 \neq 0$, and therefore the matrix $\mathbf{L}(\mathbf{x})$ can be inverted. The necessary and sufficient conditions are summarized:

$$\begin{cases} \det[\mathbf{L}(\mathbf{x})] = a^2 \neq 0 \\ r_1 + r_2 + r_3 = 3 = n \end{cases} \quad (9.10)$$

The fulfilled condition (9.11) indicates that the system (9.1) can be linearized exactly, or that the coordinate transformation can now be made.
- Step 3: Realization of the coordinate transformation.
  a) The state space **x** is transformed into a new state space **z** using (9.4).

$$\mathbf{z} = \begin{bmatrix} z_1 \\ z_2 \\ z_3 \end{bmatrix} = \begin{bmatrix} m_1^1(\mathbf{x}) \\ m_1^2(\mathbf{x}) \\ m_1^3(\mathbf{x}) \end{bmatrix} = \begin{bmatrix} g_1(\mathbf{x}) \\ g_2(\mathbf{x}) \\ g_3(\mathbf{x}) \end{bmatrix} = \begin{bmatrix} x_1 \\ x_2 \\ x_3 \end{bmatrix} \quad (9.11)$$

  b) The new state space model is calculated as follows.

$$\begin{cases} \dfrac{dz_1}{dt} = \dfrac{\partial g_1(\mathbf{x})}{\partial \mathbf{x}} \dfrac{d\mathbf{x}}{dt} \\ \qquad = L_f g_1(\mathbf{x}) + L_{h_1} g_1(\mathbf{x}) u_1 + L_{h_2} g_1(\mathbf{x}) u_2 + L_{h_3} g_1(\mathbf{x}) u_3 \\ \dfrac{dz_2}{dt} = \dfrac{\partial g_2(\mathbf{x})}{\partial \mathbf{x}} \dfrac{d\mathbf{x}}{dt} \\ \qquad = L_f g_2(\mathbf{x}) + L_{h_1} g_2(\mathbf{x}) u_1 + L_{h_2} g_2(\mathbf{x}) u_2 + L_{h_3} g_2(\mathbf{x}) u_3 \\ \dfrac{dz_3}{dt} = \dfrac{\partial g_3(\mathbf{x})}{\partial \mathbf{x}} \dfrac{d\mathbf{x}}{dt} \\ \qquad = L_f g_3(\mathbf{x}) + L_{h_1} g_3(\mathbf{x}) u_1 + L_{h_2} g_3(\mathbf{x}) u_2 + L_{h_3} g_3(\mathbf{x}) u_3 \end{cases} \quad (9.12)$$

The unknown terms in the equation (9.12) have to be calculated now.

$$L_f g_1(\mathbf{x}) = \frac{\partial g_1(\mathbf{x})}{\partial \mathbf{x}} f = \begin{bmatrix} 1 & 0 & 0 \end{bmatrix} \begin{bmatrix} -d\,x_1 + c\,\psi'_{rd} \\ -d\,x_2 - c\,T_r\,\omega\,\psi'_{rd} \\ 0 \end{bmatrix} = -d\,x_1 + c\,\psi'_{rd}$$

$$L_f g_2(\mathbf{x}) = \frac{\partial g_2(\mathbf{x})}{\partial \mathbf{x}} f = \begin{bmatrix} 0 & 1 & 0 \end{bmatrix} \begin{bmatrix} -d\,x_1 + c\,\psi'_{rd} \\ -d\,x_2 - c\,T_r\,\omega\,\psi'_{rd} \\ 0 \end{bmatrix} = -d\,x_2 + c\,T_r\,\omega\,\psi'_{rd}$$

$$L_f g_3(\mathbf{x}) = \frac{\partial g_1(\mathbf{x})}{\partial \mathbf{x}} f = \begin{bmatrix} 0 & 0 & 1 \end{bmatrix} \begin{bmatrix} -d\,x_1 + c\,\psi'_{rd} \\ -d\,x_2 - c\,T_r\,\omega\,\psi'_{rd} \\ 0 \end{bmatrix} = 0$$

The result of the coordinate transformation is then:

$$\begin{cases} \dfrac{dz_1}{dt} = -d\,x_1 + c\,\psi'_{rd} + a\,u_1 + x_2\,u_3 = w_1 \\[2mm] \dfrac{dz_1}{dt} = -d\,x_2 - c\,T_r\,\omega\,\psi'_{rd} + a\,u_2 - x_1\,u_3 = w_2 \\[2mm] \dfrac{dz_3}{dt} = u_3 = w_3 \end{cases} \qquad (9.13)$$

The following equation will be obtained from (9.13):

$$\mathbf{w} = \begin{bmatrix} w_1 \\ w_2 \\ w_3 \end{bmatrix} = \underbrace{\begin{bmatrix} -d\,x_1 + c\,\psi'_{rd} \\ -d\,x_2 - c\,T_r\,\omega\,\psi'_{rd} \\ 0 \end{bmatrix}}_{\mathbf{p}(x)} + \underbrace{\begin{bmatrix} a & 0 & x_2 \\ 0 & a & -x_1 \\ 0 & 0 & 1 \end{bmatrix}}_{\mathbf{L}(x)} \underbrace{\begin{bmatrix} u_1 \\ u_2 \\ u_3 \end{bmatrix}}_{\mathbf{u}} \qquad (9.14)$$

$$\mathbf{w} = \mathbf{p}(\mathbf{x}) + \mathbf{L}(\mathbf{x})\,\mathbf{u}$$

- Step 4: The control law, thereat $\mathbf{w}$ represents the new input vector, can be calculated by equation (9.14).

$$\mathbf{u} = \underbrace{-\mathbf{L}^{-1}(\mathbf{x})\mathbf{p}(\mathbf{x})}_{\mathbf{a}(\mathbf{x})} + \mathbf{L}^{-1}(\mathbf{x})\mathbf{w} = \mathbf{a}(\mathbf{x}) + \mathbf{L}^{-1}(\mathbf{x})\mathbf{w} \qquad (9.15)$$

Using the matrix $\mathbf{L}$ in equation (9.9), $\mathbf{L}^{-1}$ is then obtained to:

$$\mathbf{L}^{-1} = \begin{bmatrix} 1/a & 0 & -x_2/a \\ 0 & 1/a & x_1/a \\ 0 & 0 & 0 \end{bmatrix} \qquad (9.16)$$

The state-feedback control law or the coordinate transformation law (9.15) can be written in detailed form:

$$\mathbf{u} = \begin{bmatrix} u_1 \\ u_2 \\ u_3 \end{bmatrix} = \begin{bmatrix} d\,x_1/a - c\,\psi'_{rd}/a \\ d\,x_2/a + c\,T_r\,\omega\,\psi'_{rd}/a \\ 0 \end{bmatrix} + \begin{bmatrix} 1/a & 0 & -x_2/a \\ 0 & 1/a & x_1/a \\ 0 & 0 & 0 \end{bmatrix} \begin{bmatrix} w_1 \\ w_2 \\ w_3 \end{bmatrix}$$

$$= \underbrace{\begin{bmatrix} \left(\dfrac{1}{T_s} + \dfrac{1-\sigma}{T_r}\right)L_s x_1 - \dfrac{1-\sigma}{T_r} L_s \psi'_{rd} \\ \left(\dfrac{1}{T_s} + \dfrac{1-\sigma}{T_r}\right)L_s x_2 + (1-\sigma)\omega L_s \psi'_{rd} \\ 0 \end{bmatrix}}_{\mathbf{a(x)}} + \underbrace{\begin{bmatrix} \sigma L_s & 0 & -\sigma L_s x_2 \\ 0 & \sigma L_s & \sigma L_s x_1 \\ 0 & 0 & 1 \end{bmatrix}}_{\mathbf{L^{-1}(x)}} \underbrace{\begin{bmatrix} w_1 \\ w_2 \\ w_3 \end{bmatrix}}_{\mathbf{w}}$$

$$(9.17)$$

9.2.2 Feedback control structure with direct decoupling for IM

Using the state feedback or the coordinate transformation (9.17) the exact linearized IM model can be represented as in the figure 9.1. The new state model will now be the starting point for the controller design.

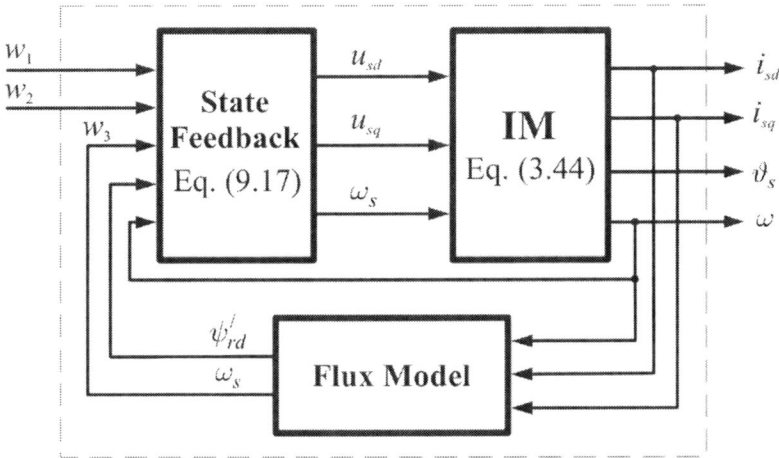

Fig. 9.1 Substitute linear process model of the IM as starting point for controller design (cf. figure 3.17)

It is not difficult for the new linear model to derive the input-output relation. After some transformations the following transfer function will be obtained:

$$\mathbf{y}(s) = \begin{vmatrix} 1/s^{r_1} & 0 & 0 \\ 0 & 1/s^{r_2} & 0 \\ 0 & 0 & 1/s^{r_3} \end{vmatrix} \mathbf{w}(s) = \begin{bmatrix} 1/s & 0 & 0 \\ 0 & 1/s & 0 \\ 0 & 0 & 1/s \end{bmatrix} \mathbf{w}(s) \qquad (9.18)$$

At more exact analysis of the equation (9.18) the following essential knowledge can be learned:

- Besides the exact linearization achieved in the complete new state space **z**, *the input-output decoupling relations are totally guaranteed.*
- The three transfer functions respectively contain only one element of integration.

Based on both these new results it seems to be possible to replace the two-dimensional current controller (figure 1.6) by a coordinate transformation and two separate current controllers for both axes *dq* (figure 9.2).

The *direct decoupling* concept in the figure 9.2 is dynamically effective for the complete state space. The two current controllers R_{isd} and R_{isq} need not have the PI characteristic, and can be designed with modern algorithms such as dead-beat control. A dynamical and nearly undelayed impression of the motor torque can be guaranteed without breaking any linearization condition.

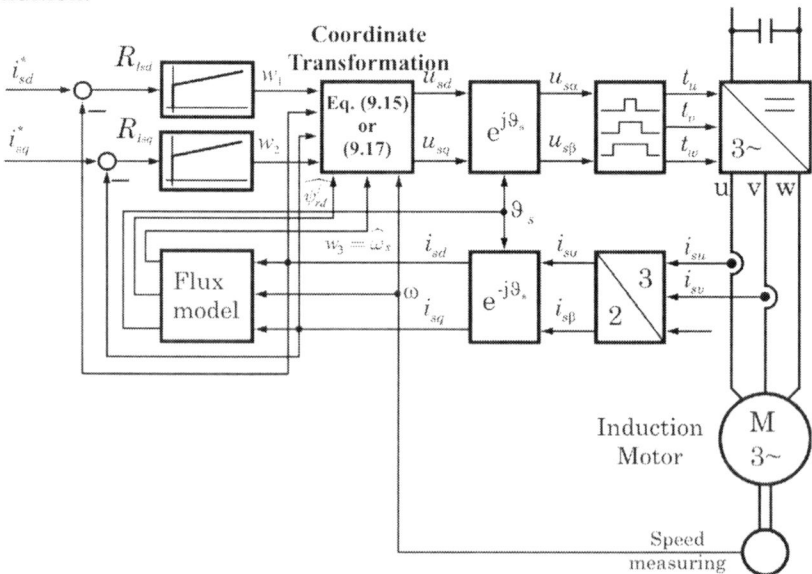

Fig. 9.2　The new control structure of the inner loop (impression of the stator current components) with direct decoupling designed using the method of exact linearization

9.3 Nonlinear control structure for drive systems with PMSM

The nonlinear process model of the PMSM was derived in the section 3.6.4 as follows:

$$\begin{cases} \overset{\bullet}{\mathbf{x}} = \mathbf{f}(\mathbf{x}) + \mathbf{h}_1(\mathbf{x})u_1 + \mathbf{h}_2(\mathbf{x})u_2 + \mathbf{h}_3(\mathbf{x})u_3 \\ \mathbf{y} = \mathbf{g}(\mathbf{x}) \end{cases} \tag{9.19}$$

$$\mathbf{f}(\mathbf{x}) = \begin{bmatrix} -cx_1 \\ -dx_2 \\ 0 \end{bmatrix}; \mathbf{h}_1(\mathbf{x}) = \begin{bmatrix} a \\ 0 \\ 0 \end{bmatrix}; \mathbf{h}_2(\mathbf{x}) = \begin{bmatrix} 0 \\ b \\ 0 \end{bmatrix}; \mathbf{h}_3(\mathbf{x}) = \begin{bmatrix} ax_2/b \\ -bx_1/a - b\psi_p \\ 1 \end{bmatrix} \tag{9.20}$$

$$y_1 = g_1(\mathbf{x}) = x_1; y_2 = g_2(\mathbf{x}) = x_2; y_3 = g_3(\mathbf{x}) = x_3$$

- Parameters: $a = 1/L_{sd}; b = 1/L_{sq}; c = 1/T_{sd}; d = 1/T_{sq}$
- State variables: $x_1 = i_{sd}; x_2 = i_{sq}; x_3 = \vartheta_s$
- Input variables: $u_1 = u_{sd}; u_2 = u_{sq}; u_3 = \omega_s$
- Output variables: $y_1 = i_{sd}; y_2 = i_{sq}; y_3 = \vartheta_s$

9.3.1 Nonlinear controller design based on "exact linearization"

The design is carried out similarly as in the case of the IM.
- Step 1: Calculation of the vector **r**.
  a) Case $j = 1$:

$$L_{h_1} g_1(\mathbf{x}) = \frac{\partial g_1(\mathbf{x})}{\partial \mathbf{x}} \mathbf{h}_1 = \begin{bmatrix} 1 & 0 & 0 \end{bmatrix} \begin{bmatrix} a \\ 0 \\ 0 \end{bmatrix} = a \neq 0 \tag{9.21a}$$

$$L_{h_2} g_1(\mathbf{x}) = \frac{\partial g_1(\mathbf{x})}{\partial \mathbf{x}} \mathbf{h}_2 = \begin{bmatrix} 1 & 0 & 0 \end{bmatrix} \begin{bmatrix} 0 \\ b \\ 0 \end{bmatrix} = 0 \tag{9.21b}$$

$$L_{h_3} g_1(\mathbf{x}) = \frac{\partial g_1(\mathbf{x})}{\partial \mathbf{x}} \mathbf{h}_3 = \begin{bmatrix} 1 & 0 & 0 \end{bmatrix} \begin{bmatrix} ax_2/b \\ -bx_1/a - b\psi_p \\ 1 \end{bmatrix} = \frac{ax_2}{b} \neq 0 \tag{9.21c}$$

From the equation (9.21) $r_1 = 1$ follows.

b) Case $j = 2$:

$$L_{h_1} g_2(\mathbf{x}) = \frac{\partial g_2(\mathbf{x})}{\partial \mathbf{x}} \mathbf{h}_1 = \begin{bmatrix} 0 & 1 & 0 \end{bmatrix} \begin{bmatrix} a \\ 0 \\ 0 \end{bmatrix} = 0 \qquad (9.22)\text{a}$$

$$L_{h_2} g_2(\mathbf{x}) = \frac{\partial g_2(\mathbf{x})}{\partial \mathbf{x}} \mathbf{h}_2 = \begin{bmatrix} 0 & 1 & 0 \end{bmatrix} \begin{bmatrix} 0 \\ b \\ 0 \end{bmatrix} = b \neq 0 \qquad (9.22)\text{b}$$

$$L_{h_3} g_2(\mathbf{x}) = \frac{\partial g_2(\mathbf{x})}{\partial \mathbf{x}} \mathbf{h}_3 = \begin{bmatrix} 0 & 1 & 0 \end{bmatrix} \begin{bmatrix} ax_2/b \\ -bx_1/a - b\psi_p \\ 1 \end{bmatrix} \qquad (9.22)\text{c}$$

$$= -\frac{bx_1}{a} - b\psi_p \neq 0$$

With the equation (9.22) $r_2 = 1$ is obtained.

c) Case $j = 3$:

$$L_{h_1} g_3(\mathbf{x}) = \frac{\partial g_3(\mathbf{x})}{\partial \mathbf{x}} \mathbf{h}_1 = \begin{bmatrix} 0 & 0 & 1 \end{bmatrix} \begin{bmatrix} a \\ 0 \\ 0 \end{bmatrix} = 0 \qquad (9.23)\text{a}$$

$$L_{h_2} g_3(\mathbf{x}) = \frac{\partial g_3(\mathbf{x})}{\partial \mathbf{x}} \mathbf{h}_2 = \begin{bmatrix} 0 & 0 & 1 \end{bmatrix} \begin{bmatrix} 0 \\ b \\ 0 \end{bmatrix} = 0 \qquad (9.23)\text{b}$$

$$L_{h_3} g_3(\mathbf{x}) = \frac{\partial g_3(\mathbf{x})}{\partial \mathbf{x}} \mathbf{h}_3 = \begin{bmatrix} 0 & 0 & 1 \end{bmatrix} \begin{bmatrix} ax_2/b \\ -bx_1/a - b\psi_p \\ 1 \end{bmatrix} = 1 \neq 0 \quad (9.23)\text{c}$$

With the equation (9.23) $r_3 = 1$ is obtained.

- Step 2: Calculation of the matrix **L**.

$$\mathbf{L}(\mathbf{x}) = \begin{bmatrix} L_{h_1} g_1(\mathbf{x}) & L_{h_2} g_1(\mathbf{x}) & L_{h_3} g_1(\mathbf{x}) \\ L_{h_1} g_2(\mathbf{x}) & L_{h_2} g_2(\mathbf{x}) & L_{h_3} g_2(\mathbf{x}) \\ L_{h_1} g_3(\mathbf{x}) & L_{h_2} g_3(\mathbf{x}) & L_{h_3} g_3(\mathbf{x}) \end{bmatrix} = \begin{bmatrix} a & 0 & ax_2/b \\ 0 & b & -bx_1/a - b\psi_p \\ 0 & 0 & 1 \end{bmatrix}$$

$$\Rightarrow \quad \det[\mathbf{L}(\mathbf{x})] = ab \neq 0$$

$$(9.24)$$

- Step 3: Realization of the coordinate transformation.

  a) The state space $\mathbf{x}$ is transformed into a new state space $\mathbf{z}$ using (9.4).

  After replacing $r_1 = r_2 = r_3 = 1$ into (9.4), the same result like (9.11) is obtained:

$$\mathbf{z} = \begin{bmatrix} z_1 \\ z_2 \\ z_3 \end{bmatrix} = \begin{bmatrix} m_1^1(\mathbf{x}) \\ m_1^2(\mathbf{x}) \\ m_1^3(\mathbf{x}) \end{bmatrix} = \begin{bmatrix} g_1(\mathbf{x}) \\ g_2(\mathbf{x}) \\ g_3(\mathbf{x}) \end{bmatrix} = \begin{bmatrix} x_1 \\ x_2 \\ x_3 \end{bmatrix} \tag{9.25}$$

  b) The new state model is similar to (9.12).

$$\left\{ \begin{aligned} \frac{dz_1}{dt} &= \frac{\partial g_1(\mathbf{x})}{\partial \mathbf{x}} \frac{d\mathbf{x}}{dt} \\ &= L_f g_1(\mathbf{x}) + L_{h_1} g_1(\mathbf{x}) u_1 + L_{h_2} g_1(\mathbf{x}) u_2 + L_{h_3} g_1(\mathbf{x}) u_3 \\ \frac{dz_2}{dt} &= \frac{\partial g_2(\mathbf{x})}{\partial \mathbf{x}} \frac{d\mathbf{x}}{dt} \\ &= L_f g_2(\mathbf{x}) + L_{h_1} g_2(\mathbf{x}) u_1 + L_{h_2} g_2(\mathbf{x}) u_2 + L_{h_3} g_2(\mathbf{x}) u_3 \\ \frac{dz_3}{dt} &= \frac{\partial g_3(\mathbf{x})}{\partial \mathbf{x}} \frac{d\mathbf{x}}{dt} \\ &= L_f g_3(\mathbf{x}) + L_{h_1} g_3(\mathbf{x}) u_1 + L_{h_2} g_3(\mathbf{x}) u_2 + L_{h_3} g_3(\mathbf{x}) u_3 \end{aligned} \right. \tag{9.26}$$

with:

$$L_f g_1(\mathbf{x}) = \frac{\partial g_1(\mathbf{x})}{\partial \mathbf{x}} f = \begin{bmatrix} 1 & 0 & 0 \end{bmatrix} \begin{bmatrix} -c\,x_1 \\ -d\,x_2 \\ 0 \end{bmatrix} = -c\,x_1$$

$$L_f g_2(\mathbf{x}) = \frac{\partial g_2(\mathbf{x})}{\partial \mathbf{x}} f = \begin{bmatrix} 0 & 1 & 0 \end{bmatrix} \begin{bmatrix} -c\,x_1 \\ -d\,x_2 \\ 0 \end{bmatrix} = -d\,x_2$$

$$L_f g_3(\mathbf{x}) = \frac{\partial g_3(\mathbf{x})}{\partial \mathbf{x}} f = \begin{bmatrix} 0 & 0 & 1 \end{bmatrix} \begin{bmatrix} -c\,x_1 \\ -d\,x_2 \\ 0 \end{bmatrix} = 0$$

After inserting of the above calculated terms into the equation (9.26), the result of the coordinate transformation is given as follows:

$$\frac{dz_1}{dt} = -cx_1 + au_1 + \frac{a}{b}x_2u_3 = w_1$$

$$\frac{dz_2}{dt} = -dx_2 + bu_2 - \left(\frac{b}{a}x_1 + b\psi_p\right)u_3 = w_2 \qquad (9.27)$$

$$\frac{dz_3}{dt} = u_3 = w_3$$

From the equation (9.27) the new input vector **w** is derived as:

$$\mathbf{w} = \begin{bmatrix} w_1 \\ w_2 \\ w_3 \end{bmatrix} = \underbrace{\begin{bmatrix} -cx_1 \\ -dx_2 \\ 0 \end{bmatrix}}_{\mathbf{p(x)}} + \underbrace{\begin{bmatrix} a & 0 & ax_2/b \\ 0 & b & -bx_1/a - b\psi_p \\ 0 & 0 & 1 \end{bmatrix}}_{\mathbf{L(x)}} \underbrace{\begin{bmatrix} u_1 \\ u_2 \\ u_3 \end{bmatrix}}_{\mathbf{u}} \qquad (9.28)$$

- Step 4: The control law or the transformation law results from the equation (9.28).

$$\mathbf{u} = \underbrace{-\mathbf{L}^{-1}(\mathbf{x})\mathbf{p(x)}}_{\mathbf{a(x)}} + \mathbf{L}^{-1}(\mathbf{x})\mathbf{w} = \mathbf{a(x)} + \mathbf{L}^{-1}(\mathbf{x})\mathbf{w}$$

$$= \begin{bmatrix} u_1 \\ u_2 \\ u_3 \end{bmatrix} = \begin{bmatrix} cx_1/a \\ dx_2/b \\ 0 \end{bmatrix} + \begin{bmatrix} 1/a & 0 & -x_2/b \\ 0 & 1/b & x_1/a + \psi_p \\ 0 & 0 & 1 \end{bmatrix} \begin{bmatrix} w_1 \\ w_2 \\ w_3 \end{bmatrix} \qquad (9.29)$$

$$= \underbrace{\begin{bmatrix} L_{sd}x_1/T_{sd} \\ L_{sq}x_2/T_{sq} \\ 0 \end{bmatrix}}_{\mathbf{a(x)}} + \underbrace{\begin{bmatrix} L_{sd} & 0 & -L_{sq}x_2 \\ 0 & L_{sq} & L_{sd}x_1 + \psi_p \\ 0 & 0 & 1 \end{bmatrix}}_{\mathbf{L}^{-1}(\mathbf{x})} \underbrace{\begin{bmatrix} w_1 \\ w_2 \\ w_3 \end{bmatrix}}_{\mathbf{w}}$$

9.3.2 Feedback control structure with direct decoupling for PMSM

Similarly to the IM and using the state feedback or the coordinate transformation (9.29) the exact linearized PMSM model can be represented as in the figure 9.3. The difference between IM and PMSM consists here in the fact that a flux model is not needed any more, because the pole flux is permanently available.

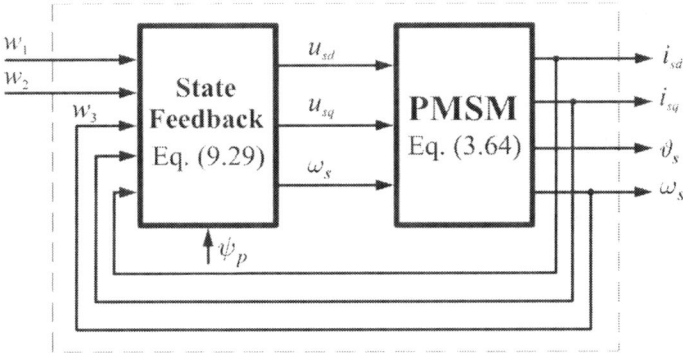

Fig. 9.3   Substitute linear process model of the PMSM

The new model in the figure 9.3 can be written in equation form as follows:

$$\begin{cases} \dfrac{d\mathbf{z}}{dt} = \begin{bmatrix} w_1 \\ w_2 \\ w_3 \end{bmatrix} = \begin{bmatrix} 0 & 0 & 0 \\ 0 & 0 & 0 \\ 0 & 0 & 0 \end{bmatrix} \mathbf{z} + \begin{bmatrix} 1 & 0 & 0 \\ 0 & 1 & 0 \\ 0 & 0 & 1 \end{bmatrix} \mathbf{w} \\[6pt] \mathbf{y} = \begin{bmatrix} z_1 \\ z_2 \\ z_2 \end{bmatrix} = \begin{bmatrix} 1 & 0 & 0 \\ 0 & 1 & 0 \\ 0 & 0 & 1 \end{bmatrix} \mathbf{z} \end{cases} \tag{9.30}$$

Similarly to the case IM with the equation (9.18), the following transfer function for the PMSM is obtained:

$$\mathbf{y}(s) = \begin{bmatrix} 1/s^{r_1} & 0 & 0 \\ 0 & 1/s^{r_2} & 0 \\ 0 & 0 & 1/s^{r_3} \end{bmatrix} \mathbf{w}(s) = \begin{bmatrix} 1/s & 0 & 0 \\ 0 & 1/s & 0 \\ 0 & 0 & 1/s \end{bmatrix} \mathbf{w}(s) \tag{9.31}$$

It can be seen easily that the same conclusions about the direct decoupling between the *dq* axes and the transfer functions of the decoupled input-output couples are valid also here. The summary then is that the control structure with direct decoupling in the figure 9.2 (of course without the flux model) can be used for the PMSM.

The new control concept in the two cases IM and PMSM with direct decoupling has some features besides the mentioned advantages which should be mentioned here:

- The transformation laws or the control laws (9.17) and (9.29) contain only static feedbacks, no time dependent components like integration and differentiation so that this nonlinear concept is very well suitable for a digital implementation.

- Because of parameter faults caused by an inaccurate parameter setting or by parameter changes during operation, stationary faults of the coordinate transformation always exist. This can be avoided by a dynamic concept using an additional parameter identification and adaptation or by additional integral parts in the transformation law.

9.4 References

Bodson M, Chiasson J (1998) Differential-Geometric Methods for Control of Electric Motors. Int. J. Robust Nonlinear Control 8, pp. 923-954

Cuong NT (2003) A Nonlinear Control Strategy based on Exact Linearization for the 3-Phase AC Drive Using Permanentmagnet-Excited Synchronous Motors. Master Thesis (in Vietnamese), Thai Nguyen University of Technology

Dawson DM, Hu J, Burg TC (2004) Nonlinear Control of Electric Machinery. Marcel Dekker, Inc., New York Basel

Duc DA (2004) A Nonlinear and Observer-based Control Strategy based on Exact Linearization for the 3-Phase AC Drive Using CSI-fed Induction Motor. Master Thesis (in Vietnamese), Thai Nguyen University of Technology

Ha TT (2003) A Nonlinear Control Strategy based on Exact Linearization for the 3-Phase AC Drive Using CSI-fed Induction Motor. Master Thesis (in Vietnamese), Thai Nguyen University of Technology

Hoa ND, Loc NX (2007) Construction of a drive test bed using DSP TMS320F2812 and Brushless DC motor, design of the non-linear controller by means of the method of the exact linearization. Bachelor Thesis (in Vietnamese), Hanoi University of Technology

Isidori A (1995) Nonlinear Control Systems. 3rd Edition, Springer-Verlag, London Berlin Heidelberg

Khorrami F, Krishnamurthy P, Melkote H (2003) Modeling and Adaptive Nonlinear Control of Electric Motors. Springer, Berlin Heidelberg New York

Krstíc M, Kanellakopoulos I, Kokotovíc P (1995) Nonlinear and Adaptive Control Design. John Wiley & Sons, Inc., New York

Nam DH (2004) A Nonlinear Control Strategy based on Exact Linearization for the 3-Phase AC Drive Using VSI-fed Induction Motor. Master Thesis (in Vietnamese), Hanoi University of Technology

Ortega R, Loría A, Nicklasson PJ, Sira-Ramírez H (1998) Passivity-based Control of Euler-Lagrange Systems: Mechanical, Electrical and Electromechanical Applications. Springer, London Berlin Heidelberg

Quang NP, Dittrich JA (1999) Praxis der feldorientierten Drehstromantriebs-regelungen. 2. Aufl., Expert-Verlag

Wey T (2001) Nichtlineare Regelungssysteme: Ein differentialalgebraischer Ansatz. B.G. Teubner Stuttgart - Leipzig - Wiesbaden.

10 Linear control structure for wind power plants with DFIM

Due to the mechanical wear at the slip-rings wound rotor induction machines are only rarely used as motors today. Regardless, they gain increasing importance for generator applications in wind and water power plants. Their main advantage in this area is that the inverter in the rotor circuit can be designed to a smaller capacity with regard to the maximum generator output. Furthermore, the DFIM can be ran in a comparatively large slip range which allows for better utilization of the available wind energy.

10.1 Construction of wind power plants with DFIM

By means of a bidirectional converter in the rotor circuit the DFIM is able to work as a generator in both subsynchronous and oversynchronous operating areas. In both cases the stator is feeding energy into the mains. Figures 10.1b and 10.1c show the energy flow in both operating areas. The rotor:

- takes energy from the mains in subsynchronous mode and
- feeds energy back to the mains in oversynchronous mode.

Independent of the mechanical speed (subsynchronous, oversynchronous) only the sign of the electrical torque determines if the DFIM is working in motor or in generator mode. According to figure 10.1a the generator mode is defined by a negative torque. The torque amplitude is equivalent to the delivered or received active power of the machine. During dynamic changes of torque or active power the adjusted power factor of the plant is required to stay uninfluenced and constant.

The front-end converter controls the active power flow into the DC link by controlling the DC link voltage amplitude. The inverter on the rotor side (figures 1.9, 1.10) is responsible for the adjustment of the desired torque. The control strategy for the reactive current in both inverters is designed to establish the desired power factor of the overall plant.

Therefore the generator is capable to work as reactive power compensator or reactive power generator as well.

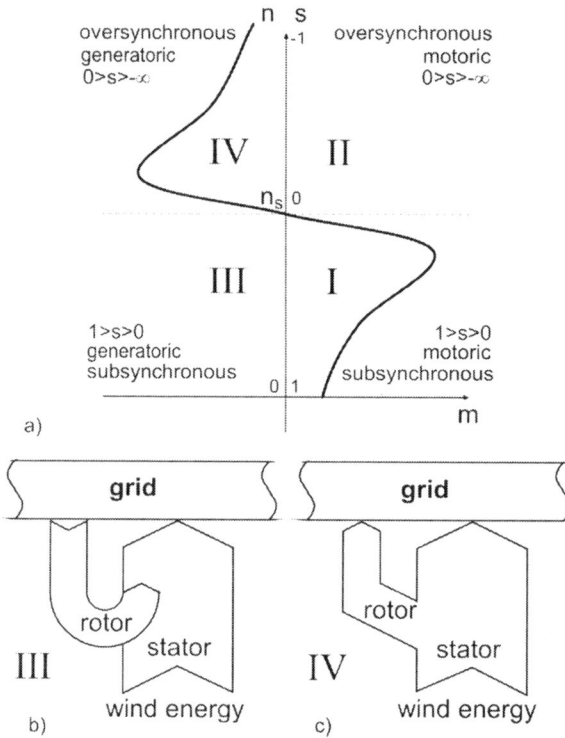

Fig. 10.1 Operating areas of the DFIM and energy flow in generator: (**a**) Operating areas of the DFIM; (**b**) Energy flow in subsynchronous mode; (**c**) Energy flow in oversynchronous mode

A number of works have been published in the recent years dealing with the problem of variable speed generators in wind or water power applications ([Suchaneck 1985], [E-Cathor 1987], [Dietrich 1990], [Pena 1995], [Tnani 1995], [Stemmler 1995]). Some of them are investigating different configurations for the converter-generator system like synchronous generator [E-Cathor 1987], oversynchronous cascade converter with natural commutation [Dietrich 1990]. In [Suchaneck 1985] a DFIM generator system for the small power range is proposed using the rotor current components as actuating variables for active and reactive power. A decoupled control of torque and rotor excitation current is presented in [Pena 1995] using stator flux orientation. Special attention is given to optimum speed tracking of the wind turbine. In [Tnani 1995] a decoupled control of active and reactive power is developed using a complete dynamic machine model for decoupling. A predictive load model is introduced to consider changing consumer conditions on the grid.

The chapter 10 deals with the often and successfully practiced control concept with decoupling between the torque m_G and the power factor $\cos\varphi$, and therewith between the active power P and reactive power Q [Quang 1997]. The design of the system is shown in the figure 10.2.

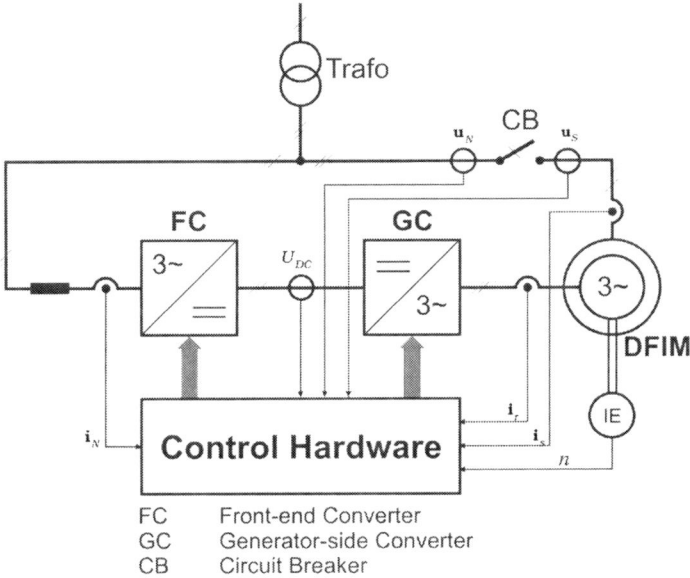

FC   Front-end Converter
GC   Generator-side Converter
CB   Circuit Breaker

Fig. 10.2 Construction of the wind power plants using DFIM

10.2 Grid voltage orientated controlled systems

The equations (1.9) – (1.15) and the representations in the section 1.3 have indicated the meaning of the grid voltage orientation in short form. The stator of the machine is connected to the constant-voltage constant-frequency mains system. Since the stator frequency is always identical to the mains frequency, the voltage drop across the stator resistance can be neglected compared to the voltage drop across the main and leakage inductances L_m and $L_{\sigma s}$. Therewith the equation (1.10) was obtained:

$$\mathbf{u}_s = R_s \mathbf{i}_s + \frac{d\psi_s}{dt} \quad \Rightarrow \quad \mathbf{u}_s \approx \frac{d\psi_s}{dt} \text{ or } \mathbf{u}_s \approx j\omega_s \psi_s \tag{10.1}$$

Equations (10.1) say that the stator flux is always lagging the stator voltage by a phase angle of nearly 90 degree. Therefore the orientation of the reference frame to the stator voltage approximately means orientation to the stator flux as well.

- For stator flux orientation: $u_{sd} = 0$, $\psi_{sq} = 0$
- For grid voltage orientation: $u_{sq} = 0$, $\psi_{sd} = 0$

The implementation of the grid voltage orientation requires the accurate and robust acquisition of the phase angle of the grid voltage fundamental wave, considering strong distortions due to converter mains pollution or other harmonic sources. Usually this is accomplished by means of a phase locked loop (PLL).

10.2.1 Control variables for active and reactive power

The torque of the DFIM is defined as:

$$m_G = \frac{3}{2}z_p\left(\psi_s \times i_s\right) = -\frac{3}{2}z_p\left(\psi_r \times i_r\right) \tag{10.2}$$

The equation (10.2) opens several ways for calculating the electrical torque m_G. Because the DFIM is controlled from the rotor circuit, the design of the torque control requires a form of the torque equation which includes the rotor current. After introducing the rotor current into the first part of (10.2) the following formula will be obtained:

$$m_G = -\frac{3}{2}z_p\frac{L_m}{L_s}\left(\psi_s \times i_r\right) = -\frac{3}{2}z_p\left(1-\sigma\right)L_r\left(\psi_s' \times i_r\right) \tag{10.3}$$

Considering the grid voltage reference frame (reference coordinate system) with $\psi_{sd} = 0$, equation (10.3) can be simplified to:

$$m_G = -\frac{3}{2}z_p\frac{L_m}{L_s}\psi_{sq}i_{rd} = -\frac{3}{2}z_p\left(1-\sigma\right)L_r\psi_{sq}'i_{rd} \tag{10.4}$$

Because the stator flux is kept constant by the constant grid voltage (refer to (10.1)) *the component i_{rd} may be considered as the torque forming current which also represents the active power P of the generator.*

The equation (10.4) needs the knowledge of the machine inductances for the torque calculation. The main inductance depends on the magnetization working point which is determined by the grid voltage amplitude. For the latter tolerances of at least +/-10% must be considered because the working point may be shifted into or out of the saturation range due to variations of the grid voltage amplitude. Therefore, using equation (1.16) for torque feedback calculation requires an accurate measuring and consideration of the L_m saturation characteristic, otherwise the requirements for the torque accuracy cannot be fulfilled. It would be better to utilize a torque model which does not require the knowledge of any machine parameters, or at least not any working point dependent ones like the main inductance. For this, the left equation (10.2) offers itself. Substituting the stator flux linkage by the first equation of (3.77) and considering grid voltage orientation with $d\psi_s/dt = 0$ it is obtained:

$$m_G = \frac{3}{2}z_p \frac{u_{sd}i_{sd} - R_s i_s^2}{\omega_s} \qquad (10.5)$$

For usual values of the stator resistance of large machines, (10.5) can be re-written with good approximation to:

$$m_G = \frac{3}{2}z_p \frac{u_{sd}i_{sd}}{\omega_s} \qquad (10.6)$$

On the other hand it was already shown in the section 1.3 that thanks to the relations (1.12), (1.14) and (1.15) *the current i_{rq} is regarded as a $\cos\varphi$ forming component or as the decisive current for the reactive power Q.*

10.2.2 Dynamic rotor current control for decoupling of active and reactive power

The control system of the DFIM consists of two parts: the generator-side control and the grid-side control (front-end converter control), in which the generator-side control was already introduced in the figure 1.9.

In section 3.4 the (continuous and discrete) state space models of the DFIM were introduced. Since the two rotor current components i_{rd}, i_{rq} play the role of P and Q control variables an inner control loop to impress the rotor current is needed.

The process model of the rotor current is represented by the figure 3.15b and by the first of both equations (3.85):

$$\mathbf{i}_r(k+1) = \mathbf{\Phi}_{11}\mathbf{i}_r(k) + \mathbf{\Phi}_{12}\mathbf{\psi}_s'(k) + \mathbf{H}_{s1}\mathbf{u}_s(k) + \mathbf{H}_{r1}\mathbf{u}_r(k) \qquad (10.7)$$

After splitting (10.7) into its real and imaginary components, the following equations are obtained in the grid voltage orientated reference frame:

$$\begin{cases} i_{rd}(k+1) = \Phi_{11}i_{rd}(k) + \Phi_{12}i_{rq}(k) + \Phi_{14}\psi_{sq}'(k) \\ \qquad\qquad + h_{11s}u_{sd}(k) + h_{11r}u_{rd}(k) \\ i_{rq}(k+1) = -\Phi_{12}i_{rd}(k) + \Phi_{11}i_{rq}(k) + \Phi_{13}\psi_{sq}'(k) \\ \qquad\qquad + h_{11r}u_{rq}(k) \end{cases} \qquad (10.8)$$

In (10.8) the stator flux and the stator voltage might be regarded as disturbances to be compensated by a feed-forward control. These values are nearly constant and therefore can be compensated exactly and fast enough by the implicit integral part of the controller, so that their feed-forward compensation may be omitted. Regardless, the compensation shall be included in the following steps to achieve a more coherent design.

The further design is identical to the description in the chapter 5. Introducing $\mathbf{y}(k)$ as output value of the vector controller $\mathbf{R}_i$, the following approach for the feed-forward compensation network may be constructed:

$$\mathbf{u}_r(k+1) = \mathbf{H}_{r1}^{-1}\left[\mathbf{y}(k) - \mathbf{\Phi}_{12}\mathbf{\psi}_s'(k+1) - \mathbf{H}_{s1}\mathbf{u}_s(k+1)\right] \qquad (10.9)$$

After inserting (10.9) into (10.7) the compensated process will be obtained:

$$\mathbf{i}_r(k+1) = \mathbf{\Phi}_{11}\mathbf{i}_r(k) + \mathbf{y}(k-1) \text{ or } z\mathbf{i}_r(z) = \mathbf{\Phi}_{11}\mathbf{i}_r(z) + z^{-1}\mathbf{y}(z)$$
$$(10.10)$$

The structure of the current controller is shown in the figure 10.3. The compensated process equation (10.10) contains a dead time of one sampling interval $\mathbf{y}(k\text{-}1)$ to account for the delay between the feedback acquisition and the voltage output (computing delay). $\mathbf{y}(k\text{-}1)$ fulfills the following equation in the z-domain:

$$\mathbf{y}(z) = \mathbf{R}_I\left[\mathbf{i}_r^*(z) - \mathbf{i}_r(z)\right] \qquad (10.11)$$

Superscript „*“: reference value (set point)

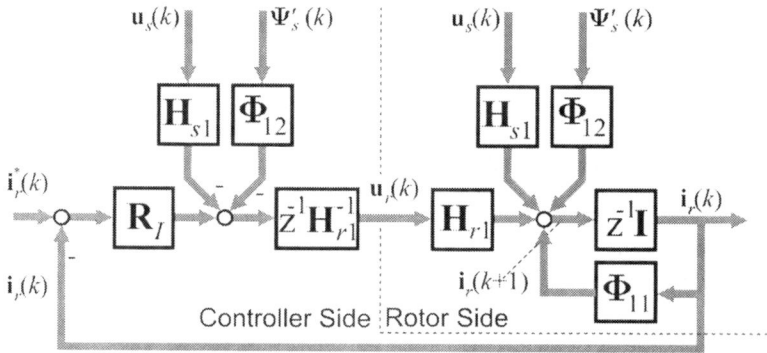

Fig. 10.3 Rotor current control structure using feed-forward compensation of stator voltage and stator flux

After substituting (10.11) into (10.10) the following transfer function of the closed current loop will be obtained:

$$\mathbf{i}_r(z) = z^{-1}\left[z\mathbf{I} - \mathbf{\Phi}_{11} + z^{-1}\mathbf{R}_I\right]^{-1}\mathbf{R}_I\mathbf{i}_r^*(z) \qquad (10.12)$$

Similarly to the chapter 5 a vector controller with dead-beat behavior for (10.12) is given by:

$$\mathbf{R}_I = \frac{\mathbf{I} - z^{-1}\mathbf{\Phi}_{11}}{1 - z^{-2}} \qquad (10.13)$$

The dead-beat behavior results in fast dynamics and accuracy, the vector design ensures good decoupling between the components i_{rd} and i_{rq}. For a less fast (and thus less noise sensitive) behavior, designs with finite adjustment times (cf. section 5.3.3) or PI-type designs may be applicable as well.

If the tracking errors are defined as:

$$x_{wd}(k) = i_{rd}^*(k) - i_{rd}(k); \quad x_{wq}(k) = i_{rq}^*(k) - i_{rq}(k) \qquad (10.14)$$

the controller equations in the time-domain including the disturbance feed-forward compensation can be written as follows:

$$\begin{cases} u_{rd}(k+1) = h_{11r}^{-1}\Big[x_{wd}(k) - \Phi_{11}x_{wd}(k-1) - \Phi_{12}x_{wq}(k-1) \\ \qquad\qquad + y_d(k-2) - \Phi_{14}\psi_{sq}' - h_{11s}u_{sd}\Big] \\ u_{rq}(k+1) = h_{11r}^{-1}\Big[x_{wq}(k) + \Phi_{12}x_{wd}(k-1) - \Phi_{11}x_{wq}(k-1) \\ \qquad\qquad + y_q(k-2) - \Phi_{13}\psi_{sq}'\Big] \end{cases} \qquad (10.15)$$

The complete structure of the generator-side control was already shown in the figure 1.9.

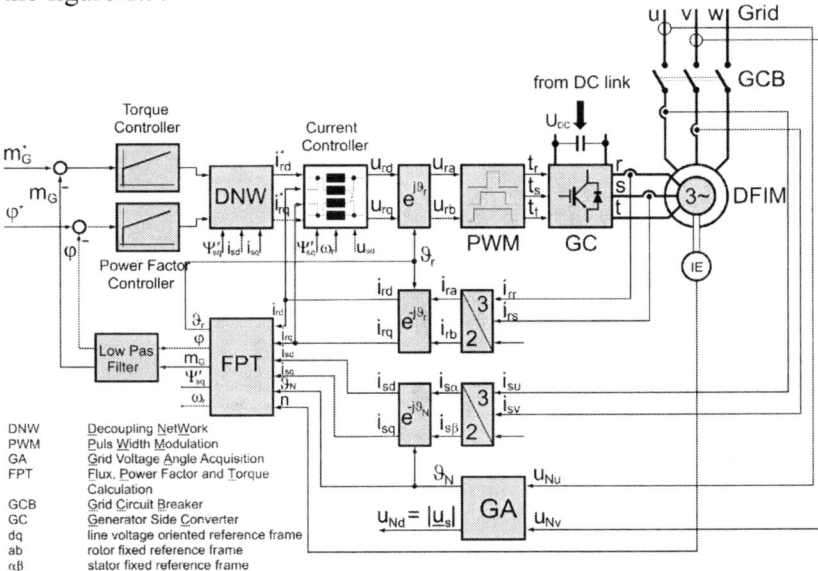

Fig. 10.4 Generator-side control structure using linear rotor current controller

10.2.3 Problems of the implementation

The decoupling structure in the figure 10.4 consists of two essential functional blocks:
- Calculation of the feedback (actual) values for torque and power factor using stator voltage, currents and speed (block FPT) and
- Calculation of the current set points (block DNW).

a) Block FPT: The following equations are realized:
- Calculation of stator flux:

$$\psi_{sq}' \approx i_{sq} + i_{rq} \tag{10.16}$$

- Calculation of torque:

$$m_G = \frac{3}{2} z_p \frac{u_{sd} i_{sd}}{\omega_s} \tag{10.17}$$

- Calculation of stator current amplitude:

$$|\mathbf{i}_s| = \sqrt{i_{sd}^2 + i_{sq}^2} \tag{10.18}$$

- Calculation of $\sin\varphi$:

$$\sin\varphi = \frac{i_{sq}}{|\mathbf{i}_s|} \tag{10.19}$$

The calculated torque and $\sin\varphi$ values are serving as feedbacks for the torque and power factor control. Both controllers are of PI type.

b) Block DNW: Using the torque and power factor controller outputs y_M and y_φ, the reference values for the inner-loop current control are calculated as follows:
- Torque forming current component i_{rd}^*:

$$i_{rd}^* = \frac{y_M}{-\frac{3}{2} z_p (1-\sigma) L_r \psi_{sq}'} \tag{10.20}$$

- $\sin\varphi$ forming current component i_{rq}^*:

$$i_{rq}^* = \psi_{sq}' - y_\varphi |\mathbf{i}_s| \tag{10.21}$$

The effectiveness of the decoupling is shown in figure 10.5. The diagrams show step responses of the torque and the power factor at both oversynchronous and subsynchronous speed. For the subsynchronous case both low pass filtered and unfiltered values of the torque and power factor are depicted. The control is optimized to the filtered values therefore the unfiltered values exhibit a stronger overshot. As already pointed out a nearly undelayed injection of the reference values is theoretically possible, however, measurement noise and current harmonics force to slow down the control dynamics. The adjusted rise time of the unfiltered torque of about 30 ms fulfills all practical requirements. During the torque transients (step size about half the nominal torque) almost no changes of cosφ are visible, the same applies for the torque during cosφ transients.

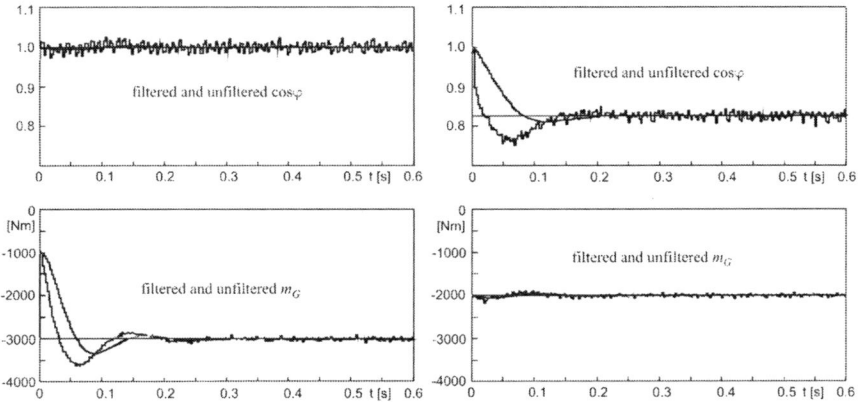

Fig. 10.5 Torque and cosφ step changes of a 620kW generator. **(left)** Torque reference -1000 → -3000Nm; **(right)** cosφ reference 1.0 → 0.825

10.3 Front-end converter current control

The front-end converter control is feeding the DC link from the grid during subsynchronous operation and feeds back energy to the supply during oversynchronous operation. This task is accomplished by means of a DC link voltage control, which is supposed to keep a constant bus voltage in all operation modes. Corresponding to the generator control the grid side contains a power factor control loop which compensates the line filter reactive-power demand and in connection with the generator-side realizes the reactive power reference for the overall plant.

The output current of the mains converter $\mathbf{i}_N$ can be split as usual into real and imaginary parts:

$$\mathbf{i}_N = i_{Nd} + j\, i_{Nq} \tag{10.22}$$

In the grid voltage orientated reference frame, the real part acts as active power producing component and the imaginary part as reactive power producing component. Therefore, i_{Nd} is a good choice to be used as control variable for the DC link voltage. Appropriately, i_{Nq} forms the control variable for the power factor control on the grid side. Similarly to the generator control, the power factor is controlled indirectly via $\sin\varphi$, being direct proportional to i_{Nq}, cf. equation (1.15). Both U_{DC} and $\sin\varphi$ controllers are of PI-type.

The commands i_{Nd} and i_{Nq} are realized by an inner-loop current control. Therefore the complete structure of the grid-side control can be given like in the figure 1.10.

As it had turned out in the previous section, the current control plays a decisive role in the control concept and for its correct realization. Therefore it shall be described in some more detail in the following section. All current control loops are designed in the grid voltage orientated reference frame.

10.3.1 Process model

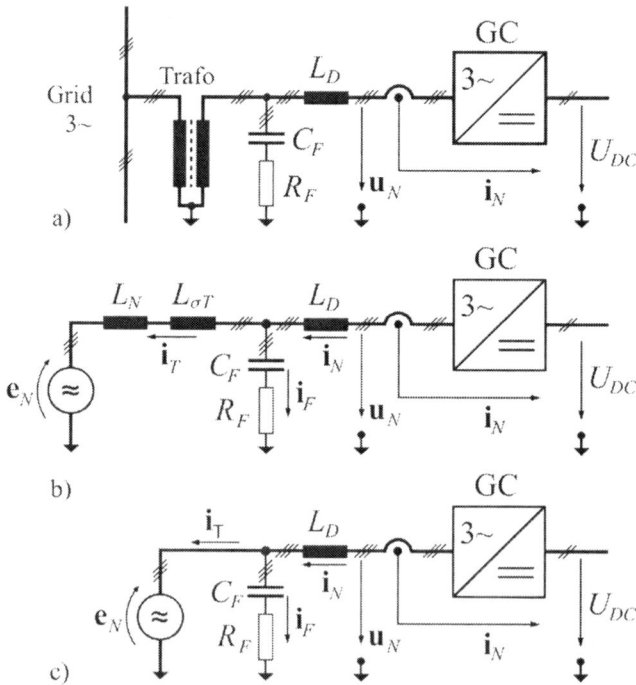

Fig. 10.6 Grid-side control plant: (**a**) Common grid-side scheme; (**b**) Equivalent circuit; (**c**) Equivalent circuit for process model derivation

The front-end converter is usually connected to the supply through the reactor L_D, RC-filter and transformer (figure 10.6a). Considering the grid inductance L_N and at the same time replacing the transformer by its leakage inductance $L_{\sigma T}$ and the grid by the voltage source $\mathbf{e}_N$ gives the equivalent circuit in the figure 10.6b. The voltage drop over the transformer and the grid inductance is considerably smaller then the one over the RC-filter and therefore may be neglected. Taking into account the resistance R_D of the inductor the equivalent circuit in figure 10.6c is finally obtained as the starting-point for the derivation of the current control process model (R_D is not drawn here).

In steady state condition the following filter equation is derived from figure 10.6c:

$$\mathbf{e}_N = \frac{1}{j\omega_s C_F}\mathbf{i}_F + R_F\mathbf{i}_F \tag{10.23}$$

In the grid voltage orientated reference frame and with:

$$\mathbf{e}_N = e_{Nd} + je_{Nq}; e_{Nq} = 0 \tag{10.24}$$

equation (10.23) can be written in component form as follows:

$$\begin{cases} e_{Nd} = R_F i_{Fd} - \dfrac{1}{\omega_s C_F}i_{Fq} \\[2mm] 0 = R_F i_{Fq} + \dfrac{1}{\omega_s C_F}i_{Fd} \end{cases} \quad \text{with} \quad \mathbf{i}_F = i_{Fd} + ji_{Fq} \tag{10.25}$$

With equation (10.25) the filter currents i_{Fd} and i_{Fq} can be calculated for the stationary condition. In steady state, these currents and the grid voltage $\mathbf{e}_N$ may be considered as constant and therefore act as constant disturbances for the control loop. They may be compensated by feed-forward control or by an integral part in the controller.

The grid current and voltage equations can be derived from the figure 10.5c:

$$\begin{cases} \mathbf{u}_N = R_D\mathbf{i}_N + L_D\dfrac{d\mathbf{i}_N}{dt} + \mathbf{e}_N \\[2mm] \mathbf{i}_N = \mathbf{i}_T + \mathbf{i}_F \end{cases} \tag{10.26}$$

After transforming (10.26) into the grid voltage orientated reference frame and substituting the current $\mathbf{i}_N$ from the first equation of (10.26) by the second equation the new voltage equation is obtained:

$$\mathbf{u}_N = R_D\mathbf{i}_T + L_D\frac{d\mathbf{i}_T}{dt} + j\omega_s L_D\mathbf{i}_T + \mathbf{e}_{Nv} \tag{10.27}$$

with:

$$\mathbf{e}_{Nv} = \mathbf{e}_N + j\omega_s L_D \mathbf{i}_F + R_D \mathbf{i}_F = e_{Nvd} + j e_{Nvq}$$

$$L_D \frac{d\mathbf{i}_F}{dt} = 0 \tag{10.28}$$

From (10.27) the continuous process model in the state space can be written in component form:

$$\begin{vmatrix} \dfrac{di_{Td}}{dt} = -\dfrac{1}{T_D} i_{Td} + \omega_s i_{Tq} + \dfrac{1}{L_D}\left(u_{Nd} - e_{Nvd}\right) \\[3mm] \dfrac{di_{Tq}}{dt} = -\omega_s i_{Td} - \dfrac{1}{T_D} i_{Tq} + \dfrac{1}{L_D}\left(u_{Nq} - e_{Nvq}\right) \end{vmatrix} \tag{10.29}$$

The time discrete state space model is given immediately by:

$$\mathbf{i}_T(k+1) = \mathbf{\Phi}_N \mathbf{i}_T(k) + \mathbf{H}_N \mathbf{u}_N(k) - \mathbf{H}_N \mathbf{e}_{Nv}(k) \tag{10.30}$$

with:

$$\mathbf{\Phi}_N = \begin{vmatrix} 1 - \dfrac{T}{T_D} & \omega_s T \\[3mm] -\omega_s T & 1 - \dfrac{T}{T_D} \end{vmatrix} ; \ \mathbf{H}_N = \begin{vmatrix} \dfrac{T}{L_D} & 0 \\[3mm] 0 & \dfrac{T}{L_D} \end{vmatrix} \tag{10.31}$$

With the process model (10.30) the controller design can be carried out now. If only the inverter output current $\mathbf{i}_N$ is measured, the filter current $\mathbf{i}_F$ must be calculated by means of (10.25) and finally the grid (or transformer) current $\mathbf{i}_T$ using (10.26).

10.3.2 Controller design

The design is similar to the rotor current controller. With the controller output $\mathbf{y}_N(k)$, the following equation can be written for the feed-forward compensation network according to (10.30) considering the computation time delay:

$$\mathbf{u}_N(k+1) = \mathbf{H}_N^{-1}\left[\mathbf{y}_N(k) + \mathbf{H}_N \mathbf{e}_{Nv}(k+1)\right] \tag{10.32}$$

After substituting (10.32) into (10.31) the compensated process model is given:

$$\mathbf{i}_T(k+1) = \mathbf{\Phi}_N \mathbf{i}_T(k) + \mathbf{y}_N(k-1) \tag{10.33}$$

Two things have to be mentioned here:

- The structure of the compensated process model is totally identical with the one of the rotor side (10.10).
- Likewise, $\mathbf{e}_N(k)$ acts as constant disturbance and therefore may be compensated automatically by an implicit integral part in the controller without any special compensation.

For control with dead-beat response and a good decoupling between the current components the controller equation can be immediately written to:

$$\mathbf{R}_{IN} = \frac{\mathbf{I} - z^{-1}\mathbf{\Phi}_N}{1 - z^{-2}}$$

(10.34)

Neglecting the feed-forward compensation of the constant disturbances the equation of the controller is obtained in the time domain as:

$$
\begin{cases}
u_{Nd}(k+1) = \dfrac{L_D}{T}\left[x_{Td}(k) - \left(1 - \dfrac{T}{T_D}\right)x_{Td}(k-1) - \omega_s T x_{Tq}(k-1)\right.\\
\qquad\qquad\qquad \left. + y_{Nd}(k-2)\right]\\[2mm]
u_{Nq}(k+1) = \dfrac{L_D}{T}\left[x_{Tq}(k) + \omega_s T x_{Td}(k-1) - \left(1 - \dfrac{T}{T_D}\right)x_{Tq}(k-1)\right.\\
\qquad\qquad\qquad \left. + y_{Nq}(k-2)\right]
\end{cases}
$$

(10.35)

with:

$$\mathbf{x}_T = x_{Td} + jx_{Tq}\ ;\ x_{Td} = i^*_{Td} - i_{Td}\ ;\ x_{Tq} = i^*_{Tq} - i_{Tq}\ ;\ \mathbf{y}_N = y_{Nd} + jy_{Nq}$$

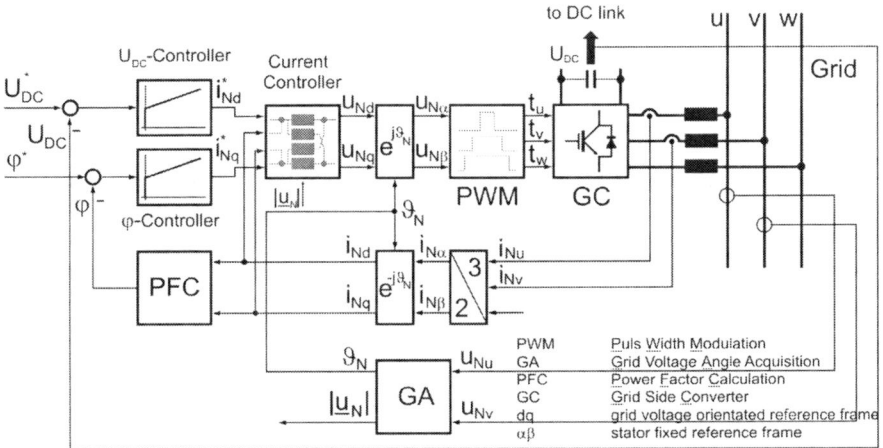

Fig. 10.7 Complete grid-side control structure

Because the control starts with immediately connected disturbance $\mathbf{e}_N$, the controller output $\mathbf{y}_N$ and its past values must be pre-initialized to avoid undesirable transients:

$$\mathbf{y}_N(k=0) = -\mathbf{H}_N\mathbf{e}_{Nv}(k=0)$$

(10.36)

The controller (10.34) - (10.35) works robust and reliable. However, if the reactor L_D is dimensioned too small, the high-dynamic behavior can cause undesirable ripples in the transformer or grid current $\mathbf{i}_T$. In this case a controller design with finite adjustment time would lead to a current waveform with a better THD coefficient. The complete grid-side control structure is shown in the figure 10.7 (cf. [Quang 1997]).

10.4 References

Arsudis D (1989) Doppeltgespeister Drehstromgenerator mit Spannungszwischen-kreis-Umrichter im Rotorkreis für Windkraftanlagen. Dissertation, TU Carolo-Wilhelmina zu Braunschweig

Dietrich W (1990) Drehzahlvariables Generatorsystem für Windkraftanlagen mittlerer Leistung. Dissertation, TU Berlin

E-Cathor J (1987) Drehzahlvariable Windenergieanlage mit Gleichstrom-zwischenkreis-Umrichter und Optimum-suchendem Regler. Dissertation, TU Carolo-Wilhelmina zu Braunschweig

Pena RS, Clare JC, Asher GM (1995) Implementation of Vector Control Strategies for a Variable Speed Doubly-Fed Induction Machine for Wind Generator System. EPE '95 Sevilla, pp. 3.075 - 3.080

Quang NP, Dittrich JA (1999) Praxis der feldorientierten Drehstromantriebs-regelungen. 2. Aufl., Expert-Verlag

Quang NP, Dittrich JA, Thieme A (1997) Doubly-fed induction machine as generator: control algorithms with decoupling of torque and power factor. Electrical Engineering / Archiv für Elektrotechnik, 10.1997, pp. 325-335

Späth H (1983) Steuerverfahren für Drehstrommaschinen: Theoretische Grundlagen. Springer-Verlag: Berlin Heidelberg New York Tokyo

Stemmler H, Omlin A (1995) Converter Controlled Fixed-Frequency Variable-Speed Motor/Generator. IPEC Yokohama '95, pp. 170 - 176

Suchaneck J (1985) Untersuchung einer doppeltgespeisten Asynchronmaschine mit feldorientierter Steuerung zur elektrischen Windenergienutzung bis 20kW. Dissertation, TU Berlin

Tnani S, Diop S, Jones SR, Berthon A (1995) Novel Control Strategy of Double-Fed Induction Machines. EPE '95 Sevilla, pp. 1.553 - 1.558

11 Nonlinear control structure with direct decoupling for wind power plants with DFIM

11.1 Existing problems at linear controlled wind power plants

An important criterion for the design of the control concept is to maintain the decoupling of active and reactive power in both steady state and dynamic mode. This requirement is fulfilled to a very high degree by using a rotor current controller (cf. chapter 10 or [Quang 1997]) to impress control variables (rotor current components) which immediately inject torque and power factor (cf. figure 10.4). This structure, in which the current controller is always based on DFIM models linearized within a sampling period, was successfully implemented in wind power plants. The linearization is made by the assumption that the sampling time T of the discretization is chosen small enough so that the rotor frequency ω_r can be regarded as constant within T. Because of this assumption the frequency ω_r is now a parameter in the system matrix of the discrete process model, and the bilinear continuous model becomes a linear time-variant system for which the well known design methods of linear systems can be used.

Nowadays most grid suppliers request ride-through of the wind turbine during grid faults (short-circuits, ref. to [Dittrich 2003]). That means the wind power plant must be able to feed reactive power into the grid to support the retaining voltage level, and the rotor frequency ω_r becomes very dynamic. In these cases, in which the linearization condition can not be fulfilled any more, a nonlinear design for the rotor current control loop would be able to deliver better results than the linear controller. Within the last few years a number of efforts had been made on this issue from both theoretical and practical [Chi 2005], [Quang 2005]. They succeeded – as in the case of the IM and PMSM – in a new control structure with direct decoupling between the dq axes using the method of the exact linearization.

11.2 Nonlinear control structure for wind power plants with DFIM

In the section 3.6.3 the nonlinear process model of the DFIM was already developed as the starting point for the controller design.

$$\begin{cases} \overset{\bullet}{\mathbf{x}} = \mathbf{f}(\mathbf{x}) + \mathbf{h}_1(\mathbf{x})u_1 + \mathbf{h}_2(\mathbf{x})u_2 + \mathbf{h}_3(\mathbf{x})u_3 \\ \mathbf{y} = \mathbf{g}(\mathbf{x}) \end{cases} \tag{11.1}$$

with:

$$\mathbf{f}(\mathbf{x}) = \begin{bmatrix} -ax_1 \\ -ax_2 \\ 0 \end{bmatrix}; \mathbf{h}_1(\mathbf{x}) = \begin{bmatrix} 1 \\ 0 \\ 0 \end{bmatrix}; \mathbf{h}_2(\mathbf{x}) = \begin{bmatrix} 0 \\ 1 \\ 0 \end{bmatrix}; \mathbf{h}_3(\mathbf{x}) = \begin{bmatrix} x_2 \\ -x_1 \\ 1 \end{bmatrix} \tag{11.2}$$

$$g_1(\mathbf{x}) = x_1; g_2(\mathbf{x}) = x_2; g_3(\mathbf{x}) = x_3$$

- Parameters:

$$a = \left(\frac{1}{\sigma T_r} + \frac{1-\sigma}{\sigma T_s}\right); b = \frac{1-\sigma}{\sigma}; c = \frac{1}{\sigma L_r}; d = \frac{1-\sigma}{\sigma L_m}; e = \frac{1-\sigma}{\sigma T_s}$$

- State variables:

$$\mathbf{x}^T = \begin{bmatrix} x_1 & x_2 & x_3 \end{bmatrix}; x_1 = i_{rd}; x_2 = i_{rq}; x_3 = \theta_r$$

- Input variables:

$$\mathbf{u}^T = \begin{bmatrix} u_1 & u_2 & u_3 \end{bmatrix}; u_1 = e\psi'_{sd} - b\omega\psi'_{sq} + cu_{rd} - du_{sd}$$

$$u_2 = b\omega\psi'_{sd} + e\psi'_{sq} + cu_{rq} + du_{sq}; u_3 = \omega_r$$

- Output variables:

$$\mathbf{y}^T = \begin{bmatrix} y_1 & y_2 & y_3 \end{bmatrix}; y_1 = i_{rd}; y_2 = i_{rq}; y_3 = \theta_r$$

11.2.1 Nonlinear controller design based on "exact linearization"

The design is made similarly as in the case of the IM.
- Step 1: Calculation of the vector **r**.
  a) Case $j = 1$:

$$L_{h_1} g_1(\mathbf{x}) = \frac{\partial g_1(\mathbf{x})}{\partial \mathbf{x}} \mathbf{h}_1(\mathbf{x}) = \begin{bmatrix} 1 & 0 & 0 \end{bmatrix} \begin{bmatrix} 1 \\ 0 \\ 0 \end{bmatrix} = 1 \neq 0 \tag{11.3}a$$

$$L_{h_2} g_1(\mathbf{x}) = \frac{\partial g_1(\mathbf{x})}{\partial \mathbf{x}} \mathbf{h}_2(\mathbf{x}) = \begin{bmatrix} 1 & 0 & 0 \end{bmatrix} \begin{bmatrix} 0 \\ 1 \\ 0 \end{bmatrix} = 0 \tag{11.3)b}$$

$$L_{h_3} g_1(\mathbf{x}) = \frac{\partial g_1(\mathbf{x})}{\partial \mathbf{x}} \mathbf{h}_3(\mathbf{x}) = \begin{bmatrix} 1 & 0 & 0 \end{bmatrix} \begin{bmatrix} x_2 \\ -x_1 \\ 1 \end{bmatrix} = x_2 \neq 0 \tag{11.3)c}$$

From (11.3) it results $r_1 = 1$.
b) Case $j = 2$:

$$L_{h_1} g_2(\mathbf{x}) = \frac{\partial g_2(\mathbf{x})}{\partial \mathbf{x}} \mathbf{h}_1(\mathbf{x}) = \begin{bmatrix} 0 & 1 & 0 \end{bmatrix} \begin{bmatrix} 1 \\ 0 \\ 0 \end{bmatrix} = 0 \tag{11.4)a}$$

$$L_{h_2} g_2(\mathbf{x}) = \frac{\partial g_2(\mathbf{x})}{\partial \mathbf{x}} \mathbf{h}_2(\mathbf{x}) = \begin{bmatrix} 0 & 1 & 0 \end{bmatrix} \begin{bmatrix} 0 \\ 1 \\ 0 \end{bmatrix} = 1 \neq 0 \tag{11.4)b}$$

$$L_{h_3} g_2(\mathbf{x}) = \frac{\partial g_2(\mathbf{x})}{\partial \mathbf{x}} \mathbf{h}_3(\mathbf{x}) = \begin{bmatrix} 0 & 1 & 0 \end{bmatrix} \begin{bmatrix} x_2 \\ -x_1 \\ 1 \end{bmatrix} = -x_1 \neq 0 \tag{11.4)c}$$

Therewith it follows $r_2 = 1$.
c) Case $j = 3$:

$$L_{h_1} g_3(\mathbf{x}) = \frac{\partial g_3(\mathbf{x})}{\partial \mathbf{x}} \mathbf{h}_1(\mathbf{x}) = \begin{bmatrix} 0 & 0 & 1 \end{bmatrix} \begin{bmatrix} 1 \\ 0 \\ 0 \end{bmatrix} = 0 \tag{11.5)a}$$

$$L_{h_2} g_3(\mathbf{x}) = \frac{\partial g_3(\mathbf{x})}{\partial \mathbf{x}} \mathbf{h}_2(\mathbf{x}) = \begin{bmatrix} 0 & 0 & 1 \end{bmatrix} \begin{bmatrix} 0 \\ 1 \\ 0 \end{bmatrix} = 0 \tag{11.5)b}$$

$$L_{h_3} g_3(\mathbf{x}) = \frac{\partial g_3(\mathbf{x})}{\partial \mathbf{x}} \mathbf{h}_3(\mathbf{x}) = \begin{bmatrix} 0 & 0 & 1 \end{bmatrix} \begin{bmatrix} x_2 \\ -x_1 \\ 1 \end{bmatrix} = 1 \neq 0 \tag{11.5)c}$$

With the equation (11.5) $r_3 = 1$ is obtained now.
- Step 2: Calculation of the matrix $\mathbf{L}$.

$$\mathbf{L}(\mathbf{x}) = \begin{bmatrix} L_{h_1} g_1(\mathbf{x}) & L_{h_2} g_1(\mathbf{x}) & L_{h_3} g_1(\mathbf{x}) \\ L_{h_1} g_2(\mathbf{x}) & L_{h_2} g_2(\mathbf{x}) & L_{h_3} g_2(\mathbf{x}) \\ L_{h_1} g_3(\mathbf{x}) & L_{h_2} g_3(\mathbf{x}) & L_{h_3} g_3(\mathbf{x}) \end{bmatrix} = \begin{bmatrix} 1 & 0 & x_2 \\ 0 & 1 & -x_1 \\ 0 & 0 & 1 \end{bmatrix} \tag{11.6}$$

Therewith it is: $\det\left[\mathbf{L}(\mathbf{x})\right]=1\neq 0 \quad \forall\mathbf{x}$

- Step 3: Realization of the coordinate transformation.

a) The state space $\mathbf{x}$ is transformed into a new state space $\mathbf{z}$ using the equation (9.4). After replacing $r_1 = r_2 = r_3 = 1$ into (9.4), the same result like (9.11) for the case IM and (9.25) for the case PMSM will be obtained:

$$\mathbf{z}=\begin{bmatrix}z_1\\z_2\\z_3\end{bmatrix}=\begin{bmatrix}m_1^1(\mathbf{x})\\m_1^2(\mathbf{x})\\m_1^3(\mathbf{x})\end{bmatrix}=\begin{bmatrix}g_1(\mathbf{x})\\g_2(\mathbf{x})\\g_3(\mathbf{x})\end{bmatrix}=\begin{bmatrix}x_1\\x_2\\x_3\end{bmatrix} \tag{11.7}$$

The equations (9.11), (9.25) and (11.7) show that the new state variables are identical to the old ones, and therefore the physically fundamental properties of the system "electrical machines IM, PMSM and DFIM" remain unchanged after the transformation of state coordinates.

b) The new state model is the same as (9.12) and (9.26):

$$\begin{cases}\dfrac{dz_1}{dt}=\dfrac{\partial g_1(\mathbf{x})}{\partial \mathbf{x}}\dfrac{d\mathbf{x}}{dt}=L_f g_1(\mathbf{x})+L_{h_1}g_1(\mathbf{x})u_1+L_{h_2}g_1(\mathbf{x})u_2+L_{h_3}g_1(\mathbf{x})u_3\\[2mm]\dfrac{dz_2}{dt}=\dfrac{\partial g_2(\mathbf{x})}{\partial \mathbf{x}}\dfrac{d\mathbf{x}}{dt}=L_f g_2(\mathbf{x})+L_{h_1}g_2(\mathbf{x})u_1+L_{h_2}g_2(\mathbf{x})u_2+L_{h_3}g_2(\mathbf{x})u_3\\[2mm]\dfrac{dz_3}{dt}=\dfrac{\partial g_3(\mathbf{x})}{\partial \mathbf{x}}\dfrac{d\mathbf{x}}{dt}=L_f g_3(\mathbf{x})+L_{h_1}g_3(\mathbf{x})u_1+L_{h_2}g_3(\mathbf{x})u_2+L_{h_3}g_3(\mathbf{x})u_3\end{cases} \tag{11.8}$$

with:

$$\begin{cases}L_f g_1(\mathbf{x})=\dfrac{\partial g_1(\mathbf{x})}{\partial \mathbf{x}}f(\mathbf{x})=\begin{bmatrix}1&0&0\end{bmatrix}\begin{bmatrix}-ax_1\\-ax_2\\0\end{bmatrix}=-ax_1\\[4mm]L_f g_2(\mathbf{x})=\dfrac{\partial g_2(\mathbf{x})}{\partial \mathbf{x}}f(\mathbf{x})=\begin{bmatrix}0&1&0\end{bmatrix}\begin{bmatrix}-ax_1\\-ax_2\\0\end{bmatrix}=-ax_2\\[4mm]L_f g_3(\mathbf{x})=\dfrac{\partial g_3(\mathbf{x})}{\partial \mathbf{x}}f(\mathbf{x})=\begin{bmatrix}0&0&1\end{bmatrix}\begin{bmatrix}-ax_1\\-ax_2\\0\end{bmatrix}=0\end{cases}$$

These terms have to be inserted into the equation (11.8), and the result of the transformation will be obtained to:

$$\frac{dz_1}{dt} = -ax_1 + u_1 + x_2 u_3 = w_1$$

$$\frac{dz_2}{dt} = -ax_2 + u_2 - x_1 u_3 = w_2 \qquad (11.9)$$

$$\frac{dz_3}{dt} = u_3 = w_3$$

From the equation (11.9) the new input vector $\mathbf{w}$ is given as follows:

$$\mathbf{w} = \begin{bmatrix} w_1 \\ w_2 \\ w_3 \end{bmatrix} = \underbrace{\begin{bmatrix} -ax_1 \\ -ax_2 \\ 0 \end{bmatrix}}_{\mathbf{p(x)}} + \underbrace{\begin{bmatrix} 1 & 0 & x_2 \\ 0 & 1 & -x_1 \\ 0 & 0 & 1 \end{bmatrix}}_{\mathbf{L(x)}} \underbrace{\begin{bmatrix} u_1 \\ u_2 \\ u_3 \end{bmatrix}}_{\mathbf{u}} \qquad (11.10)$$

- Step 4: The control law results from (11.10) as follows.

$$\mathbf{u} = \underbrace{-\mathbf{L}^{-1}(\mathbf{x})\mathbf{p(x)}}_{\mathbf{a(x)}} + \mathbf{L}^{-1}(\mathbf{x})\mathbf{w} = \mathbf{a(x)} + \mathbf{L}^{-1}(\mathbf{x})\mathbf{w}$$

$$\begin{bmatrix} u_1 \\ u_2 \\ u_3 \end{bmatrix} = \underbrace{\begin{bmatrix} \left(\dfrac{1}{\sigma T_r} + \dfrac{1-\sigma}{\sigma T_s}\right)x_1 \\ \left(\dfrac{1}{\sigma T_r} + \dfrac{1-\sigma}{\sigma T_s}\right)x_2 \\ 0 \end{bmatrix}}_{\mathbf{a(x)}} + \underbrace{\begin{bmatrix} 1 & 0 & -x_2 \\ 0 & 1 & x_1 \\ 0 & 0 & 1 \end{bmatrix}}_{\mathbf{L}^{-1}(\mathbf{x})} \underbrace{\begin{bmatrix} w_1 \\ w_2 \\ w_3 \end{bmatrix}}_{\mathbf{w}} \qquad (11.11)$$

11.2.2 Feedback control structure with direct decoupling for DFIM

Similarly to the cases IM and PMSM and using the state feedback (11.11), the exactly linearized DFIM model is represented in the figure 11.1. It can be easily recognized that also here – like for IM – only the submodel of the rotor current is linearized exactly. In the case IM it was the submodel of the stator current. Both equations (9.11) and (11.7) express this clearly.

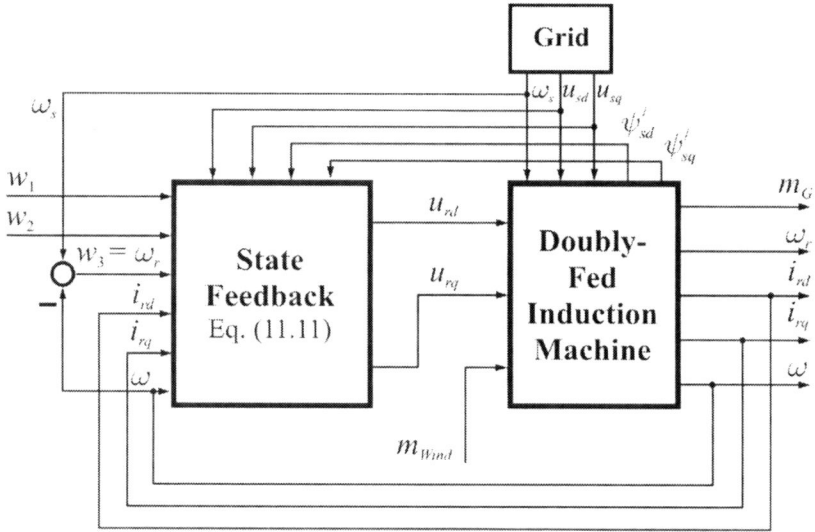

Fig. 11.1 Substitute linear process model of the DFIM

The new linear process model of the DFIM is identical to the model expressed by the equation (9.18) and the comments to (9.18). Using the coordinate transformation (11.11) or the structure in the figure 11.1, the new generator-side control scheme can be derived as in the figure 11.2.

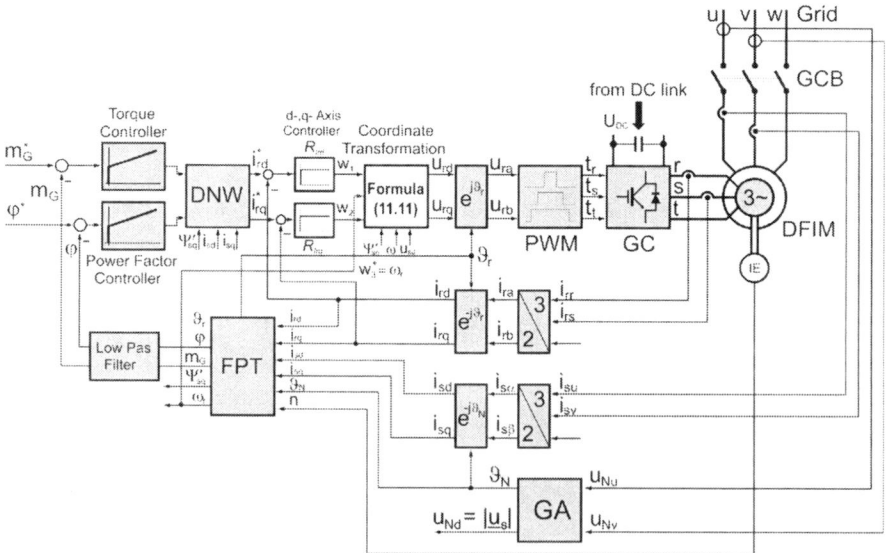

Fig. 11.2 Generator-side control scheme using exact linearization by state coordinate transformation and two separate axis controllers to impress current components

To verify the capabilities of the newly proposed nonlinear control scheme, a series of simulations (cf. [Quang 2005]) have been performed. Results are shown below in figure 11.3 for the case of step changes in the grid voltage of different amplitudes. This special test procedure has been chosen due to the following considerations:

- Generator systems are required to stay operational during grid voltage faults (fault ride-through, FRT). The control system should be capable to maintain this operability as far as possible to avoid falling back to hardware protection circuitry to ensure FRT.
- From the control point of view, a grid voltage step change poses a strong disturbance to the system where its qualities can clearly be revealed.

In the simulations, the results for a linear control system according to figure 10.4 and the nonlinear scheme outlined above are compared for 3 different voltage steps to 70%, 50% and 25% retained grid voltage. Both control schemes had been implemented into an otherwise identical converter-generator system of a 2500 kW wind power plant. Fore sole comparison of the control systems, hardware protection and FRT features had been excluded deliberately.

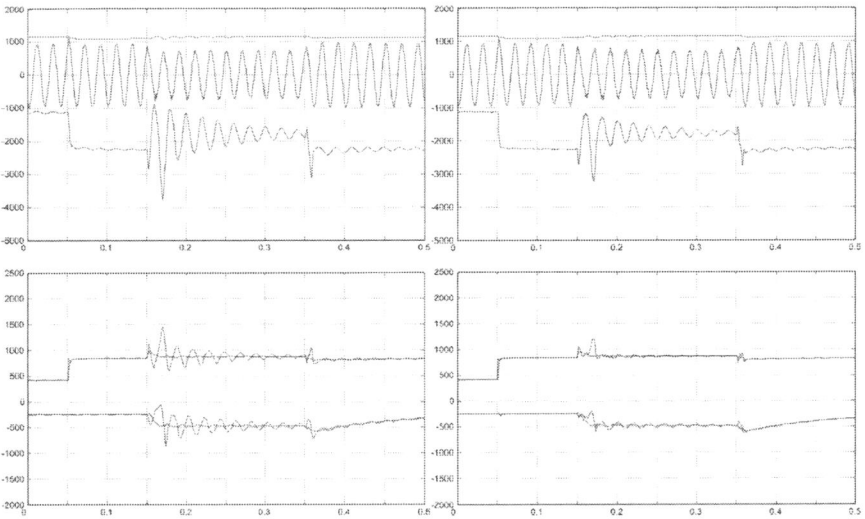

Fig. 11.3a   Grid voltage drop to 70% retaining voltage, 2500 kW converter-generator system: (**left**) linear control scheme, (**right**) nonlinear control scheme, (**top**) Speed [rpm], grid voltage [V], torque [10 Nm], (**bottom**) rotor current d (torque) [A], rotor current q (flux) [A]

322 Nonlinear control structure for wind power plants

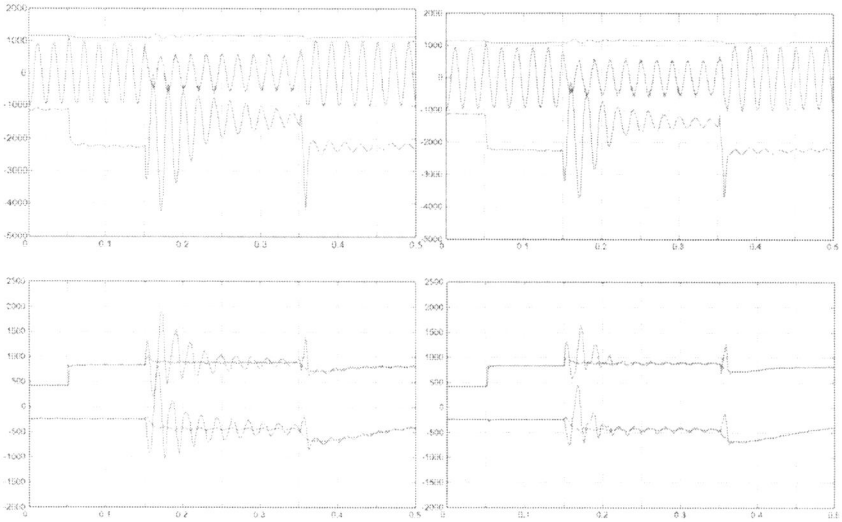

Fig. 11.3b Grid voltage drop to 50% retaining voltage, quantities like figure 11.3a

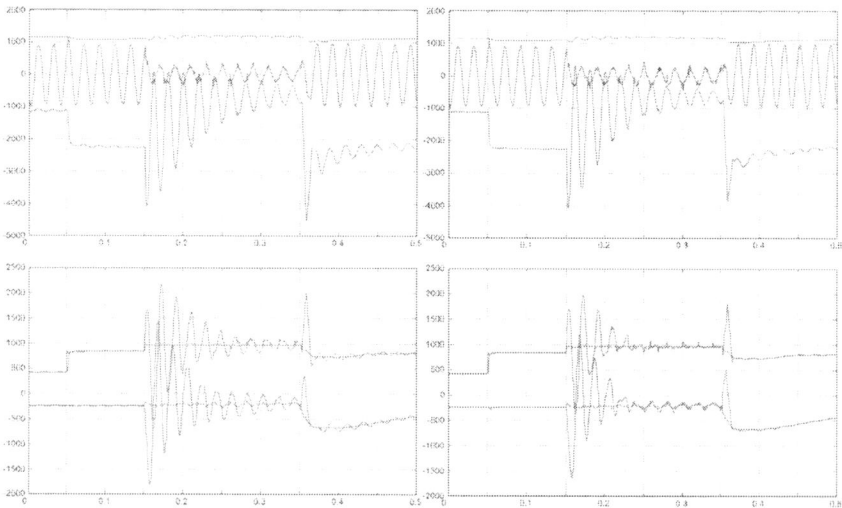

Fig. 11.3c Grid voltage drop to 25% retaining voltage, quantities like figure 11.3a

For the examination of the control behaviour two different time spans must be distinguished: Immediately after the voltage step, the system will lose its controllability due to limitation of the rotor voltage. The performance here depends on how effective the limited voltage is assigned to the respective d- and q-components. After the voltage limitation is overcome and the system controllability regained, the performance can be judged from how fast the rotor currents are forced to follow their set points again.

The results show that the new direct decoupling concept clearly outperforms the linear control in both aspects:

- Smaller oscillation amplitudes of stator and rotor currents occur in the first milliseconds after the fault instant while the rotor current controllers work in limitation mode. This means practically, that the system may cope with more serious fault events without triggering hardware protection functions.
- The system control functionality is regained very fast after the controllers return to linear operation, resulting in short recovery time from disturbances and continuation of defined control behaviour.

These results of simulation were also confirmed by a practical laboratory implementation [Lan 2006]. It should be mentioned also, that the differences between linear and nonlinear control with respect to torque behaviour are considerably less pronounced. This had to be expected, since the nonlinear decoupling has been developed with respect to rotor current, where the performance difference is clearly visible. To extend the nonlinear approach also to flux and torque control will be a matter of further research.

11.3 References

Bodson M, Chiasson J (1998) Differential-Geometric Methods for Control of Electric Motors. Int. J. Robust Nonlinear Control 8, pp. 923-954
Chi PL, Tuan NQ (2005) Design of a Nonlinear Control Algorithm with Direct Decoupling for Wind Power Plant Using DFIM and Validation by Simulation with MATLAB & Simulink and PLECS. Bachelor Thesis (in Vietnamese), Hanoi University of Technology.
Dawson DM, Hu J, Burg TC (2004) Nonlinear Control of Electric Machinery. Marcel Dekker, Inc., New York Basel
Dittrich JA, Stoev A (2003) Grid Fault Proof Doubly-Fed Induction Generator System. EPE Toulouse
Khorrami F, Krishnamurthy P, Melkote H (2003) Modeling and Adaptive Nonlinear Control of Electric Motors. Springer, Berlin Heidelberg New York
Krstić M, Kanellakopoulos I, Kokotović P (1995) Nonlinear and Adaptive Control Design. John Wiley & Sons, Inc., New York
Lan PN (2006) Linear and Nonlinear Control Approach of Doubly-Fed Induction Generator in Wind Power Generation. Dissertation, TU Dresden
Ortega R, Loría A, Nicklasson PJ, Sira-Ramírez H (1998) Passivity-based Control of Euler-Lagrange Systems: Mechanical, Electrical and Electromechanical Applications. Springer, London Berlin Heidelberg

Phuoc ND (2005) Combining of Exact Linearization and Model Reference Techniques for Design of Adaptive GAS Controller and Application to Adaptive Control of Induction Motor. Intern. Conf. CD Proc. of 11th European Conf. on Power Electronics and Applications, 11-14 Sept., Dresden

Quang NP (2004) Nonlinear Control Structures: New Application Perspective in 3-Phase AC Drives. Proc. of the 8th Intern. Conf. on Mechatronics Technology ICMT, November 8-12, Hanoi, pp. 213 – 219

Quang NP, Dittrich JA (1999) Praxis der feldorientierten Drehstromantriebsregelungen. 2. Aufl., Expert-Verlag

Quang NP, Dittrich A, Lan PN (2005) Doubly-Fed Induction Machine as Generator in Wind Power Plant: Nonlinear Control Algorithms with Direct Decoupling. Intern. Conf. CD Proc. of 11th European Conf. on Power Electronics and Applications, 11-14 Sept., Dresden

Quang NP, Dittrich A, Thieme A (1997) Doubly-Fed Induction Machine as Generator: Control Algorithms with Decoupling of Torque and Power Factor. Electrical Engineering / Archiv für Elektrotechnik, 10.1997, pp. 325-335

Wey T (2001) Nichtlineare Regelungssysteme: Ein differentialalgebraischer Ansatz. B.G. Teubner Stuttgart - Leipzig - Wiesbaden.

12 Appendices

12.1 Normalizing - the important step towards preparation for programming

So far, the algorithms with their control variables and parameters are given in the originally derived physical form. An implementation or a programming in this form is mainly impossible. Before this practical implementation can start, all algorithms have to be normalized and scaled if necessary, e.g. for fixed point and partly also for floating point processors. The purpose of the *normalization* consists of transferring these variables and parameters into a unity-less form and thus to prepare them for programming. The *scaling* is primarily necessary to increase the numerical accuracy which is of great importance for the use of fixed point processors.

This important step towards preparation for programming is demonstrated on two examples.

a) Example 1:
The first equation of (3.55), which can be written in detail as follows, serves as the first example:

$$\begin{cases} i_{sd}(k+1) = \Phi_{11}\, i_{sd}(k) + \Phi_{12}\, i_{sq}(k) + \Phi_{13}\, \psi'_{rd}(k) + h_{11}\, u_{sd}(k) \\ i_{sq}(k+1) = -\Phi_{12}\, i_{sd}(k) + \Phi_{11}\, i_{sq}(k) - \Phi_{14}\, \psi'_{rd}(k) + h_{11}\, u_{sq}(k) \end{cases}$$

$$(12.1)$$

From (12.1) the following can be noticed:

- The variables like currents $i_{sd}, i_{sq}, \psi'_{rd}$ and voltages u_{sd}, u_{sq} have to be normalized.
- From the parameters only Φ_{11}, Φ_{13} are already unity-less. All others have to be normalized.

For the normalization of the currents, the maximum inverter current I_{max} is often chosen. For the normalization of the voltage, the maximum value, which is $2U_{DC}{}^{1)}/3$, is chosen. The quantity U_{DC} is itself variable and, with respect to the hardware, has to be normalized by U_{max} while measuring. The equation (12.1) is totally identical with the following:

$$
\left|
\begin{aligned}
\frac{i_{sd}(k+1)}{I_{max}} &= \Phi_{11}\frac{i_{sd}(k)}{I_{max}} + \Phi_{12}\frac{i_{sq}(k)}{I_{max}} + \Phi_{13}\frac{\psi'_{rd}(k)}{I_{max}} \\[2mm]
&\quad + h_{11}\frac{2U_{max}}{3I_{max}}\left(\frac{U_{DC}(k)}{U_{max}}\right)\left|\frac{u_{sd}(k)}{\frac{2}{3}U_{DC}}\right| \\[4mm]
\frac{i_{sq}(k+1)}{I_{max}} &= -\Phi_{12}\frac{i_{sd}(k)}{I_{max}} + \Phi_{11}\frac{i_{sq}(k)}{I_{max}} - \Phi_{14}\frac{\psi'_{rd}(k)}{I_{max}} \\[2mm]
&\quad + h_{11}\frac{2U_{max}}{3I_{max}}\left(\frac{U_{DC}(k)}{U_{max}}\right)\left|\frac{u_{sq}(k)}{\frac{2}{3}U_{DC}}\right|
\end{aligned}
\right.
\tag{12.2}
$$

In the equation (12.2) new symbols are introduced and replaced:

$$
i^N_{sd,sq} = \frac{i_{sd,sq}}{I_{max}}; \; \psi'^N_{rd} = \frac{\psi'_{rd}}{I_{max}}; \; u^N_{sd,sq} = \frac{u_{sd,sq}}{2U_{DC}/3}
$$

$$
h^N_{11} = k_u U^N_{DC}; \; k_u = h_{11}\frac{2U_{max}}{3I_{max}}; \; U^N_{DC} = \frac{U_{DC}}{U_{max}}
\tag{12.3}
$$

Superscripts N: normalized quantities

The parameters Φ_{12}, Φ_{14} are frequency dependent. The value f_{max} is used for the normalization of the frequencies. Using (3.54) it can be written then:

$$
\Phi_{12} = \omega_s T = 2\pi f_s T = (2\pi f_{max} T)\left(\frac{f_s}{f_{max}}\right) = k_{f1} f_s^N
$$

$$
\Phi_{14} = \frac{1-\sigma}{\sigma}\omega T = \frac{1-\sigma}{\sigma}2\pi f T = \left(\frac{1-\sigma}{\sigma}2\pi f_{max} T\right)\left(\frac{f}{f_{max}}\right)
\tag{12.4}
$$

$$
= k_{f2} f^N
$$

In (12.4) the symbols mean:

$^{1)}$ U_{DC}: DC link voltage of the inverter

$$f_s^N = \frac{f_s}{f_{max}}; f^N = \frac{f}{f_{max}}; k_{f1} = 2\pi f_{max}T; k_{f2} = \frac{1-\sigma}{\sigma}2\pi f_{max}T$$

(12.5)

The equation (12.1) can now be rewritten in the normalized form, in which the constants k_u, k_{f1} and k_{f2} as well as the constant parameters Φ_{11}, Φ_{13} have to be calculated only at the beginning, i.e. at the initialization of the system.

$$\begin{cases} i_{sd}^N(k+1) = \Phi_{11} i_{sd}^N(k) + \Phi_{12} i_{sq}^N(k) + \Phi_{13} \psi_{rd}^{/N}(k) \\ \qquad\qquad +h_{11}^N u_{sd}^N(k) \\ i_{sq}^N(k+1) = -\Phi_{12} i_{sd}^N(k) + \Phi_{11} i_{sq}^N(k) - \Phi_{14} \psi_{rd}^{/N}(k) \\ \qquad\qquad +h_{11}^N u_{sq}^N(k) \end{cases}$$

(12.6)

The original equation (12.1) exists now in programmable form without loss of its physical meaning. *The voltages represent the degree of modulation in this normalized form* in which the variable DC link voltage U_{DC} is considered by the parameter h_{11}^N, which has to be updated on-line.

To achieve the most possible numerical accuracy with fixed point or integer arithmetic, the normalized quantities (represented in hexadecimal form) are shifted to the left (multiplied with 2) as much as possible without producing overflow. For normalized currents, voltages and frequencies which accept only values smaller than one, the multiplication factor can be e.g. 2^{15} for 16 bit fixed point processors. This process is commonly described as the *scaling*. The multiplication factor of 2^{15} is the *scaling factor* which at the same time means the number of digits behind the comma. For parameters, which by their nature are already greater than one, the scaling has to be carried out in a way that on the one hand the maximum word length is used, but on the other hand overflow is avoided simultaneously. In principle, this problem does not exist any more with the use of floating point processors and only appears for conversions between data types again, e.g. between integer numbers and signed floating point numbers.

b) Example 2:
A further typical example is shown by normalizing and scaling of the quantities within the equation (12.6) for the calculation of the angular velocity ω.

$$\vartheta(k+1) = \vartheta(k) + \omega(k)T$$
$$= \vartheta(k) + 2\pi f(k)T \qquad (12.7)$$

If the values 2π and f_{max} are chosen as normalizing quantities for the angle and frequency, and it is considered that:

- the frequency has to be signed (e.g. positive for right and negative for left rotation), and
- the angle has to be unsigned (i.e. only forwards counting 0, π, 2π, 3π, 4π...)

then e.g. 2^{15} is possible to be used as scaling factor for the frequency at 16 bit word length, and 2^{16} for the angle. From (12.7) it will be obtained:

$$\left[2^{16}\frac{\vartheta(k+1)}{2\pi}\right] = \left[2^{16}\frac{\vartheta(k)}{2\pi}\right] + \left[2^{15}\frac{f}{f_{max}}\right](2f_{max}T) \qquad (12.8)a$$

or

$$\left[2^{16}\vartheta^N(k+1)\right] = \left[2^{16}\vartheta^N(k)\right] + \left[2^{15}f^N(k)\right]k_{f3} \qquad (12.8)b$$

In the equation (12.8)b it means:

- $2^{16} \times \vartheta^N$ the unsigned integer calculation quantity for the angle,
- $2^{15} \times f^N$ the signed calculation quantity for the frequency

12.2 Example for the model discretization in the section 3.1.2

A system of second order following (3.5) is given with:

$$\mathbf{A} = \begin{bmatrix} a & \omega \\ -\omega & a \end{bmatrix} \quad \mathbf{B} = \begin{bmatrix} b & 0 \\ 0 & b \end{bmatrix} \quad \mathbf{C} = \begin{bmatrix} 1 & 0 \\ 0 & 1 \end{bmatrix} \qquad (12.9)$$

a) *Method* 1: Series expansion with truncation after the linear term (Euler). The use of (3.14) provides:

$$\mathbf{\Phi} = \begin{bmatrix} 1+aT & \omega T \\ -\omega T & 1+aT \end{bmatrix}; \quad \mathbf{H} = T\begin{bmatrix} b & 0 \\ 0 & b \end{bmatrix} \qquad (12.10)$$

b) *Method* 2: Series expansion with truncation after the quadratic term. The use of (3.14) provides again:

$$\mathbf{\Phi} = \begin{bmatrix} 1+aT+\left[(aT)^2-(\omega T)^2\right]\big/2 & \omega T(1+aT) \\ -\omega T(1+aT) & 1+aT+\left[(aT)^2-(\omega T)^2\right]\big/2 \end{bmatrix}$$

$$(12.11)a$$

$$\mathbf{H} = bT \begin{bmatrix} 1 + aT/2 & \omega T/2 \\ -\omega T/2 & 1 + aT/2 \end{bmatrix} \tag{12.11)b}$$

c) Method 3: Euler discretization in a suitable coordinate system.

The method is applicable if the system matrix **A** owns the symmetry properties of the special block diagonal structure (12.9) which is often the case at the modeling of three-phase machines. In this case, state and input quantities can be understood as complex variables.

$$\mathbf{x} = x_x + j x_y \tag{12.12}$$

$$\overset{\bullet}{\mathbf{x}}(t) = (a - j\omega)\,\mathbf{x}(t) + b\,\mathbf{u}(t)$$

At first, the continuous system is viewed in a coordinate system which circulates with the frequency $-\omega$ with respect to the target coordinate system, i.e. $\mathbf{x} = \mathbf{x}^{\omega} e^{-j\omega t}$ (to the topic "transformation of coordinates" cf. chapter 1). Considering the product rule, the following is obtained for the time derivative:

$$\overset{\bullet\,\omega}{\mathbf{x}}(t) = a\,\mathbf{x}^{\omega}(t) + b\,\mathbf{u}^{\omega}(t) \tag{12.13}$$

The discretization using Euler method leads to the following discrete state equation:

$$\mathbf{x}^{\omega}(k+1) = (1 + aT)\,\mathbf{x}^{\omega}(k) + bT\,\mathbf{u}^{\omega}(k) \tag{12.14}$$

For the inverse transformation into the target coordinate system, the equation (12.14) has to be subjected to the counter-rotation, i.e.

$$\mathbf{x}^{\omega}(k) = \mathbf{x}(k)e^{j\vartheta(k)}, \; \mathbf{x}^{\omega}(k+1) = \mathbf{x}(k+1)e^{j\vartheta(k+1)} \tag{12.15}$$

Thereat, the discrete transformation angle ϑ results by Euler discretization as follows:

$$\vartheta(k+1) = \vartheta(k) + \omega T \tag{12.16}$$

The state equation is now in the target coordinate system:

$$\mathbf{x}(k+1) = e^{-j\omega T}\left[(1 + aT)\mathbf{x}(k) + bT\,\mathbf{u}(k)\right] \tag{12.17}$$

or resolved with discrete transfer matrices:

$$\mathbf{\Phi} = (1 + aT)\begin{bmatrix} \cos\omega T & \sin\omega T \\ -\sin\omega T & \cos\omega T \end{bmatrix}; \; \mathbf{H} = bT\begin{bmatrix} \cos\omega T & \sin\omega T \\ -\sin\omega T & \cos\omega T \end{bmatrix} \tag{12.18}$$

d) Method 4: Substitute function using the Sylvester-Lagrange substitute polynomials.

The eigen values of the continuous system matrix **A** will be:

$$\lambda_{1,2} = a \pm j\omega$$

It follows then for equations (3.19) to (3.22):

$$M(\lambda) = (\lambda - \lambda_1)(\lambda - \lambda_2)$$

$$M_1 = (\lambda - \lambda_1), \quad M_2 = (\lambda - \lambda_2)$$

$$m_1 = (\lambda_1 - \lambda_2), \quad m_2 = (\lambda_2 - \lambda_1) \tag{12.19}$$

$$R(\lambda) = \frac{\lambda_1 e^{\lambda_2 T} - \lambda_2 e^{\lambda_1 T}}{\lambda_1 - \lambda_2} + \lambda \frac{e^{\lambda_1 T} - e^{\lambda_2 T}}{\lambda_1 - \lambda_2}$$

Therewith the substitute function $\mathbf{R(A)}$ can be given as:

$$\mathbf{R(A)} = \mathbf{\Phi} = \frac{\lambda_1 e^{\lambda_2 T} - \lambda_2 e^{\lambda_1 T}}{\lambda_1 - \lambda_2} \mathbf{I} + \frac{e^{\lambda_1 T} - e^{\lambda_2 T}}{\lambda_1 - \lambda_2} \mathbf{A} \tag{12.20}$$

and finally the system matrix $\mathbf{\Phi}$:

$$\mathbf{\Phi} = e^{aT} \begin{bmatrix} \cos \omega T & \sin \omega T \\ -\sin \omega T & \cos \omega T \end{bmatrix} \tag{12.21}$$

The input matrix $\mathbf{H}$ is calculated by direct integration of $\mathbf{\Phi}$ according to (3.12):

$$\mathbf{H} = \frac{b}{a^2 + \omega^2} \begin{bmatrix} e^{aT}(a \cos \omega T + \omega \sin \omega T) - a & e^{aT}(a \sin \omega T - \omega \cos \omega T) + \omega \\ -e^{aT}(a \sin \omega T - \omega \cos \omega T) + \omega & e^{aT}(a \cos \omega T + \omega \sin \omega T) - a \end{bmatrix}$$

$$\tag{12.22}$$

12.3 Application of the method of the least squares regression

The method of the least squares regression is often used for the optimization of control loops or the identification of the system parameters. The goal is normally to find an approximate function $y(x)$ in the form of a polynomial of nth order

$$y(x) = a_0 + a_1 x + a_2 x^2 + \ldots + a_n x^n \tag{12.23}$$

from a set of m experimental measurement pairs $[y_i, x_i]$, $(i = 1,2,3,\ldots,m)$ and by the prerequisite, that the loss function (cf. [Rojiani 1996]):

$$S = \sum_{i=1}^{m} [y_i - y(x_i)]^2 \tag{12.24}$$

is minimized. A typical application example is the off-line identification of the main inductance L_s (cf. section 6.4.4, figure 6.18) in dependence on the magnetizing current i_μ. As an approach for $L(i)$[1] a polynomial of 4th order, thus n = 4, is very suitable.

$$L(i) = a_0 + a_1 i + a_2 i^2 + a_3 i^3 + a_4 i^4 \tag{12.25}$$

The task is now to determine the coefficients a_0, a_1, a_2, a_3 and a_4 from m pairs $[L_i, i_i]$ with (i=1,2,3,...,m). To minimize the loss function, at first (12.25) has to be inserted into (12.24), and then the partial derivations

$$\frac{\partial S}{\partial a_0}; \frac{\partial S}{\partial a_1}; \frac{\partial S}{\partial a_2}; \frac{\partial S}{\partial a_3}; \frac{\partial S}{\partial a_4} \tag{12.26}$$

have to be set to zero. Thereby a system with (n+1)=5 linear equations results (cf. [Rojiani 1996]):

$$
\begin{bmatrix}
m & \sum_{i=1}^{m} i_i & \sum_{i=1}^{m} i_i^2 & \sum_{i=1}^{m} i_i^3 & \sum_{i=1}^{m} i_i^4 \\
\sum_{i=1}^{m} i_i & \sum_{i=1}^{m} i_i^2 & \sum_{i=1}^{m} i_i^3 & \sum_{i=1}^{m} i_i^4 & \sum_{i=1}^{m} i_i^5 \\
\sum_{i=1}^{m} i_i^2 & \sum_{i=1}^{m} i_i^3 & \sum_{i=1}^{m} i_i^4 & \sum_{i=1}^{m} i_i^5 & \sum_{i=1}^{m} i_i^6 \\
\sum_{i=1}^{m} i_i^3 & \sum_{i=1}^{m} i_i^4 & \sum_{i=1}^{m} i_i^5 & \sum_{i=1}^{m} i_i^6 & \sum_{i=1}^{m} i_i^7 \\
\sum_{i=1}^{m} i_i^4 & \sum_{i=1}^{m} i_i^5 & \sum_{i=1}^{m} i_i^6 & \sum_{i=1}^{m} i_i^7 & \sum_{i=1}^{m} i_i^8
\end{bmatrix}
\begin{bmatrix}
a_0 \\ a_1 \\ a_2 \\ a_3 \\ a_4
\end{bmatrix}
=
\begin{bmatrix}
\sum_{i=1}^{m} L_i \\
\sum_{i=1}^{m} i_i L_i \\
\sum_{i=1}^{m} i_i^2 L_i \\
\sum_{i=1}^{m} i_i^3 L_i \\
\sum_{i=1}^{m} i_i^4 L_i
\end{bmatrix}
\tag{12.27}
$$

The system (12.27) can be merged into the following form:

$$\mathbf{A}[5,5] * \mathbf{a}[5] = \mathbf{b}[5] \tag{12.28}$$

thereat $\mathbf{A}[5,5]$ and $\mathbf{b}[5]$ are given by the measurement pairs, following (12.27). If C is used as programming language, then the calculation can be realized by the following program section as an example, where $\mathbf{A}[5,5]$ and $\mathbf{b}[5]$ are summarized in a matrix $\mathbf{A}[5,6]$ with $\mathbf{b}[5]$ as the 6th column. In this example it is assumed, that „MeasNum" is the number of the measurement pairs and „PolyOrder" the order of the approached polynomial. Here it holds:

MeasNum = 10; PolyOrder = 4

During measuring the current is increased by 0,1×ImNominal step by step from 0,1×ImNominal to ImNominal (e.g. nominal magnetizing

---

[1] For simplification L_s is replaced by L, and i_μ by i

current). The 10 measured inductance values are saved in the field $L[0]...L[9]$.

```
/* Calculation of the sums or the elements of the matrix A[ ][ ] */
s[0] = (float)MeasNum;
for (i = 1; i <= 2*PolyOrder; i++)
  {  s[i] = 0.0;
     for (j = 1; j <= MeasNum; j++)
        s[i] += pow(ImNominal * (float)j/10.0, i);
  }
/* The sums are assigned to the elements of the matrix A[ ][ ] */
for (i = 0; i <= PolyOrder; i++)
   for (j = 0; j <= PolyOrder; j++)
      A[i][j] = s[i][j];
/* Calculation of the vector b[ ] as the 6th column of the matrix A[ ][ ] */
A[0][PolyOrder+1] = 0.0;
for (i = 0; i < MeasNum; i++)
   A[0][PolyOrder+1] += L[i];
for (i = 1; i <= PolyOrder; i++)
   {  A[i][PolyOrder+1] = 0.0;
      for (j = 0; j < MeasNum; j++)
         A[i][PolyOrder+1] += L[j] * pow(ImNominal * (float)(j+1)/10.0, i);
   }
```

After calculation of $\mathbf{A}[5,5]$ and $\mathbf{b}[5]$ the linear equation system (12.27) or (12.28) can be relatively simply solved by using the *Gauss elimination method*. The first step of the method is the *forward elimination*, which can be summarized (cf. [Sedgewick 1992]) as follows:

The first variable in all equations, with exception of the first one, has to be eliminated by addition of suitable multiples of the first equation to each of the other equations. Then the second variable in all equations, with exception of the first two, has to be eliminated by addition of suitable multiples of the second equation to each of the equations from the third up to the last one (now named as the N-th). Then the third variable in all equations, with the exception of the first three, has to be eliminated etc. To eliminate the i-th variable in the j-th equation (for j between $i+1$ and N), the i-th equation must be multiplied with a_{ji}/a_{ii} and subtracted from the j-th equation.

The described procedure is too simple to be completely right: a_{ii} (now named as *pivot element*) can become zero, so that a division by zero could arise. This can be avoided because any arbitrary row (from the $(i+1)$-th to the *N*-th) can be swapped with the i-th row, so that a_{ii} in the outer loop is different from zero. For swapping it is the best to use the row for which the value in the *i*-th column is the greatest with respect to the absolute amount. The reason is, that in the calculation considerable errors can arise if the pivot value, which is used to multiply a row by a factor, is very small. If a_{ii} is very small, a_{ji}/a_{ii} can become very big. This process, called the partial pivoting, is realized in the example of the L_s identification by the following program section.

```
/* Forward elimination: partial pivoting */
for (i = 0; i <= PolyOrder; i++)
   {  max = i;
      for (j = i+1; j <= PolyOrder; j++)
         if (fabs(A[j][i]) > fabs(A[max][i])) max = j;
      for (k = i; k <= PolyOrder+1; k++)
         {  Temp = A[i][k];                    /* Temp: temporary variable */
            A[i][k] = A[max][k];
                     A[max][k] = Temp;  }
      for (j = i+1; j <= PolyOrder; j++)
         for (k = PolyOrder+1; k >= i; k- -)
            A[j][k] -= A[i][k] * A[j][i] / A[i][i]
}
```

After the step of the forward elimination is completed, the field below the diagonal of the modified matrix **A**[5][5] contains only zeros. The step of the *backwards insertion* can be executed now to calculate the coefficients a_0, a_1, a_2, a_3 and a_4.

```
/* Backwards insertion: Calculation of a0, a1, a2, a3, a4,
   then storage in a[0]...[4] */
for (j = PolyOrder; j >= 0; j- -)
   {  Temp = 0.0;
      for (k = j+1; k <= PolyOrder; k++) Temp += A[j][k] * a[k];
      a[j] = (A[j][PolyOrder+1] - Temp) / A[j][j];
   }
```

With the calculated coefficients a_0, a_1, a_2, a_3 and a_4, the magnetizing curve $L(i)$ in form of a polynomial (cf. equation (12.25)) is now available.

12.4 Definition and calculation of Lie derivation

An unexcited system (input vector $\mathbf{u} = \mathbf{0}$) is defined (cf. [Wey 2001, appendix B], [Phuoc 2006, section 5.1.2]) as follows:

$$\frac{d\mathbf{x}}{dt} = \mathbf{f}(\mathbf{x}) \qquad (12.29)$$

A scalar function $v(\mathbf{x})$ is given. The derivation of this scalar function along the freely moving state trajectory (along the vector field $\mathbf{f}(\mathbf{x}) \in \mathbb{R}^n$) of the unexcited system (12.29):

$$\mathbf{x}(t) = \mathbf{\Phi}_t^f(\mathbf{x}_0) \qquad (12.30)$$

can be given as follows:

$$L_f v(\mathbf{x}) = \sum_{i=1}^{n} \left[\frac{\partial v(\mathbf{x})}{\partial x_i} f_i(x_1, \cdots, x_n) \right] \qquad (12.31)$$

with:

$$\frac{\partial v(\mathbf{x})}{\partial \mathbf{x}} = \left[\frac{\partial v}{\partial x_1}, \quad \frac{\partial v}{\partial x_2}, \quad \cdots, \quad \frac{\partial v}{\partial x_n} \right] \qquad (12.32)$$

Using (12.32), the Lie derivation $L_f v(\mathbf{x})$ can also be formulated as a scalar product (a scalar function):

$$L_f v = \frac{\partial v}{\partial \mathbf{x}} \mathbf{f} \quad \Rightarrow \quad L_f v(\mathbf{x}) = \frac{\partial v(\mathbf{x})}{\partial \mathbf{x}} \mathbf{f}(\mathbf{x}) \qquad (12.33)$$

The function $L_f v(\mathbf{x})$ returns the quantitative change of $v(\mathbf{x})$ along the trajectory (12.30). The figure 12.1 illustrates this fact.

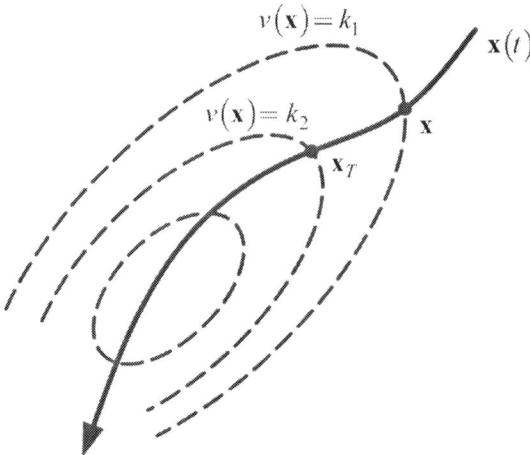

Fig. 12.1 Derivation of the scalar function $v(\mathbf{x})$ along the state trajectory $\mathbf{x}(t)$

The dashed curves in the figure 12.1 represent the sets of points inside $\mathbb{R}^n$ at which the function $v(\mathbf{x})$ has the same values. The dashed curve with the point $\mathbf{x}$ contains the set of points which fulfills $v(\mathbf{x}) = k_1$, and the curve with $\mathbf{x}_T$ the set of points fulfilling $v(\mathbf{x}_T) = k_2$. In this case, the speed of the quantitative change of $v(\mathbf{x})$ along $\mathbf{x}(t)$, from point $\mathbf{x}$ to the point $\mathbf{x}_T$, will be:

$$L_f v(\mathbf{x}) = \lim_{T \to 0} \frac{k_2 - k_1}{T} = \lim_{T \to 0} \frac{v(\mathbf{x}_T) - v(\mathbf{x})}{T} = \lim_{T \to 0} \frac{v\left(\boldsymbol{\Phi}_T^f(\mathbf{x})\right) - v(\mathbf{x})}{T}$$

$$= \frac{\partial v}{\partial \mathbf{x}} \lim_{T \to 0} \frac{\boldsymbol{\Phi}_T^f(\mathbf{x}) - \mathbf{x}}{T} = \frac{\partial v(\mathbf{x})}{\partial \mathbf{x}} \frac{d\mathbf{x}}{dt} = \frac{\partial v(\mathbf{x})}{\partial \mathbf{x}} \mathbf{f}(\mathbf{x})$$

$$(12.34)$$

The Lie derivation has the following properties:
- The multiple Lie derivation of $v(\mathbf{x})$, at first along the vector field $\mathbf{f}(\mathbf{x})$ and then along $\mathbf{g}(\mathbf{x})$, can be written as follows:

$$L_g L_f v(\mathbf{x}) = L_g \left[L_f v(\mathbf{x}) \right] = \frac{\partial \left[L_f v(\mathbf{x}) \right]}{\partial \mathbf{x}} \mathbf{g}(\mathbf{x}) = \frac{\partial}{\partial \mathbf{x}} \left[\frac{\partial v(\mathbf{x})}{\partial \mathbf{x}} \mathbf{f}(\mathbf{x}) \right] \mathbf{g}(\mathbf{x})$$

$$(12.35)$$

- Let $w(\mathbf{x})$ be an additional scalar function, then the following relation is valid:

$$L_{wf} v(\mathbf{x}) = \left[L_f v(\mathbf{x}) \right] w$$

$$\text{because} \quad L_{wf} v(\mathbf{x}) = \frac{\partial v(\mathbf{x})}{\partial \mathbf{x}} (w\mathbf{f}) = \underbrace{\left[\frac{\partial v(\mathbf{x})}{\partial \mathbf{x}} \mathbf{f}(\mathbf{x}) \right]}_{L_f v(\mathbf{x})} w(\mathbf{x}) \qquad (12.36)$$

- Let k be an integer number, then the k-fold Lie derivation of $v(\mathbf{x})$ along $\mathbf{f}(\mathbf{x})$ can be recursively calculated as follows:

$$L_f^k v(\mathbf{x}) = \frac{\partial \left[L_f^{k-1} v(\mathbf{x}) \right]}{\partial \mathbf{x}} \mathbf{f}(\mathbf{x}) \quad \text{with} \quad L_f^0 v(\mathbf{x}) = v(\mathbf{x}) \qquad (12.37)$$

12.5 References

Rojiani KB (1996) Programming in C with numerical methods for engineers. Prentice-Hall International, Inc.

Sedgewick R (1992) Algorithmen in C. Addison-Wesley

Phuoc ND, Minh PX, Trung HT (2006) Nonlinear control theory. Publishing House of Science and Technique, Hanoi (in Vietnamese)

Wey T (2001) Nichtlineare Regelungssysteme: Ein differentialalgebraischer Ansatz. B.G. Teubner Stuttgart - Leipzig - Wiesbaden.

Index

Printed in the United States
125098LV00002B/43-60/P